ANALECTA HUSSERLIANA

THE YEARBOOK OF PHENOMENOLOGICAL RESEARCH

VOLUME LIII

For sequel volumes see the end of this volume.

THE REINCARNATING MIND, OR THE ONTOPOIETIC OUTBURST IN CREATIVE VIRTUALITIES:

Harmonisations and Attunement in Cognition, the Fine Arts, Literature

Phenomenology of Life and the Human Creative Condition (Book II)

Edited by

ANNA-TERESA TYMIENIECKA

The World Phenomenology Institute

Published under the auspices of
The World Institute for Advanced Phenomenological Research and Learning
A-T. Tymieniecka, President

KLUWER ACADEMIC PUBLISHERS

DORDRECHT / BOSTON / LONDON

Library of Congress Cataloging-in-Publication Data

Phenomenology of life and the human creative condition / edited by Anna-Teresa Tymieniecka.
p. cm. -- (Analecta Husserliana ; v. 52-54)
Papers presented at the Second World Phenomenology Congress, Sept. 12-18, 1995, Guadalajara, Mexico.
"Published under the auspices of the World Institute for Advances Phenomenological Research and Learning."
Includes index.
Contents: bk. 1. Laying down the cornerstones of the field -- bk. 2. The reincarnating mind, or, The ontopoietic outburst in creative virtualities -- bk. 3. Ontopoietic expansion in human self -interpretation-in-existence.
ISBN 0-7923-4445-6 (hardback : alk. paper : bk. 1). -- ISBN 0-7923-4461-8 (hardback : alk. paper : bk. 2). -- ISBN 0-7923-4462-6 (hardback : alk. paper : bk. 3)
1. Phenomenology--Congresses. 2. Husserl, Edmund, 1859-1938--Congresses. 3. Life--Congresses. 4. Creative ability--Congresses. I. Tymieniecka, Anna-Teresa. II. World Institute for Advanced Phenomenological Research and Learning. III. World Congress of Phenomenology (2nd : 1995 : Guadalajara, Mexico) IV. Series.
B3279.H94A129 vol. 52-54
[B829.57]
142'.7--DC21 97-2276

ISBN 0-7923-4461-8

Published by Kluwer Academic Publishers,
P.O. Box 17, 3300 AA Dordrecht, The Netherlands.

Sold and distributed in the U.S.A. and Canada
by Kluwer Academic Publishers,
101 Philip Drive, Norwell, MA 02061, U.S.A.

In all other countries, sold and distributed
by Kluwer Academic Publishers,
P.O. Box 322, 3300 AH Dordrecht, The Netherlands.

Prepared with the editorial assistance of Robert S. Wise

Printed on acid-free paper

Printed in the Netherlands.

TABLE OF CONTENTS

PART THREE
INFUSION OF THE AESTHETIC SENSE INTO HUMAN EXPERIENCE

PART FOUR
FINE ARTS AND THE CONVEYING OF THE AESTHETIC EXPERIENCE

PART FIVE
THE EDUCATION OF THE EYE AND THE EXPERIENCE OF ART

ACKNOWLEDGEMENTS

As this volume goes to press we have to remember all who have been instrumental in its origination and preparation. We owe thanks to those who made our Guadalajara Second World Phenomenology Congress, the work of which was destined here, possible. First of all we thank Dr. Santiago Mendez Bravo, Rector and Dr. Cesario Hernández Hernandez, Vice-Rector of the Universidad del Valle de Atemajac – and Dr. Xavier Rodríguez, Chairman of the Philosophy Department – in Guadalajara for having received the Congress with warm hospitality. We also thank Dr. Sergio X. Vazquez and Dr. Miguel Jarquín Marín who extended the hospitality of the Instituto de Terapia Gestalt Region Occidente (INTEGRO) and led the Organization Committee which prepared the ideal conditions for the conference. Their contributions to the program in featuring workshops in the particular methods of therapy practiced at the INTEGRO have enriched it with stimulating philosophical ideas. We thank them too for opening the INTEGRO doors for the founding session of the newest affiliated society of the Institute, the *Sociedad Iberoamericana de Fenomenologia*, of which Dr. Jorge García Gomez of Long Island University is the first Secretary General.

Dr. Tomasz Panz, the General Secretary of our Institute did a wonderful job in promptly and precisely – as in his laboratory work – handling all matters concerning the organization of the program and the correspondence with participants, while Assistant Editor Robert J. Wise, Jr., contributed to the smooth running of the Congress by his gracious reception and registration of the participants, a reprise of his role at the First Congress of Phenomenology at Santiago de Compostela. He was aided in this by our Administrative Assistant, Louis T. Houthakker, who handles our day-to-day membership concerns.

The wonderfully generous gesture on the part of our Mexican hosts at INTEGRO in arranging housing for numerous foreign participants with their colleagues merits not only thanks but first of all admiration for the warm hearts of our Mexican hosts, which made this Congress an unforgettable experience of camaraderie and friendship under the banner of the phenomenology of life.

A-T. Tymieniecka (ed.), Analecta Husserliana, Vol. LIII, ix–x.

In this volume, which pursues further inquiries focused upon the "creative virtualities" of the Human Condition and the great lines of their actualization in bringing forth the specifically human significance of life, there are two special sections on the phenomenology of life in the fine arts and in aesthetics. These are the fruits of sessions organized by Professor Patricia Trutty-Coohill, and we owe her our warmest thanks for her contribution.

We are indebted to Mr. Robert Wise, Miss Isabelle Houthakker and Mr. Mitchell Seagrave for the copy-editing and proofreading of this volume. Special thanks are due to Mr. Wise, who takes considerable pains to smooth the numerous translations into idiomatic English.

Of course, the authors who made this book possible cannot be overlooked. The book is their own work!

A-T. T.

PHENOMENOLOGY WORLDWIDE

The Reincarnated Mind or The Ontopoietic Outburst in Creative Virtualities: Harmonisation and Attunement in Cognition, the Fine Arts, Literature, Aesthetics

Phenomenology which has oscillated between pure consciousness and at the other extreme, the lived body, has appeared in the last decades to be so rarefied in its approaches and ideas, that one would believe it to be evaporating altogether, ceding its place to its Avatars. Yet the present collection, one of four books presenting the work of the Second World Congress of Phenomenology belies such an assumption. On the contrary, in this collection, as in the other three books, we witness a new worldwide philosophical community emerging from the debris of the classical phenomenological movement – one which vibrates with new ideas and intuitions.

This strong manifestation of the ideas taking shape and already presented in several of our books in their germinal stage for more than a decade now, proceeds from a novel, a strikingly novel, philosophical inception: the creative virtualities by which humans differentiate between the human universe and the world humans share with animals and plants.

The result here is a new mind, which we want to present here in its various aspects. This mind is new, not because it has changed into the "emperor's new clothes" and thus stands denuded, stripped to its bare essentials, as with some of the new mind conceptions that reduce it to the organic level, but because in its sense that all is disclosed when we gain access to the genetic source it forms the human creative forge.

Here is the new pivot that phenomenology of life has established – the creative act of the human being which operates the crystallisation of the Human Creative Condition in innumerable rays of novel rationalities, novel with respect to the vital rationalities of life, and within which the Human Condition emerges in its pristine state. Now we can differentiate adequately the main functions of the human processing of life's significances as it transforms them imaginatively and projects new ones, unprecedented among long-reigning rationalities of *bios*. Knowledge, mind, artistic activity, are by no means new fields for philosophical reflection. What we take to be a striking advance beyond hitherto

A-T. Tymieniecka (ed.), Analecta Husserliana, Vol. LIII, xi–xii.

propounded philosophical doctrines of the mind, is our not giving any priority to the intentional conceptualising work of the mind, but deriving from that the modalities of the rest of the mind's activity. On the contrary, this strikes at the Gordian knot of all philosophical fallacies with respect to the differentiation of human faculties and their respective sources and engines, by bringing forth the Human Creative Condition with its creative virtualities as the fulcrum of all which is specifically human – of humanness as such. Thus we may assign to the human functioning its proper source within the creative orchestration of all human forces and virtualities of natural *bios*, animality, and the bringing to recognition of each of them in its own right.

I have written at length about the nature of the Human Creative Condition and the creativity of the human being as such (see my *Logos and Life*). In this collection, originally presented at the world congress, we gather studies which evidence the various functions stemming from and orchestrated by the work of the creative virtualities of the human being as they are crystallised in various specifically human functions. It is in the service of the specifically human ontopoietic unfolding, or interpretation-in-existence, human creativity as such, that the human mind manifests itself. Operative in orchestrating these functions, the creativity is at its peak in artistic endeavors which are as Kant would say the exemplary works of the human being. I have often declared that the aesthetic sense as a specifically creative virtuality comes first in establishing the creative orchestration of human functioning, that is specifically the human mind as such. Therefore, aesthetic and artistic creativity comes first and last in this collection. In between we situate the other basic constituents of the creative orchestration of human functioning, such as emotion, imagination, experience as such, judgement, infusion of aesthetic states of mind, language as instrument of aesthetic communication, etc.

Aesthetics at its highest level, as the life of the human spirit in the fine arts, and its cultivation round off a collection rich in insights and reflections.

A-T. T.

PART ONE

BRINGING FORTH SOME BASIC POINTS OF HUMAN CREATIVITY

The congress in session.

J. J. VENTER

ROAD-SIGNS TO "CREATIVITY"

1. INTRODUCTION: "CREATION" AND WORLD VIEW

> The archetype of the creator is the Divine Being (...). But in the divine creative act something is made to exist where nothing existed before. Since human beings are not able to make something out of nothing, the human act of creation always involves a reshaping of given materials, whether physical or mental. The "something new," then is a form made by the reconstitution of, or generation from, something old (Barron, 1969, p. 10).

Barron's description of creativity has world view implications. He uses the idea of a divine being producing something where nothing existed before, but in focusing on human creation, he seems to use "reconstitution" and "generation" as synonyms of "creativity." In fact, all through his book, he proposes a direct parallel between (natural) generative processes (such as making a baby) and creativity (such as creating a new literary genre, or innovating a management practice).

This terminological fluidity is something of recent date. Ancient world views can be distinguished from one another as "creationist" versus "generationist" – in accordance with the views on what happened "on the first occasion" (Egyptians) or "in the beginning" (Hebrew tradition).

In an organistic representation, that which came later would be considered as related to that which was earlier by descent (as in a tribe) – they are members of the same family, which developed according to an autonomous process. Individual human subjectivity is difficult to conceive of, initially, within this representation – the *process* is encompassing, and where individuality is forced to the forefront (e.g. during the Greek Sophistic era and afterwards), it may be viewed as sacrilege (as happened with Socrates), and may call forth forms of mysticism (rejection of one's own individuality, as in Egyptian Neo-Platonism), or elitism (certain individuals are considered more expressive of the fullness of the cosmic process, and therefore more privileged – autonomous! – than others).

A-T. Tymieniecka (ed.), Analecta Husserliana, Vol. LIII, 3–27.

In a creationist representation, the creatures are considered to be dependent upon their creator, and subject to his structuring commands – their situation is one of heteronomous dependence. Individuals are here related to their creator, and therefore considered as accountable to him, in terms of the heteronomy of their creatureliness. Although subjectivity and personal responsibility is strongly developed in this representation, the strong sense of heteronomy prevents a subjectivistic autonomism. The human subject could in this context only be considered a "creator" in a metaphorical or in an analogical sense – the latter became the approach of medieval thinkers who combined biblical creationism with Greek organistic thinking, thus representing the living cosmos as God's "fellow" in creative activity.

There are presently an abundance of studies on human creativity/innovation. Interestingly, this happens at a time when strict Judaeo-Christian creationism is on the decline in favour of organistic evolutionism. "Creativity" is presently allocated a source in the encompassing process, which seems to reduce the individual human subject to a(n embarrassing) moment in the process (or a cross-point in the network) as in the quote from Barron, above.

Human beings, says Barron, can only "create" the new from the old. But Rationalism, in its extreme form, accepted the autonomy of the individual rational subject (as in Kant): the Divine Creator is replaced by self-determining man, constructing a rational order for himself, imposing this order onto nature according to his plan (as Kant admits and Heidegger highlights), and forges practical life into the mould of reason supreme. Given autonomous man, *"reality" was reduced to history* (progressive man producing the new world from the old), which (finally) lead to a fusion of the idea of the autonomous rational human subject with the idea of the divine creator (as in Metaphysical Idealism). It was all in the context of a world considered predictably logical and logically predictable, the individual subject was already dissipating into a universal subject; "world-process" was replacing subjectivity.

Evolutionism popularised the idea of a creative *process*; scientific and technical innovation increased a sense of human creativity; the dynamics of growth, profit, and innovation demanded more and more novelty, originality. At the same time the attention was drawn to irrational factors as determinants of man's life. The individual, autonomous, subject with his rational creativity imposing the law onto reality is no more. The tendency is on the one hand to highlight the unexpected, the original, the unique (the collapse of logicality as identical with order), and on the other once again to reduce the individual subject to a moment in an encompassing process, or a cross-point in a network – a mystical decentering; a new embarrassment about own individuality (already prefigured in rationalism's universal subject). Universal "unreason" in some forms of Irrationalism, replaces Rationalism's universal reason. In Post-Modernism the individual subject finally disappears and only the network of "unreason" remains. "Creativity," shifted by Rationalism from the supra-historically divine to the historically human, cannot be overlooked anymore (as in ancient organism); therefore it has to be located somewhere in the all-encompassing process, in which the individual subject may then appear (paradoxically) as both a created and a self-creative moment of the family-descendancy. Or, in the words of Barron:

> The problem of psychic creation is a special case of the problem of novelty in all of nature. By what process do new forms come into being? The specification of the conditions under which novelty appears in human psychical functioning is the task to which the psychology of creativity addresses itself. In doing so, it links itself to the general scientific enterprise of describing the evolution of forms in the natural world (1969, p. 9).

What Barron has in mind is a process of replacing old forms by new ones: baby-making and thinking are near relatives in "creativity" terms.

Given its history, present-day conceptions of creativity are not merely the products of an irrationalist fad. They are rather the remnants of an idea from the Judeo-Christian tradition (rooted in the Genesis account), but moulded by modern Rationalism, and fused into the "foreign" body of organistic thinking by twentieth-century irrationalism.

We are, therefore, confronted by basically two "opposing" conceptions of human "creativity." One takes the individual human subject as the initiator of novelty – it is moulded on the idea of a personal God who creates ex-nihilo, and may either ascribe divine powers to the individual human subject, or conceive of human creativity as something analogous to divine creativity. The other retains the idea that creative power transcends the individual human subject and ascribes the initiating of novelty to some universal "divine" entity, or to some universal spirit, which is supposed to determine individual innovative work in some mystical way.

In the following pages I shall attempt to analyse some representative twentieth-century conceptions of creativity (those of Berdyaev, Jung, and Tymieniecka), aiming at crystallising a conception of creativity which may accommodate different nuances from these conceptions, without accepting the deterministic implications of the second one, but also rejecting the modernist idea of the human individual as an autonomous divinity.

2. BERDYAEV

Berdyaev (as early as 1914) understands human creativity *as freedom in (in the love of) God* (1995, p. 11). Liberation for him means freedom from subjection to diabolic (illusory) necessity – one form of which is the Church's doctrine of freedom as "liberation from sin." Berdyaev developed his own conception of original sin (and its opposite: creative freedom):

> For original sin is slavery, bondage of the spirit, subjection to diabolic necessity, the incapacity to affirm oneself as a free creative agent; man loses himself by his assertion in the necessity of the world and not in the freedom of God. The way of liberation from "the world" for the creation of a new life is at the same time the way of liberation from sin, the overcoming of evil, the gathering of spiritual forces for life which is divine (1955, p. 12).

The "sinfulness" of human nature provides no ground to leave man in the lower depths of the world, to deliver man to the results of his sins; for the latter leads to a palsied indifference to good and evil; in fact to a confusion between God and the devil, Christ and the Antichrist. Berdyaev wants a manful liberation of the spirit from this "degeneration" of Christian obedience and humility. The creative act is power, conquest, liberation from egoistic and self-conceited concentration on one's own salvation. In Christ man is already united with the divine nature; any sense of isolation is sin against the divine calling and God's need of man. To be a person and a creator means experiencing everything as universal, by experiencing the whole world within oneself (1955, p. 13). Berdyaev's "soteriology" does not seem to count on grace; it is rather rooted in the divinity of man:

> Doubt of man's creative power is a self-conceited reflection, a morbid egotism. Humility and doubting modesty in places where there should be daring confidence and decision are always disguised metaphysical pride, reflective retrospection and egotistic isolation, born of fear and terror. Times are coming into the life of humanity when it must help itself, conscious that the absence of transcendent aid is not helplessness; because man can discover limitless aid immanent within himself if he dares to reveal in himself, by

> the creative act, all the power of God and the world, the true world, freed from the illusory world (1955, p. 13).

Degenerate Christianity is not the only obstructor of the *universal love* which creates. Science and therefore also the philosophies pretending to be "scientific," such as transcendentalism and positivism, are stuck in adaptation to the necessity of "this world." The "scientific" is really the invalid transfer of the criteria valid for science only, to other spheres of spiritual life, thus elevating science to the supreme criterion of the whole life of the spirit, e.g. culture, religion, philosophy (1955, p. 21ff). Science is the condensed economic description of the given world, for the sake of its self-preservation; a method of partial orientation for man lost in the dark forests of this world's life. Philosophy can never be scientific, for this would mean accepting bondage to necessity. Philosophy is the very opposite:

> (. . .) philosophy is a general orientation to the whole of being and not a partial orientation to the constitutions of being. Philosophy seeks the truth, not truths. Philosophy loves wisdom. Sophia moves all true philosophy. At the summit of philosophic consciousness, Sophia enters into man. In its bases and principles (. . .) science may be dependent upon philosophy but never can the reverse be true. A philosophy of science may be admitted but scientific philosophy is inadmissible. By its essence and by its purpose philosophy has always striven to be the liberation of the human spirit from its bondage to necessity. (. . .) Philosophy is a knowing way out of the given world: it is vision which surpasses the necessity of the world (1955, 27–28).

It is exactly the philosopher which opens up the possibilities for creativity – for discovering the (internally reflected) universe, captured in holistic (time encompassing) intuition, conversing with the oecumenical, universal reason, which is the real principle of creativity. The universal validity of this may not be recognised from the lower level, and leaves the philosopher isolated, but creativity may not be lowered for the sake of a larger common ground and general acceptance (1995, p. 35).

The ontology (man as divine microcosmos, macrocosmos, universal reason), the intuition focussed on the supratemporal: the expectation of Sophia at the summit; the philosopher as creator; the concept of this world as the uncreated evil opposite of God – all these express the elitism of Gnosticism.

Does this Gnosticism provide any avenues of *practicability*?

Creativity is mystery in Berdyaev. Genius, in opposition to talent, is the acceptance of another world and a burning impulse towards it (p. 174); the practical life of a genius, however, is a sacrificial, monastic one in the "world," a "maladjustment" which reveals the creative secrets of being as the "other world" (as the "universal"), while talent is viewed as simply successful functional directedness at cultural canonicity (p. 178). Natural sex is played out against the creativity of the cosmic androgyny; family life in this world against the unbounded love which knows no law – the androgyny which unifies male and female in a higher God-like being (p. 223); symbolic art against theurgic art which unifies all arts and the beauty of nature in creating new being (p. 247ff); and ethics of obedience against one of creative transcendence (p. 272); old magic and old mysticism, bound to nature and God respectively, against new magic and new mysticism, which will express mono-pluralism and immanent-transcendence (cf. pp. 318–319).

Finally, *cultural* creativity comes to expression in the post-redemptive (apocalyptic) epoch in which creative man meets Christ returning:

> We cannot merely passively await the coming of Christ, we must be up and go toward Him. The sense of the apocalyptic will lead to a new religious life only if it becomes actively creative instead of passively expectant. The coming of Christ, in which the absolute Man will be fully revealed in all his power and glory, is connected with man's creative act, with an anthropological revelation. Man's Christological nature will be revealed in man's creative act. The Coming Christ will come only to a humanity which courageously accomplishes a Christological self-revelation, that is, reveals in its own nature divine power and glory (1955, p. 336).

Berdyaev wants it all: God in the world, and the world in God; transcendence, yet inner-worldly monasticism; rejection of humanism, yet affirming the divinity of man; man orienting himself in "this world" by scientific adaption, yet philosophical, Gnostic freedom in the love of God; manful personhood yet mystic transcendence. And yet, fundamentally, he sells us introspective, mystical transcendence as *the* fountain of creativity, from which all truth springs:

> Man's consciousness of himself as the centre of the world, bearing within himself the secret of the world, and rising above all things of the world, is a prerequisite of all philosophy: without it one could not dare to philosophize. (. . .) Man is a small universe – that is the basic truth for knowing man, and the basic truth which precedes every act knowing (. . .) this exclusive self-consciousness of man is not one of many truths acquired as the result of philosophizing; this is a truth which precedes every creative act of philosophic perception (1955, pp. 58–59).

The kind of occult elitism displayed by Berdyaev gives *creativity* a content very reminiscent of D. H. Lawrence's romantic fascism. It appears to say: all men are (or can be) divine creators, but some (those who recognise themselves as microcosms) are more divine than others. On them there rests no obligation to descend to the common (lowly) creatures. A person who pretends to be God knows no normative boundaries. "Creativity" in this context, therefore, need not know about other people.

Berdyaev's and Jung's conceptions of "creativity" give us some insight into the crisis of "canonicity," "normativity," and other similar guiding concepts. The movement of world pictures in the direction of the organistic is more often than not accompanied by a new kind of spiritualism, according to which the "divine" overlaps to a large extent (or totally) with the "earthly." The "divine," of course, cannot be subject to any order or law – thus: whether it expresses itself in an individual (as in Berdyaev) or in a universal (as in Jung), the entity in which it is expressed is subject only to its own decisions or processes. This shift follows on the collapse of Rationalism's identification of "logicality" with "order." Thus discourse about "canons" (e.g. in literature), or "norms," or "legitimacy" has landed in serious problems. The wind of "creativity" now goes where it wants, and nobody knows whither it comes or goes.

3. C. G. JUNG

There seems to be some kinship between Berdyaev's views on creativity and those of Jung. Both of these men have their roots in the organistic, esoteric, mystical tradition. Both of them presuppose some universal, spiritual source of creativity. But there is also a clear difference between the two: Berdyaev uses the micro-macro-cosmos schema to defend the individual creativity of the elite human subject, by representing God and the living universe as expressed in elite man. In this way he can defend mono-pluralism, and transcendent-immanentism, which pays (at least) lip-service to the autonomous subject of early irrationalist humanism. Jung, on the other hand, tends to a universalist explanation of the creativity: the collective unconscious and the archetypes, as universal determinants, initiate the creative process, changing the course of historical events.

Jung, in a few essays written between 1930 and 1950, distinguished between psychological and visionary works of art (1978, p. 87ff) – the first being consciously psychologically self-explanatory; the second really interests the psychologist, because it demands explanation. In the course of his analyses of creativity, he seems almost to forget about the first type. He warns about causal explanations of the psychic process of creation:

(. . .) because the creative urge which finds its clearest expression in art is irrational and will in the end make a mock of all our rationalistic undertakings. All conscious psychic processes may well be causally explicable; but the creative act, being rooted in the immensity of the unconscious, will forever elude our attempts at understanding (1978, p. 87).

The gist of Jung's elucidation of creativity resides in his distinction between the conscious and the unconscious. *Psychological* art draws its materials from conscious life; the material of *visionary* art is unfamiliar:

It is something strange that derives its existence from the hinterland of man's mind, as if it had emerged from the abyss of prehuman ages, or from a superhuman world of contrasting light and darkness. It is a primordial experience which surpasses man's understanding and to which in his weakness he may easily succumb (1978, p. 90).

Jung explicitly rejects Freudian repression as an explanation for this type of art; it is not a substitute for personal experience, or the product of a twisted mind. He insists that visionary art has a deeper meaning, even if the artist himself would deny it. Vision expresses a deeper experience than human passion – the latter falls within the sphere of conscious experience; whilst our visionary intuition points to things unknown, hidden, "by their very nature (. . .) secret" (1978, p. 94). The visionary artist is not the only one to be in touch with this night side of life; so also is the prophet and the seer, and it has been known to man from time immemorial. In the vision the imagery of the collective unconscious shows traces of primitive levels of development in dreams and mental disturbances (as can also be found in the anatomical structure; cf. 1978, p. 97).

The creative act has its source in the collective unconsciousness, which will restore the balance whenever an epoch becomes one-sided. It is the herald of a new era (1978, pp. 98–99; cf. also p. 103). It therefore does not reside in the will of the artist; it is not explicable from his idiosyncrasies (which will rather detract from the artistic value of his work).

Art is a kind of innate drive that seizes a human being and makes him its instrument. The artist is not a person endowed with free will who seeks his own ends, but one who allows art to realize its purposes through him. As a human being he may have moods and a will and personal aims, but as an artist he is "man" in a higher sense – he is "collective man," a vehicle and moulder of the unconscious life of mankind. (. . .) He is sometimes (. . .) fated to sacrifice happiness (. . .) he is everywhere hemmed round and prevailed upon by the Unconscious, the mysterious god within him; so that ideas flow to him – he knows not whence; he is driven to work and to create – he knows not to what end; and is mastered by an impulse for constant growth and development – he knows not whither (1978, pp. 101–102).

The compensatory drive of the collective unconscious awakens the relevant archetypes (such as that of the Wise Old Man in times of transition), whenever society deviates from the right track. The archetypal images "instinctively" arise to restore the balance (1978, p. 104). The artist expressing it thus meets the psychic need of his society; to understand him the reader must allow the work to shape him as it did the artist. This is a mystical communication which removes the isolating barriers of conscious life:

He has plunged into the healing and redeeming depths of the collective psyche, where man is not lost in the isolation of consciousness and its errors and sufferings, but where all men are caught in a common rhythm which allows the individual to communicate his feelings and strivings to mankind as a whole. This re-immersion in the state of *participation mystique* is the secret of artistic creation and of the effect which great art has upon us, for at that level of experience it is no longer the weal and the woe of the individual that counts, but the life of the collective. That is why every great work of art is objective and impersonal, and yet profoundly moving (1978, p. 105).

Human subjectivity as conscious life seems to have become a total embarrassment – it is man lost in isolation. Art is activated by the unconsciousness, which also supplies the material from its collective, spiritualistic, evolutionary memory, and which provides the only real communication by submerging the individual into the collective rhythm of universal, objective, unconscious, experience. Practicing art and appreciating it, here in fact means immersion, being seized upon, engulfed. (No subject; no centre.)

"*Working* creatively" would be impossible in this view. Jung can only advise us to open up . . . in case the time is ripe for a: NEW AGE! I am not sure that Jung has the right to use the term "creativity" – what he conceives of is an organistic birth process (as in ancient Egypt). There is no sign here of creative endeavour (initiative) in the individual human subject.

4. A-T. TYMIENIECKA

Tymieniecka attempts to overcome the multiplicity of perspectives on the place and role of man in the world, breaking through the vicious circle implied by the cognition/reality antithesis. Hers is an attempt to uncover the *basic unity* of the different experiential perspectives (eidetic, emotive, aesthetic, intentional) and the corresponding regional ontologies (of the social, aesthetic, ontic, transcendental, etc.) for once isolated, these individual fragments do not give us enough clues to their interconnectedness (1988, pp. 3–5). It is, in fact, the continuation of the phenomenological attempt to access the Archimedean point:

First, the constructivism of the human givenness is foremostly not only that of the development of his life-course, but also secondly, the human being is not merely (as Husserl proposed) a meaning bestowing agent, the maker of his life-world; what comes first is that *his very life in itself is the effect of his self-individualisation in existence through inventive self-interpretation of his most intimate moves of life*. Thirdly, although the aspect of man's evolution as a living being has already been acknowledged by Max Scheler, yet Scheler misses the critical point: he believes that self-constructivism through invention and creativity closes life's possibilities. We intend to show, to the contrary, that it expands them into possible worlds of life. In brief, I propose that the access to the Archimedean point from which, alone, the unity of all possible perspectives on man's experience can be explained, and the key to the Human Condition be obtained, lies in the creative act of the human being which makes him human – the creative act of man where the differentiating factors of the macrocosm of life differentiate (1988, pp. 5–6).

Note: it is creativity which makes the human being human; it is the creative act which provides access to the Archimedean point from which the unity of perspectives can be explained and the key to our human condition obtained. Note further that inventive self-interpretation of the individual is constitutive for self-individualization, which is his very life. *In other words: creativity is what distinguishes the human from other life-forms; inventive self-interpretation individualizes, and discloses possible worlds of human expression.*

Tymieniecka wants to go deeper than the Husserlian question of the origin of order in the context of human (intentional?) involvement with the world – *she wants to reach for the origin of the forms of this involvement; that is the origin of life itself.*

The human agent bestows meaning on his orbit, but it is not cognition which discloses the origin of order; access to the latter can only be found in "*the essential ways in which the human being enacts and delineates the enactment of his life*" – this is how the human being (through response, deliberation, selection, choice, imaginative invention, planning, etc.) performs his self-interpretation in existence (and cognition is inseparably part of this). At the base of it all is the *creative act* of man – this is how Tymieniecka hopes to "overturn" phenomenology (1988, pp. 6–7).

The "orbit"-metaphor suggests that orderly life revolves around a centre as origin. This centre is the creative act, rooted in what is called "the Logos of life." It remains somewhat unclear whether this Logos of life is a universal impulse (somewhat like Bergson's *élan vital*), or a kind of "law" (like the Stoic *Logos*), or even an expression of the *telos* (which she often refers to). This is of some importance: one of the major contemporary issues about "creativity" concerns the role accorded to "universals" or "wholes" (laws, processes, et cetera), as opposed to individual human subjects.

Tymieniecka focuses in on the question – the basis for a new critique of reason – "what is the primeval *Logos of life* from which the creative endeavour draws its disconnected, yet somehow cohering, manifold and multiple rationale" (1988, p. 16)? This can be analysed into two specific questions: "(i) What are the crucial and decisive sources of sense? (She distinguishes three factors of 'sense': the moral, the poetic and the intellectual.) (ii) What is the thread of coherence among the discreet fragments which binds them together in the self-interpretative game, in which to 'live' means delineating a self-individualized path, endowing each step with significance?" (1988, 16). The search for the sources of sense and their coherence is an attempt to transcend the Husserlian constitutive consciousness in its universal functions, and to push through to the real individual in his actual existence.

Tymieniecka draws the attention to the undeserved privileging of the rational on the one hand, the logos in Western thought, and to the unjustified rejection of the quest for rational principles by present-day irrationalists on the other. Instead of the one-sided opposition between reason and anti-reason, she proposes a triptychal approach: (i) The "poetic" logos ("which presides over the creative work of man proper, and through which man establishes the cornerstones of his human experience"); (ii) the opposite extreme, the "spiritual" anti-logos (which upturns man's creative aspirations in seeking to discover "ultimate reality" by dissolving the ties projected by the poetic logos); (iii) the middle tableau which unifies the two extremes in a plurirational operational system using a skeleton of structures, regulations and principles (1988, pp. 16–17).

It is exactly the middle tableau – the discursive modality of the intellect – which allows us to reconstruct the *mechanisms* of man's self-interpretation in existence, "with its central spine, the *creative apparatus*," and to locate the parameters of human existence: from the expanse of meaningfulness to transcendence (at the two extremes) (1988, p. 17). One could, therefore, say that she distinguishes a foundational, non-rational Logos (the "poetic"), which is near the centre of the orbit; the non-rational anti-logos (which is rooted in transcendence, outside the orbit of life). And somewhere at orbit level a multiplicity(!) of rationalities integrate the two extremes and make them intelligible.

It seems as if Tymieniecka really still wants to take "rationality" seriously and attempts to avoid the trap of irrationalism. On the other hand she does also take account of the irrational factors in human life, as well as their interconnectedness with rationality. The way in which she conceives of the interconnections, however, leads to a dialectical conception of rationality – an autonomous centre bound together with a heteronomy-imposing transcendence. This apparent dialectical tension reappears at other levels in Tymieniecka's thought, and binds her conception to, yet separates it from, that of "non-spiritual" rationalism.

From Eros to Logos

The key, I think, to understanding Tymieniecka's conception of human creativity, lies in the idea of a regulative *telos*. The entelechy inherent in the development of the real individual enables us to account not only for the individual's emergence, but for his life cycle and constructive attainment (1988, p. 22). In fact, although she also refers in this context to individual autonomy, the entelechy seems to presuppose a heteronomy (and the dialectical play mentioned above, reappears):

> When I approached the subject in the Leibnizian fashion (. . .) "Why is there something rather than nothing?," I noted that conceived in the cosmological perspective, the real autonomous individual does not contain in the framework of his being his sufficient reason for being: his individual telos is custom-made to deal within temporal, contingent, and natural-existential conditions, therefore it is subjected to various circumstances and is partly predetermined by the total life process. The reasons ordering his origin (. . .) remain foreign to him. Asked to account for them, his *telos* refers to the concrete *universal order of things*. This in turn, refers to the architectonical plan of the objective constitution of the universe of things and beings. Once there, the thread of the cosmological

analysis, that is to say, of *conjecture* – established on the basis of the eidetic and intentional data, breaks down before leading us to a terminus (1988, p. 23).

Thus the origin of the telos of the life-process as well as of the individual is part of a deeper "why?" which escapes cosmological analysis – somewhat in the fashion of Blondel's autonomous philosophy pointing beyond itself to a heteronomy.

Importantly, for Tymieniecka, a spontaneity, an impulse, on the level of passion, musters all our energies, directs them at its own telos, as foreshadowed in the vision of a totally new universe (transcending established categories and demanding new *types* of objects). The creative act emerges within "the human-being-in-conflicts," in rebellion "against the interpretation of the real that the present phase of the constituted world gives us" (1988, pp. 24–26). The renewal of forms is a product, on the one hand, of "blind" elemental Nature striving, but also (more importantly), of the capability of man and his *desiderium naturale*, to transcend present objectivity:

(. . .) the *creative function*, guided by its own *telos*, generated the *Imaginatio creatrix* in man, as the means, *par excellence*, of specific human freedom; that is, freedom to go beyond the framework of the *life-world*, the freedom of man to surpass himself. (. . .) While recognizing the creative function as a basis of the fundamental human condition, we shall discover, then, that *to be human means to be creative* (1988, p. 26).

The telos which brings about the creative function (i) allows the human being to escape the closed horizon of Nature (i.e. purely survival-oriented behaviour) and to let the Imaginatio Creatrix emerge; (ii) breaking the bonds of the intentional system (natural determination), it guides the generation of new forms as fruits of human invention (creative possibility); (iii) allows us to focus in on the pre-intentional dimension of human functioning at its subliminal source (i.e. the root-unity of individual human life) (1988, pp. 28–29).

To understand creativity, Tymieniecka contends, we need more than beaten tracks and schemas remaining identical, for we are in search of "the laws of the unrepeatable, the unique" (1988, p. 35). This she attempts to do by working from the universal condition of to-be-in-the-world towards the unique (the conflict issuing from the aspiration to surpass this world) – "from Eros to Logos, whose union presides over the passage from a present life-world to possible worlds" (1988, p. 39). Husserl, according to her, in his attempt to overcome the dichotomy of consciousness versus reality (by conceiving of the mind as "constitutive," that is generative of the self and of the human universe; i.e. by letting knowing and being arise in the same genesis of the constitutive consciousness, yet *limiting itself to the discovery of rules and laws observed in comparison to the world as it actually is*), missed out on the individual, unforeseeable, "unrepeatable" powers which arise as a function of an original work, and blows up the habitual frames of the world (1988, pp. 37–38). Briefly stated, the contrast seems to be: *constitutive consciousness submerges the individual in the universal; creativity rather expresses the transformations of the established universal by the individual.*

Conflict and Unity; Transcendence

"Creativity," in terms of the first panel of the triptych, is bound up with Tymieniecka's conception of life as conflict – a conflict between the superior vocation of man and the natural laws which govern his existence, between the spirit and the senses (i.e. between "creative man" and "natural man") (1988, p. 40).

Searching for the point of contact between the extremes, Tymieniecka, though accepting on the one hand that creation is placed at the level of conscious functions superior to those of sensation itself, argues on the other hand that the process of creation is strictly and inevitably bound to the process of transformation of the forms of sensibility and affectivity, establishing a new relationship between mind and body, man and nature, etc. (1988, p. 73). Universe and mind meet in the body as intermediary, through sensation.

[My problem with this approach does not lie in the attempt to find unity (the search for the Archimedean point as such), but in the assumption of a conflict between two extremes. It is, I think, modern rationalism (reason challenging the world/nature in the court of its judgment (Kant), imposing its own order on "reality" according to a plan (as Heidegger characterises it)), which supposed and practised alienation between consciousness and nature, mind and mechanical body (Descartes). The basis of this is the doctrine of the autonomy of man-identified-with-reason, which became exploitative man.]

The conflict between man and nature for Tymieniecka, is a *historical* one. Although there is an initial spontaneity at every stage of human creation, such a spontaneous experience is completely individual and unshareable. But this is at its very emergence complemented by a universally human system of functioning (1988, p. 134) – it shows itself not as strictly personal, but as part and continuation of humanity's collective effort (p. 136). Although the new turns given to history are the results of personal genius breaking the chain of collective heritage, the creative effort cannot simply ignore the collective heritage: we must engage in a critically reflective dialogue with the "universal" spirit.

> Thus, in order to reach behind the established *ways of our own response, molds for spontaneity of volition, cognition, emotion, and to overcome the totality of human tradition* that has infiltrated with pre-ordained forms the very pattern of our functions, *we cannot abandon, discard, or neglect the collective heritage.* We must, on the contrary, enter into a dialogue with it and pierce not only the reservoir of knowledge and the forms of thought, but even the pre-established forms of generation down to the cyclic pattern of the natural life and the ultimate web of the temporal and spatial rules which have pre-organized our experience within their molds (1988, p. 145).

This critical, reflective dialogue pushes through to *limiting conceptions*, to the *unconditioned* which is independent from formal structures as screens: it is both *the reality of life* and the *creative endeavour* of man – it is the Archimedean point from which the new possibilities are disclosed:

> The radicality of clearing the ground instead of its naive acceptance – like that we show in our practical empirical life attitudes – consists in a critical scrutiny which stops at nothing in order to wring from our most personal, intimate source a point of contact with the "ultimate" (1988, p. 150).

It's according to Tymieniecka, the quest of artist, inventor, scholar and philosopher, to transcend the discovered and established forms of human consciousness "and come to grips with reality" (which we can approach, but not define). Crucial, I think, will in this context be what is meant by the "ultimate" and how this is related to the creative process and reality.

"Natural man" for Tymieniecka, will limit himself to assuring the proper run of the course of natural unfolding, and enjoys the stability and security itself; *but "creative man" divorces himself in aim and purpose from nature, setting for himself aims which are meant to gratify intelligence alone.* This is already the case in sense perception, which is interpretative and less predictable than physical work upon "matter"; more so in the case of cognition of inner states, abstract ideas and theories. We cannot account for the emergence of the *creative vision*, and it cannot be commanded and retained at will; at the other end is the *creative work* which does not enter the chain of natural generation, but is elevated above this, both transient and enduring (1988, pp. 153–154).

Thus, somewhere, in the centre of one's being, an impulse arises which tends upwards, out of the material world and towards the "purely" intellectual. Will it reach past this, or again be met by something higher?

Tymieniecka's presentation of the creative process leaves us with the same mysteriousness as most other analyses do – at least as regards its source and its end. Particularly important, however, is that this process is self-expanding to a summit (both individually and culturally), after which decay sets in, and the central question of the third panel of the triptych (about the meaning of it all), comes to the fore. This is how she summarises the creative process.

Human self-individualization is the product of the creative orchestration which allows the living being to invent new avenues for his life. "*Creative orchestration*" here refers to a configuration of the orchestrative system, established by the creative factor in such a way that flexibility and the possibility of choices are introduced into the system, and planning for interpretative transformation becomes possible; and to introducing these into "stuffy" life-situations. And the "orchestrative system" consists of the three constructive faculties (imagination, intellect, and will), and the four sources of meaningfulness (the vital, the poetic, intellectual/structural, and moral senses) – this is the vehicle of culture/civilization/history. In brief *it is the human being taking on a specific form*. It concerns more than (just) the increasing complexity of the vital unfolding; it also touches on

> the *complete series* or incalculably rich forms that are being invented and created by the human being for establishing life's meaningfulness, which are to be experienced, lived, discoursed upon, appreciated, rejected, etc. (1988, p. 164).

New communal forms in human life do not appear from "nowhere." The self-interpretative creative system needs models/prototypes (in the absence of which, culture wanes). Tymieniecka supposes a peculiar mode of the creative orchestration in which the faculties are intensely activated. The *will* acquires dominion over the intellectual faculties in a uniquely goal-oriented process, by which it distances itself tending towards an aim which *does not belong* to the established meaningfulness of the individual's life-course, and unfolding new meanings. Something *alien* is brought into the life-world, something which is not immediately understood – which becomes an exemplary work, a model, a new type of meaningfulness, being "a guidepost for the creative advance of man's self-interpretation in existence" (1988, pp. 165–166).

The investigation into the creative process leads, according to Tymieniecka, to the main question of Western philosophy, "the issue of the unity of all." Rejecting an all-embracing schema or ultimate principle, she proposes an *infinitely expanded cohesion of life*, in which all functional modalities are "unified" by the focus from which they proceed – the creative process itself. Their "independence" from one another is no more than an objectified status which is not given, but as Husserl states, *projected* by the intentional power of human consciousness. Tymieniecka, therefore, approaches these modalities,

> in their PRIMOGENITAL FUNCTIONAL PROFICIENCY: that is, they will appear to us as specific factors of the creative functional system in the specific significant role which they play in its architectonic organization (1988, p. 168).

The creative process shows this unity in three phases:

(a) The creative *vision*. This creates a unity of interplay of all human faculties, but it does not stand on its own with regard to the later phases, for the creative vision is itself transformed by its embodiment (given the limitations of the media used and creative intent (referent) and the changes in personality of creative man).

(b) The *construction* in objective form. Here we see that the differences in the realm of being (body, soul, intellect, and spirit), disappear, all being unified in aim.

(c) The *transition*. When the objectification takes place, this is done with reference to vision, material and the creative intent – that to which the work refers; its "message." It is exactly in the transition that, through an entrusting of a spiritual message, the artist/creator reaches the common springs of the material and spiritual. The gap between subject and object disappears; what springs from the darkest corners of subjectivity, becomes objective, whilst the process itself "recreates" subjectivity.

What remains, then, is not only the question as to the source of the creative endeavour, but also more seriously, the question of the "meaningfulness" of creativity as such, in the context of life's divergences and limited span:

> In fact the creative process makes its way through the innumerable differentiations among the interpretative elements which are objectified, and so appear to be so alien, so divergent or incongenial or contradictory or opposed to each other that they could

> not, as they stand, enter into a combination. Indeed, unless there be some subterranean or innermost factor present to all of them – or to their virtual interplay – that of the Great Game of Creation, the gist of its workings would remain a complete mystery. The creative orchestration may (. . .) work *de facto* with all the available sources of significance; it may be assumed to be the forge of all the *multiple rationales* of human world. (. . .) Our query would not, however, be complete without asking: How do these signifying factors stand in relation to each other and *in virtue of what* do they interplay with each other (. . .) (1988, pp. 172–173).

The creative process does, therefore, not receive its unity from some "Neo-Platonistic" metaphysical oneness, but is rather the playful expression of many different factors, somehow tuned to one another, like a symphony orchestra.

Spiritual Meaning

Thus we arrive at the *third panel* of Tymieniecka's triptych – the important question of the deepest inspiration of our cultural activities: "spirituality." For Tymieniecka, the spiritual is at the very heart of human life. It permeates culture, and therefore is probably for her the deepest unifying factor of all factors involved in the orchestration, for it is the way in which meaning is attributed to life:

> Moreover, we know that the authentic spiritual element has played an essential role in this world through the long centuries. By its transformation of human life through the uplifting of the personal spirit, by its permutation of customs, and by its inspiration of achievements, the authentic spiritual element remains the living soul of all true culture. With its fluid contours, diffuse contents directly inaccessible to reason; and its insistent practice of cutting against the good sense of the everyday world, the spiritual life does not pass the test of rational justification for its existence (1988b, p. 28).

Analogously to the (Husserlian) constitutive act, Tymieniecka characterises the spiritual as (not a state of mind but) an *act*. Such acts occur in the ordinary flow of life, and are organised into a unique experience. *The spiritual act exhibits its presence in immanent perception in absolute evidence, accompanied by a conviction of a presence, which penetrates consciousness everywhere.* It differs, however, significantly from the constitutive act, by the absence of a directly intended object, but rather uses images, hooking onto the rational, but *aims* at transcendence "beyond all objectivity." *The spiritual act is realised in the "natural" life in the meaning attributed by the individual to his own life, and through him, to communal life.* It is not limited to the moral; it also touches on the aesthetical, the modes of receptivity, and concentrates the whole individual in its transcending act (1988b, pp. 29–33).

Although it does not work through structuring objectivity, the spiritual act nevertheless contains a *message*, which is personal (subjective); it is proclamatory, it communicates:

> It does not construct anything but in its quality of "livedness," interpreted in relation to its two poles of reference, it launches an *appeal.* The spiritual act proclaims a "truth," not of logical understanding, but of faith, of a "range" that surpasses the objective frame. (. . .) It thereby knows itself to be the transforming agent of the subject's *modality of being*. The surge of the spiritual act marks the "re-nascence" or even the "re-nascence" of a new modality of being for the subject (1988b, p. 35).

In its aim to transcend objectivity, the spiritual act reveals the limits of the world, and an infinite horizon opens up. Thus the wheel has come full turn: the spiritual act attains its complete form in the attainment of the telos initially mentioned.

In summary: Tymieniecka's conception of creativity is actually identical with her philosophical "anthropology" – for it is creativity, the individualising self-interpretation, which makes a human being *human.* Creativity is rooted in the Logos of life, which finds expression in the poetic logos, and meets up with the spiritual anti-logos in the middle tableau of multiple rationalities. In the creative process the naturally and historically given is transcended through the autonomy of the individual struggling with nature and in critical dialogue with history.

Of the three conceptions discussed, hers is the most balanced, allowing for individual responsibility, yet recognising a spiritual "beyond"; it gives credit to universality yet celebrates individuality. But the balance is attained by a dialectical interplay in which the relationships are conceived of as polar opposites – this may lead to contradictions.

5. CONCLUSION

With hindsight, one can explain the abundance of studies about creative persons, their behavioral patterns, characters and personalities, by the emergence of Irrationalism. Early Irrationalism pointed to the original (the logically unexpected), while retaining the individual subject as "creator" – even creator of a "world" (like Sartre). Although the accent presently (in later Irrationalism) seems to shift from the creative subject to processes and networks, the term "creativity" is probably so embedded into our discourse, that we cannot discard it. Tongue in the cheek, one could say it seems to indicate what brings us all together in conferences, namely adding something to that "sein sum Tode" which our employers call "curriculum vitae."

On the other hand one can admit with Dresden, who meticulously studied works considered "creative," and the self-reflections of "creative persons," as well as psychological research about these, that we really do not know what "creativity" is (Dresden, 1987, 207ff). Yet on the other hand we do know that there are decisive changes in different areas of human culture, which are characterised as "original," or "new," or "creative."

The conceptions analysed above seem to coincide in associating "creativity" with "the production of something from (almost) nothing," "originality," "transcending nature or the present situation," and "spirituality." "Creativity," therefore, implies a deviation from established ways not simply through the production of something materially new (like a better carving knife), but something *which changes the established contextual relationships in which it functions; it brings about new "meaning."* Thus, *the wider the span of relationships affected by an innovation, the more "creative"* it would seem to be.

But there are also differences among the conceptions analysed – the most important probably being their views on *the role of the individual human subject*:

(a) We can, like Jung, eliminate the individual subject as the initiator of renewal, and ascribe innovation to the "collective unconscious," or some other process. But this, of course, has dangerous implications: what kind of accountability/responsibility can be attributed to that somebody (or group) through whom the collective unconscious develops a unique new method of torture? Jung develops his conception of "creativity" with reference to "visionary" art which may be fairly innocent (such as James Joyce's *Ulysses*), and assumes that the collective unconscious will always reset the balance, which of course implies that we have to accept Jung's view of the onesidedness of an era. In fact, to follow Jung would mean denying the hard work preceding and following any mysterious "Aha! – Erlebnis," as well as the playing with "ideas" testified to by persons like Poincaré and Einstein (Ghiselin, 1952, 33–44). And as long as we keep on playing the curriculum vitae game, we would in practice deny what Jung says.

(b) We could, like Berdyaev, stretch the *homo mensura* tradition to its very limits: concentrating reality in the human subject considered divine. Berdyaev stresses freedom (through the love of God) from natural necessity (explicitly not freedom from "sin"), but we must remember that for him the human subject is "divine" – it has godlike *power*. We do not need to stress the gnostic elitism he connects with it; the products of the assumption of *homo-mensura* are all around us (and we are afraid of them). It almost enforces a conflictual conception of the relationship between man and world.

(c) The third option would be in line with Tymieniecka's approach, which allows for a human subject, in the context of a life-process, eking out an individuality, in the interplay of vision and embodying of the vision. In this context the idea of transcending the natural course of events or the established, guided by a telos and a sense of the meaning of life (which she calls the "spiritual"), may be a defensible option. I think it is unfortunate that she retains the conflictual conception of the relationship between man and nature.

Given the role ascribed to the individual aspect in Berdyaev and Tymieniecka's conceptions, they associate "creativity" with "individual freedom," "intelligence," "genius," the "person," "breaking with the past" (Tymieniecka by dialogue and Berdyaev elitistically), etcetera.

I would, however, like to specifically take issue with the idea of an *autonomous* human subject. "Transcendence" of the established, constituted, cannot simply be an autonomous act, for "autonomy" implies a denial of the limitations of the human subject, whether taken individually or collectively, and can only lead to the destructive attempts at controlling the world, which we have Modernity to thank for. I therefore propose a kind of "heteronomy" which recognizes the individual human subject in its interactive relationships with its environment, while holding it responsible and keeping it accountable, instead of Tymieniecka's dialectical play with autonomy (versus "telos" and "spirituality").

Barron's description of creative products are helpful in this regard:

> The important defining properties of these creative products are their originality, their aptness, their validity, their adequacy in meeting a need, and (. . .) "esthetic fit" or "elegance." The emphasis is on whatever is fresh, novel, unusual, ingenious, divergent, clever, and apt (Barron, 1969, p. 20).

"Fresh," "ingenious," "divergent," "apt," "meeting a need," "valid," "elegant" – all these adjectives can probably be used for initiatives considered in a very negative light, for example a new way of psychological torture. The question is whether we want to use the adjective, "creative," for such a *negative* initiative. Analysing some such "negativity," for example the behaviour of a gang of criminals, may help us somewhat further.

A gang of criminals, however "antisocial" they may be, cannot escape certain conditions:

(a) It needs a minimum of organisational structure.
(b) It cannot do without some social status relationships, such as leadership.
(c) It presupposes a kind of "legal" structure, such as agreements on how the booty is to be divided.
(d) Some minimal internal morality, such as loyalty amongst themselves vis-a-vis the police, is required.
(e) They will need codes, signs – a specialised language – enabling them to communicate without their intentions being detected.
(f) As a gang, they cannot do without an economy – risk and input is weighed against possible output.
(g) Apart from social power, some physical power instruments (weapons, tools) are part and parcel of being a gangster – and new ways have to be found to avoid security measures.
(h) Logicality: of course; contradictory instructions will not do.
(i) Of course, the other conditions called "natural," such as the emotional ability to handle the stress, the need for health, food, shelter, apply to them too.

My assumption is that the hardened criminal would want to escape these conditions but cannot. Neither can those of us who do our best to avoid being characterised as "criminal." This demands an expansion of our concept of what is "natural" – we are all subject to minimal conditions of organisation, social status, legitimacy, loyalty, signification, efficiency, as well as the emotional and physical laws. "Natural" is therefore not limited to physico-biological aspects, but encompasses "mental" conditions as well.

These natural conditions can be enslaving, or they can be the platform for creative renewal. The gangster serves only his own interest – the minimal conditions for functioning as a gangster, such as cooperating in an organisation, accepting a specific status, honouring agreements, practising loyalty, risking your life for financial gain, are burdensome. But the same will apply if one lives the well intended philosophy of social darwinist "naturalism": if excellence is equated with competitive elimination of the other, then all these conditions become bonds and burdens, sharing becomes dangerous, for the other may appropriate "my" novelty before I can put my signature onto it.

How can these conditions serve as a platform for a jump in the opposite direction, for non-enslavement: for Berdyaev's "creative freedom?"

It is significant that different circumscriptions and institutional forms of these conditions do occur. Plato defines "justice" in terms of his view of man and social status ("remaining where you belong"); Aristotle's hierarchical interpretation of the "golden mean" is expressed as "treating equals equally and unequals unequally"; and the Romans wanted "to give each his due." *We may be unable to avoid legalities, but we are able to change their contents. We cannot avoid the question of efficiency, but if it becomes a terror (as Lyotard calls it), we can change its format.*

The conditions are outlines which have to be colored in, by interpreting them, to use Van Peursen's term, as "a system of rules." "Creativity" in this context means "the changing of a rule system," or as Tymieniecka characterises it: "transcending objectivity."

We have in the "new" South Africa institutionalised a new set of rules, interpreting human rights in a reconciliatory rather than completely competitively individualist way. But the new rule system is already having a stagnating effect, for newly defined rights are used polaristically. What inspirational power can carry us over this obstacle?

Horkheimer somewhere says that giving your time to something is to give it *love*; brute power is always in a hurry. Berdyaev mentions "love" and "wisdom." Heidegger might have come up with "care." Why should I relinquish some of my rights (willingly accept discrimination against myself), if not for the sake of "love?" Why should I share my ideas about improving nature conservation before publishing them, if not for the sake of "love" of nature and mankind? Why should I avoid marketing an incomplete product to remain ahead in a competitive economy if not for the sake of "love" of the consumer and his quality of life?

"Love" in this sense does not mean an affection, but rather assuming *co-responsibility, sharing* in *caring*. Heavy words, interconnected. "Co-responsibility" refers to dialogical interaction and accepting the duty to be accountable. "Sharing" opens up the dialogical interaction from another angle; that of recognition of one's own limitations, and the openness, sensitivity for the other. "Caring" focuses in on the interaction from a third guarding post: that of protecting, recovering, disclosing and empowering – not leaving the other in the lurch.

Love, in this sense of assuming responsibility *with-and-for* the other, sets us free from the enslavement which the natural conditions exert upon the selfish, the egoist, the violent, the ideologue. It distinctly denies the autonomy or divinity of the subject, although it accepts individual initiative. It goes against Berdyaev's and D. H. Lawrence's elitism (which elevates the creative individual above his fellows as a God); it accepts collective activity as a fountain of creativity, without eliminating individual responsibility.

The conditions sketched above are disclosed for creativity by love. A specialistic focus may dislodge innovation in a specific area, all too often with destructive enslaving consequences elsewhere. *It is only when love attains its full sense, relating the different natural conditions to one another in a positive way (i.e. where individuals, disclosing natural possibilities, recognize their own limitations in interactive dialogical co-responsibility), that we are able to change rule systems which have become calcified and obstructive*; that we are able to transcend the established and the given. I.e. it is love which guides innovation to take as many as possible relevant relationships into account, and therefore increases the scope of relevant relationships responsibly taken into account in the innovating process. In this conception, *the innovator assumes responsibility for the extension of relationships affected, as well as the ways in which they are affected.*

Changing of rule systems does not imply "autonomy" for an individual or a collective. "Autonomy" means nothing else than control and mastering – which implies enslavement, exploitation and destruction. To recognize and be guided by the heteronomy of love, is to practice Tymieniecka's spiritual disclosure and living out its telos. This opens up avenues for the oppressed; allows the powerless to gain access to legitimate power; provide food for the hungry; for:

> it is prepared to open up key symbols and instruments to the foreigner, the alien; it searches for innovative modes of living to allow animals to remain in the fields; it induces one to work hard at healing broken relationships at a time when enduring relationships are in jeopardy. *Creativity is responsible self-sacrifice.* (. . .)

Potchefstroom University, South Africa

REFERENCES

Barron, F., *Creative Person and Creative Process* (New York/Toronto/London: Rinehart and Winston, 1969).

Berdyaev, N., *The Meaning of the Creative Act*. Tr. by D. A. Lowrie (London: Gollancz, 1955).

Dresden, S., *Wat is Creativiteit?* (Amsterdam: Meulenhof, 1987).

Ghiselin, B., *The Creative Process* (New York/Scarborough/London: New American Library/New English Library, 1952).

Jung, C. G., *The Spirit in Man, Art and Literature*. Tr. by R. F. C. Hull (Princeton: Princeton University Press, 1978) (Bollingen Series XX).

Maritain, J., *Creative Intuition in Art and Poetry* (London: Harvill Press, 1953).

Tymieniecka, A-T., *Logos and Life: Creative Experience and the Critique of Reason. Analecta Husserliana*, XXIV (Dordrecht/Boston/London/Tokyo: Kluwer Academic Publishers, 1988(a)).

Tymieniecka, A-T., *Logos and Life: The Three Movements of the Soul, or the Spontaneous and the Creative in Man's Self-Interpretation-in-the-Sacred. Analecta Husserliana*, XXV (Dordrecht/Boston/London: Kluwer Academic Publishers, 1988(b)).

PATRICIA TRUTTY-COOHILL

HOW I WENT UP TO IMAGE PHENOMENOLOGY AND CAME DOWN ENTANGLED . . .

I wanted to make an emblem of phenomenology for the flyers I was sending out. Something simple and clear, something graphic, like the Renaissance "Festina lente," make haste slowly.[1] But what could symbolize that for us *sein, ist in der Welt sein* – the acceptance of matter, things, others, as most real and even edifying?

One thinks of Alcyoneus, the son of Ge, the Earth, who gained strength when he touched his mother, but he is not portrayed very favorably in art. In the Great Frieze of the Altar of Zeus at Pergamum,[2] he reaches for earth to no avail because Athena has him caught by the hair. I decided against Alcyoneus. Having the personification of phenomenology slain by Wisdom hardly seems appropriate.

There is the Giant Antaeus, the great wrestler, who is invincible so long as some part of him touches Earth, from whom he draws his strength. Sadly for us though, the familiar images of Antaeus[3] do not cast him in a favorable light. In Pollaiuolo's painting, Hercules locks his arms around the giant's waist, lifts him, thus keeping him away from the earth, the source of his strength. Hercules crushes him until Antaeus finally weakens and dies. It would hardly be appropriate to have the symbol of phenomenology crushed by the symbol of courage!

But all that speculation was not a total loss. *Touching* might provide the key.

In the Christian tradition one image of touching comes to mind: The doubting of Thomas. For me the most effective image is an Ottonian ivory from the tenth or eleventh century in the National Museum in Berlin.[4] "Then saith he to Thomas, Reach hither thy finger and behold my hands; and reach hither thy hand, and thrust it into my side: be not faithless, but believing." But to identify phenomenology with Thomas's doubting and *needing* to be convinced that the object exists, hardly seems appropriate. On the other hand to identify phenomenology with Christ who allows, indeed, invites contact, who opens his body, accepting man's touch so that he can be believed – that would be hubris.

More appropriate would be the Buddha in the mudra of calling the earth to witness.[5] The story is that when the historical Buddha was on

A-T. Tymieniecka (ed.), Analecta Husserliana, Vol. LIII, 29–33.

the point of proving his Buddha perfection, the gods of the earth warned him that he would be attacked by demons. Buddha calmed them, saying that he would suppress these evil-doing gods by his power alone, through his Enlightened knowledge. The demon king appeared and challenged him to put his words into action. The Buddha, pointing to the ground with his finger, called upon the gods of the earth, who rose up and killed the demons.[6] While the iconography is generally appropriate, perhaps designation of the non-phenomenologists as demons goes a little too far.

So the search continued – until I read Anna-Teresa Tymieniecka's discussion of creative function in philosophy and art, disciplines that "progress . . . along two contrastingly formulated yet equally essential quests" – the one to know, the other "to give expression to the knowledge." She writes that as this double interrogation attempts to understand the sense of human existence, "it interlaces and intertwines within the radius of interrogation."[7] Both quests intend to "synthesize provisorily this infinitely vast field" and "to discover its first and last principles. Its synthesizing endeavor in inventing the SIGNIFICANT KNOTS OF THE SPECIFICALLY HUMAN REALITY and its principles leads to the discovery of relevant units of sense and (its expression) with the CIPHER."[8]

My problem in a nutshell – and she had even pointed the way. Instead of imaging the basic *sein, ist in der Welt sein*, I could image its effect! After all, the image is not just for the whole study, but for an organization devoted to it, an organization which does really experience and cope with being and being in the world. Tymieniecka's key words, "interlace," "intertwine," "significant knots," "radius of interrogation" led in one direction, toward Celtic design, and in particular, to book illumination. So I turned to *The Book of Kells*, Ms. 58. A. I. 6, housed at Trinity College Library, Dublin. It was produced in the late eighth to early ninth century, probably in Iona and Kells. Whole pages measure 13″ × 9 1/2″. I chose one detail, from folio 130 r, the beginning of the Gospel of St. Mark, because it showed men "intertwined" within a "radius," making, I would hope, a "significant knot."[9]

The small design shows a group of three men in profile, with eyes full front, within a field bounded by a spiral. Their individual bodies are knotted and they hold each other's wrists. This graphic well suits the group Professor Tymieniecka has organized. There is one problem

The Book of Kells, Ms. 58 A. I. 6. f. 130 r. The manuscript was produced in the late eighth to early ninth century, probably in Iona and Kells. Whole pages measure 13 × 9 1/2 inches. The manuscript is housed at Trinity College Library, Dublin.

with this design: it contains *three* men whereas the passage from Tymieniecka that inspired this choice, speaks of *two* kinds of knowledge: art and philosophy. When translating an idea into a graphic, one must understand the effect of certain factors, especially number. Two will set up a dichotomy, tension, and frequently stasis. Tymieniecka's intention, on the other hand, is to emphasize the progress of the two methods that attempt to "understand the sense of human existence." To have only two men would make the image static.[10] The trinitarian approach adds dynamism to the composition – a dynamism which can be interpreted in this case as life movement.

A major iconographic difficulty thus disposed of, it is possible to launch a full-scale semiotic analysis of the composition in terms of Professor Tymieniecka and the World Institute for Phenomenological Research and Learning.

Within a circular framework, three figures are shown in profile. The limbs of each individual are interlaced. Hair flows out of each head, meandering to encircle the legs of another and to enframe his body. The torso and legs are outlined by a double band of the same thickness as the entangling hair, as if the strand were an extension of their surfaces. The long strand under (attached to)? the lower figure makes a hairpin bend and turns into a frame that spirals to enclose all the figures. Although the black field might not be immediately grasped as Tymieniecka's "infinite vast," one might, in applying the design to our purpose, interpret the blackness as very deep and the spiral which bounds it as a convention for suggesting continuous movement throughout that infinite and vast field.

In all great compositions, meaning can be taken from the whole and from certain focussed events. So it is with the interlace of men on folio 130 recto in the *Book of Kells*. There is the sense of the whole, for the spiral of life is generated from the entanglements of individuals . . . or vice-versa. There is a sense of focussed event, for central to the image is the motif of the great hands that reach out through the entanglement of self to touch, to grasp. And so, their great eyes are opened and they see.

NOTES

[1] For "Festina lente" see James Hall, *A History of Ideas and Images in Italian Art*, New York: Harper & Row, 1983, p. 274.

[2] *Athena and Alcyoneus*, from the east side of the Great Frieze of the Altar of Zeus at Pergamum, *c*. 180. B.C. Marble, height 2.3 m. Berlin, Pergamonmuseum, Staatliche Museen. See Richard Tansey and Fred S. Kleiner, *Gardner's Art through the Ages*, Fort Worth, Philadelphia: Harcourt Brace, 1995, p. 175.

[3] The small panel, 16 × 10.5 cm is generally attributed to Antonio Pollaiuolo (1426–1498), its in the Uffizi in Florence. See Leopold D. Ettlinger, *Antonio and Piero Pollaiuolo*, London: Phaidon, 1978, cat. no. 10, plate 92.

[4] This small ivory panel was made in either Trier or Echternach, and has been variously dated to the tenth or eleventh century. See Tansey and Kleiner, *op. cit.*, p. 374.

[5] This Thai Buddha was made in the 14th century in the Sukhothai high style. Bronze, 94.9 cm high, in the Collection of H.R.H. Prince Chalermbol Yugala, Bangkok. See Sherman Lee, *History of Far Eastern Art*, Englewood Cliffs: Prentice Hall, 1994, p. 144.

[6] As reproduced in E. Dale Saunders, *Mudra, A Study of Symbolic Gestures in Japanese Buddhist Sculpture*, New York: Pantheon, Bollingen Series, 1960, p. 81.

[7] Anna-Teresa Tymieniecka, *Poetica Nova*, Reidel, 1982, p. 15.

[8] *Ibid.*, p. 16.

[9] The actual image is taken from a source which presents such designs in a very graphic manner, suitable for black and white reproduction, *A Coloring Book of Ancient Ireland*, Santa Barbara: Bellerophon, 1978.

[10] One thinks of the dualistic *yang-yin* symbol in which the opposites both contain their opposites. The two halves are separated by a diametrical line, that curves so that each side makes room for the other. Beautiful symbol, but it is already associated with other systems. What is more, there is no sense of "entanglement" and wilful reaching out, realizations we associate with phenomenology.

SLAWOMIR MAGALA

ART AS MORAL GAMBLE

Phenomenological Aspects of Creative Responsibility

The idea that art is an activity reminiscent of a moral gamble is not a new one. Quite a number of philosophers have been pointing out that in undertaking an artistic activity one cannot be certain of the outcomes (e.g., B. Williams, Th. Nagel). The problem, although noticed, did not figure prominently in philosophical, ethical and aesthetical consideration. The reason for this relative neglect is to be found in a fairly tacit consensus that philosophical analysis of the "context of justification" of a work of art merits much more attention than its "context of discovery" (reasons of expediency, no doubt; it is much simpler to tackle those art works which have survived than to speculate on many more of which very few will). Thus the infantile disorder of "scientificity" in the philosophy of art (and, to a large extent, art itself) has made us wait for the full-blown analysis of a moral gamble involved in the artistic activity. Let us begin our analysis with some intuitive remarks of Ingarden on the ontological basis of responsibility (Ingarden, 1987), and then proceed to a processual analysis of the situation of an artist and a connoisseur as a moral gambler.[1]

Ingarden distinguishes four types of contexts in which we speak of responsibilities, which do not begin in dreams (pace Delmore Schwartz, albeit the problem of dreams did intrigue Ingarden in a number of ways). First, we can say that somebody *is* responsible for something, second, that somebody *assumes* responsibility, third, that someone is *held* responsible and fourth, that somebody *acts* in a responsible way. It is the fourth type of context that is most relevant to our discussion – as it is for Ingarden, who notices that a copyright for particular acts forms the basic criterion of responsibility. Responsible action is, according to Ingarden, a concretization of general value, a successful attempt to enforce an actualization of a value. The copyright label (needless to say Ingarden did not use this term, for which the undersigned is to be held responsible) matters in so far as an actor not only actualizes a value, but also somehow stresses this value's value (*Wertigsein*, as the author of "Das literarische Kunstwerk" says in the German original of his study on the ontological bases of responsibility), in so far as he vouch-

A-T. Tymieniecka (ed.), Analecta Husserliana, Vol. LIII, 35–47.

safes for it. The concept of responsibility is thus linked to the actualization, "concretization" of value, i.e., to intervention into its mode of existence. As Ingarden succinctly puts it: "The agent is responsible for destroying or creating this real or reality-based ontological fundamental of a value's concretization" (Ingarden, 1987, p. 167). This general approach towards responsibility is compatible with the aesthetic view held by the author of *Untersuchungen zur Ontologie der Kunst* on the mode of existence of a work of art and the contribution made by an actor, i.e., an artist:

> If we distinguish between the work of art (the picture) and the aesthetic object (the concretization of the work of art constituted in the aesthetic experience), then we must also draw a distinction between two different categories of value that are correlative of one another, between artistic and aesthetic values. The former occur in the work of art itself, the latter appear in concreto only in the aesthetic object. (Ingarden, 1989, p. 231)

Thus for Ingarden the test of the artistic cuisine is in the eating of the pudding of a work of art: experiential grasping of the latter in immediate apprehension results either in holding the artist in question responsible for a legitimate work of art or in questioning the links between artistic attempts and aesthetic results. In other words, beauty is, in a sense, in the eye (but also in the mind and in the cultivated taste) of the beholder, thus presupposing the ascription of responsibility to the owner of the copyright to the object of beauty. This approach raises the following problems which are relevant from the point of view of modern artistic processes and modern ways of cultivating the aesthetic taste. First, how do we select the artistic objects which we allow to trigger our aesthetic experience. Second, why do the selection criteria change if there is no progress in the arts. Third, do we have to gamble in supplying artistic objects, experiencing them and evaluating them into the aesthetic ones, and if so, why.

Let us imagine that we are facing a novel by Samuel Beckett, *Dream of Fair to Middling Women*, a novel by Thomas Pynchon, *Gravity's Rainbow* and a novel by William Burroughs, *Naked Lunch*. All three of them are novels which may offer us an aesthetic pleasure if we decide to read them. However, being mortal, we have less than infinity to read them and thus we have to drop two so that one can survive to reveal her beauty (if a novel, like a boat, is of a feminine gender) to our beholding eye. I have narrowed our options by limiting this choice to novels only (as a homage to Ingarden, whose study *The Literary Work of Art* focusses on literature at the expense of music and painting).[2]

The first novel was left unpublished until the author died and his fame prompted the owners of the typescript to publish it against his explicit wishes. The second is a mature work of an acknowledged master of arcane and difficult art, often thought to surpass the late James Joyce's "Finnegan's Wake" in historical density and avant-garde linguistic virtuosity.

The third is an acknowledged masterpiece of a countercultural and subcultural artist attempting to depict the reality of drug addicts, of the marginal, the low, and the hopeless. Which one should we choose and why?

It is easy to notice that our choice depends on the way in which our taste has been cultivated. If we were brought up in the faith in ruling institutional culture, then the fact that Samuel Beckett had been awarded a Nobel Prize for literature means that we should prefer him above his fellow writers. Cultivating taste in good faith in the validity of our institutions as reliable judges of aesthetic quality thus prompts us to look forward to the experience of the earliest novel by the author of the Nobel Prize-winning *Waiting for Godot*. In a sense we are hedging our bets by relying on competent judges serving in acknowledged institutions – even though it might be argued that the first juvenile novel fails to meet the standards Beckett imposed on his own products later on in his life. However, we might argue that our taste has been cultivated in opposition to the ruling institutions, that we have gone to the elitist circles of the few subtle minds praising the arcane, the obscure, the complex, the difficult – in other words that our taste has been shaped by an opposition to the public's preferences, even if the public in question be an educated one. Having thus established a preference for works of art inaccessible to the broader circle of potential readers we have little difficulty in selecting the novel of greatest formal complexity and the one which most readers would find too difficult to read.

But having chosen Beckett or Pynchon we are still keeping our faith in cultural or countercultural institutions, although our canons differ and so do the articles of faith. Nevertheless a faith it is and in arguing for Beckett rather than Pynchon or the other way round we may justify our choice with a formally identifiable aesthetic value. This would hardly be possible in the case of selecting Burroughs. "Naked Lunch" is difficult to evaluate from the point of view of cultural or countercultural standards of cultivating good taste. It was written as a celebration of human freedom in escaping the iron cage of ideal values, both cultural and countercultural. A number of writers, for instance Henry Miller, William Burroughs, Jean Genet, or Louis Ferdinand Céline, to mention but a few, have deliberately blurred the borderlines of the aesthetic experience. They have demonstrated the futility of cultivating taste within institutions and counterinstitutions by radically exploding the repertoires of accepted features of legitimate works of art.

One may protest by saying that even the most radical expansion of the repertory of features a work of art can have in order to provoke an aesthetic experience does not yet mean that the very principle of cultivating taste had changed. Perhaps the principle remains and what changed is simply a belief in established institutions and counterinstitutions. We do not believe in any monopoly of taste, granted either by tradition or revolution and we reserve a right to consider even those options which from the point of view of the taste arbiters would be inadmissible. Hence Burroughs and *Naked Lunch.* He would not have qualified from the viewpoint of the aesthetic standards of the Swedish Royal Academy. He would not have qualified in point of the aesthetic standards of the admirers of *Finnegan's Wake.* But even by picking *Naked Lunch* up and neglecting thousands of other similarly extravagant and shocking literary products we do demonstrate a certain "elitist" reference to the rules of sensemaking. Why Burroughs and not, let us say, Bukowski? Why, *Tropic of Cancer* and *Sexus* and not *The Story of O* or *Vinnetou*? Because within the cultural process of making sense of some literary works of art enough readers have decided to share their analyses of their experiences, thus demonstrating to the broader public the potential merits of objects which would otherwise remain obscure and would not lead to the acknowledged aesthetic experiences. The shock value, the value of novelty have certainly helped to trigger these first critical responses of the newly-born aesthetic converts, but they were free to do so because of the acknowledged, even if tacit, defense of the realm of taste against all attempts to limit it. The cult of novelty and the possibility of going beyond the limits of "good" taste in cultivating and changing it are certainly the basic mechanisms safeguarding the "gambler's freedom." As another philosopher once observed:

> The spiritual concreteness of the creative new (the ontological new, the new of reality, not a mere change of previously existing element) must prevail fundamentally over mechanism, habit, causality – conditionality, extrinsic construction based on given elements, abstract will, arbitrariness, or any defect of spontaneity or plasticity. (Vivante, 1980, p. 114)

We may therefore venture a hypothesis that all other things being equal a choice of Burroughs testifies a preference for novelty in artistic form *and* aesthetic value and confirms our basic freedom to select artistic objects triggering our aesthetic experience over and beyond the established matrixes of "good taste." It testifies to our willingness to gamble. Please note that when we select Pynchon or Beckett we are also gambling, since it is quite likely that we shall fail to incite an aesthetic experience. But at least we will have forestalled adverse criticism by hedging our bets, i.e., by trying to select according to an institutional guarantee of established or countercultural value. Having chosen pure novelty we are able to demonstrate our freedom, but we are certainly taking higher risks than those taken by the readers of Beckett or Pynchon. Thus if we are to decide whether to invest money in a large edition of any of the three novels, we can lose money on all of them, but only in the case of risking money on Burroughs can we face serious charges of carelessness.

Our case, however, is not hopeless if the only tribunal to pronounce judgement upon us is the tribunal of philosophical imagination. Ingarden hints at this possibility when he admits, albeit in a footnote, that:

> It is, of course, quite a different matter if the viewer detaches himself, as it were, from what he has learned from aesthetic experience, and does not merely lose it from his field of vision, but also forgets it, and then completely blindly and coolly passes purely intellectual judgements on the aesthetic worth of an object. (Ingarden, 1989, p. 251)

What for Ingarden was an exceptional situation deviating from the rule that judgements are to be meted out meticulously and within a definite frame of idealizational and institutional reference has become a commonplace and a rule for critics of the end of the twentieth century. We do change the criteria for selecting artistic objects and passing aesthetic judgements upon them knowing full well that there is no progress in the arts: the question posed by Rousseau was relevant but incorrect and his answer correct but irrelevant.[3] But change we must, for the following reasons: a) we feel that the reality of the social processes we participate in and the quality of our relationships to others in these processes do change and require a repertory of responses for which the existing institutions and philosophies are insufficient; b) we do not want to stop gambling since that would mean the end of history and the pushing away of a philosophical ladder while we are strongly motivated to continue playing our games (i.e., gambling); and c) gambling on values, including aesthetic and ethical ones, confirms our freedom, increases our repertory of responses and maintains the continuity of the game.

Let us sum up our considerations. We select artistic objects which we subsequently allow to trigger our aesthetic experience either from the viewpoint of an established set of cultural values, a set of counter-cultural values or from the point of view of a new mix of values and institutions or from the point of view of their substitutes (a secret, condemned community of drug addicts, or cybergenetic mutants in the case of Burroughs). Second, even if there are no particular changes in the sets of values we are relying on, we are still free to exercise our faculty of judgement even if it means deliberate gambling on an outcome that is even more uncertain than the others.[4] This brings us to the third question: do we have to gamble in selecting artistic objects in order to pronounce aesthetic judgements, and if so, why.

If we follow Ingarden in accepting the individual concretization of a schematic intentionality of an artistic object as the basis of aesthetic experience (of which an aesthetic judgement is an outcome but one that becomes a "lex" which "retro agit"), then the answer is bound to be positive. An artistic object is a candidate whose author wants him to be admitted to the Popperian "third world" of cultural products.[5] It will be there even if "falsified" as a candidate (i.e., even if it is proven that no aesthetic concretizations of individual experience can be traced back to it). However, if it is admitted as an object genuinely "capable" of triggering, enhancing or even generating aesthetic experiences, its position in the relative and historically mutable hierarchy of objects in the third world will change. And since the hierarchy of values which is being used in all intersubjective acts of judgement has to be anchored in this Popperian "third world," most of the artists, critics and conoisseurs of art will try – other things being equal – to rely on higher-ups in the hierarchy of artistic objects and to report their experiences as linked to higher-ups or objects deserving of being moved up in the above-mentioned hierarchy. *Ars longa* (at least the "ars" I admire), and *vita brevis* (then why waste it on inferior "triggers" of experience). But since everybody is in on this game, the ultimate outcome is bound to be random, unforeseen, emergent, coincidental, like the posters envisioned by William Burroughs in "Blade Runner": "FREEDOM TO MUTATE" and "ORGONE ENERGY NOT RADIATION" (Burroughs, 1979, p. 67).

There are two significant consequences of the admittance of this democratic, egalitarian, anti-hierarchical procedure for selecting artistic objects for a potential aesthetic experience for single human individuals. First, those who have a vital stake in having their object accepted try to aggregate human individuals through the anticipation of a common aesthetic denominator (let us call it the Hollywood principle). On the other hand, tolerance and liberalization of admittance criteria for artistic objects means that recognition at the gate-keeping point becomes increasingly difficult if not impossible: all that is solid art melts into an unspecific mix, an abstract expression of collective intentionality for triggering aesthetic judgements (let us call it the credit card principle).

The Hollywood principle of anticipating a common aesthetic denominator is being followed in order to lower the risk of negative selection of an artistic product by the largest possible number of individuals. Successful application of this principle means that numerous individuals decide to select the product which promises them a chance of experiencing an aesthetic "pleasure." Since the number of individuals has to be large, the common denominator which would appeal to all of them is bound to be fairly simple and non-specific. The appearance of Rambo on the American cultural scene[6] meant that a common aesthetic denominator had been found in a cross-breeding of the countercultural ideal of a hippie with the conservative ideal of a country-loving tough Marine. Rambo feeds on natural foods and wears a bandanna on long hair, but serves in special forces and obeys his superiors. We can see that the search for a common aesthetic denominator is being conducted by mixing the ingredients of the product after measuring their expected symbolic appeal. The quality of the aesthetic experience remains beyond the reach of those who do the mixing – it is irrelevant to them. Once a potential viewer, reader, "client" pays to acquire the artistic product, all that matters is that he is not disappointed (i.e., that all the promised symbolic rewards are there – countercultural myths of closeness to nature, entertaining suspense and surprising action, etc.), or that he is at least ready to try the formula again. A similar argument can be made with respect to "Schindler's List," where a story of a small-scale German business adventurer is set against the background of Nazi genocide. A common aesthetic denominator is an expectation of profound interest in the possibility of "poetry after Auschwitz," as Adorno had once put it. The very possibility of a fascinating, "entertaining" Hollywood product which tells of one of the two most inhuman systematic genocides of our century introduces an appealing thrill (although in this particular case the mixing of the historical and mythological dimensions does provoke some resistance; the market success of Rambo films far surpasses that of "Schindler's List").

The principle of anticipating a common aesthetic denominator means that ingredients are gathered in many areas of culture, not necessarily in art alone. This generalized search does not remain without consequences for the mode and locus of existence of aesthetic concretization: we are systematically trained to attempt triggering aesthetic experience with landscapes, products of industrial design, photographs, concepts of potential works of art, public buildings packaged in plastic foil, etc. The situation is not unlike the gradual replacement of gold with paper money – the value of rare metal has been generalized in purely symbolic objects and they, in turn, have been generalized into purely symbolic, "contractual" means of exchange:

> Bankers would accept gold and store it for safekeeping, and in return for the gold they issued paper certificates to the depositors of gold. The certificates then could be used as a medium of exchange, just as the gold itself was. The certificate was a kind of substitute for the gold. It had complete credibility as an object of value, because at any point, it was exchangeable for gold. Commodity money had thus been replaced by contract money. A stroke of genius occurred when somebody figured out that we can increase the supply of money by issuing more certificates than we have gold. [. . .] The next stroke of genius came when somebody figured out [. . .] – we can forget about the gold and just have the certificates. With this change we have arrived at fiat money and this is the situation we are in today. (Searle, 1995, p. 43)

The process of the ephemeralization of art is similar to the process of disconnecting bank notes from gold reserves in that human individuals are being socialized (their standards of taste are being influenced) so that their aesthetic experience can be triggered with a much broader range of products. The term "much broader" has two meanings. First, the class of accepted artistic objects is broader – the novels of William S. Burroughs are included along with the novels of Thomas Mann. Second, artistic objects have lost their monopoly on triggering aesthetic experiences – reality itself has found in a "para-aesthetic" situation (Golaszewska, 1984, pp. 234–250). Let us consider the broadening of the class of artistic objects capable of triggering aesthetic experience in the latter of these two senses.

Golaszewska discusses four types of the "aesthetization" of reality, i.e., of our being brought to perceive the real world as if it were made of artistic products. First, there is the purely intentional, emotional, individual "imaginative comment" on reality which enables us to focus on significant structural features of our environment. Second, there is the discovery of the harmony and beauty of a natural or artificial environment (high mountains, big city). Third, there is the profound revelation of a meaning of the world which constitutes/reveals an outline of values present in the world (Golaszewska says that this type of aesthetization results in the making of "an axiological object"). Fourth, there is the exhortation to action (rituals, ceremonies, gestures).

The first and the second type of aesthetization coexist with the traditional concept of art. The fourth has been discovered by the dictatorships of the present century. It is the third type of aesthetization which makes the concept of art as moral gamble possible. Is the king naked? Can even the subtlest critics distinguish between aesthetically pleasing and aesthetically incompetent renderings of a symphony? What if non-aesthetic loyalties prevail? Nobody wanted to publish Samuel Beckett's prose before he won a Nobel Prize for a theatre play which got staged by sheer coincidence.

With the conceptual artists we have reached the stage of the professionals' refusal to deliver material objects which could serve as "concretization" sets for aesthetic experience. Conceptual artists were, in fact, saying that designing, inventing, conceiving, thinking a work of art and communicating this scheme, this intellectual design, this thinking procedure to the potential amateurs of aesthetic experiences will bring about the same effects as the production of material art objects. Golaszewska correctly observes that aesthetization is being identified with artistic creativity and a "reader," "viewer" or "receiver" of aesthetic experience becomes a co-creator or even a creator, while: "aesthetization as a process becomes much more important than its results" (Golaszewska, 1984, p. 243).

Golaszewska rejects the possibility of reducing the aesthetization processes to more general axiologizing ones.[7] However, she notices that the mode of participating in the process of aesthetization has changed: the role of the reader, to paraphrase Eco, becomes more important, and the role of a creator *sensu stricto* becomes more mediated and modest. Golaszewska has introduced the term "perceptor" of a work of art. One of the possible explanations of this changed, increasingly important role of the "perceptor" in modern aesthetization processes can be found in the radically anti-authoritarian trend of the present century's cultural processes (of which the aesthetization and the axiologizing processes are important parts). Ingarden had a premonition of this trend when finishing his Oslo lectures (Ingarden, 1974), since having commented on the idea of an "immanent perception" (and having concluded that it is open both to the danger of *regresus ad infinitum* and *petitio principii*), he focussed on Husserl's thesis that "reality is essentially non-autonomous." According to Ingarden, Husserl meant that reality depended in a crucial way on pure consciousness – and that this was exactly its fundamental mode of being, as opposed to the mode of being of pure consciousness.

One may venture a hypothesis that Ingarden's focus on a non-autonomous nature of reality manifests his uneasiness about the necessity of continuing Husserl's analyses of "experiencing" – without any external guarantees of the scientificity which both Husserl and Ingarden still hoped for in the twenties and thirties. For this is the real difference between the founders of phenomenology and their students (Golaszewska studied under Ingarden): for Husserl and for Ingarden[8] there is no gambling in constructing the world and ascribing values to it. Even if the process is complex (aesthetic concretization as elucidated by Ingarden), it is theoretically possible to become "an ideal perceptor" chasing "an ideal artist." Hence no gambling has to be involved and the gambling that actually takes place is due to the insufficient competence of artists and perceptors. The present perception of the existential situation of both an artist and a perceptor is different. It is not only insufficient knowledge (a deficiency which can, in principle at least, be remedied), which is at play, but also a continuous changing of rules of the game by participants in the processes of aesthetization. Hence not only Mann, but also Kafka, not only Kafka, but also Beckett, not only Beckett, but also Pynchon, not only Pynchon, but also Burroughs, not only Burroughs. . . . With modern art, participants in the aesthetic experience, including artists and "perceptors," are allowed to gamble on unexpected choices, increased networks, but also on rules of the game, on the level of abstraction of the symbolism involved, on the directness of authenticity conveyed, on the limits of acceptability involved, on the making and unmaking of axiological and aesthetic objects and of the contexts in which they are being experienced.

In his *Sentimental Journey* (through France and Italy), Lawrence Sterne compared the French and the English claiming that the French were more amiable and sociable – they resembled shining coins, which have often been used and whose value was well known. The English, on the other hand, were rough and less sociable, resembling rare old coins which have been circulating so infrequently that nobody knew what they were worth. It is quite possible that modern aesthetic experiences are part and parcel of processes so broad and varied that some of our cherished works of art can appear like those rare old coins to the collectors of others. The only feeling of stability comes from our awareness that as long as we participate in the aesthetization processes – there is art rather than nothing.

Erasmus University,
Rotterdam

NOTES

[1] One can ask if the analysis in question should not begin with a discussion of Heidegger's concept of the role of a philosopher as a "shepherd of being" with an artistic, linguistic and philosophical "German shepherd" to work with. It has been my deliberate decision not to do so, since I think that Heidegger's stream of post-Husserlian analysis begs the question precisely at this juncture. First, because Heidegger appears to have trusted poetic intuition, second because he applied this trust in 1933. Therefore Ingarden offers a much more promising starting point.

[2] This fundamental study in the aesthetics of literature had been first published in German as *Das literarische Kunstwerk: eine Untersuchung aus dem Grenzgebiet der Ontologie, Logik und Literaturwissenschaft* (1931). The Polish translation *O dziele literackim: Badania z pogranicza ontologii, teorii jezyka i filozofii literatury* appeared in 1960, and the English one *The Literary Work of Art: With an Appendix on the Functions of Language in the Theater* in 1973.

[3] Rousseau asked if there is progress in the arts, sciences and morality, and the question was quite relevant to his contemporaries blinded by the superstitious Enlightenment. However it was also incorrect, since by asking it he only confirmed the distinction between the material irreversibility of time and causality and the essential reversibility ("convertibility") of mental and cultural processes. A correct question would have demanded an explanation of the desire to have, experience and contribute to progress as a metaphysical midwife of evolutionism. The answer he gave to the question was basically correct: while progress in science and technology, although slow and ambiguous, can be observed from a number of points of view, it would be hard to repeat these observations in the arts and the situation in ethics is far from clarity, too. However, this answer is also wholly irrelevant when it is a potential methodology for judging change that is sought.

[4] My most unusual encounter with the Burroughs option took place in March 1994 in the Polish, northern Sudety Mountains. When rain had started in Karpacz, I decided to walk up towards the well-known skiing lodge there overlooking a mountain lake. My calculations proved correct; having walked above the rain cloud level I found myself quietly ploughing through snow towards the lodge. Rain kept the other clients away and

enjoying my meal I started chatting with a twenty-year-old girl, a student of sociology from nearby Wroclaw. Having few clients on this day, she was reading William Burroughs' *Junky* and enthusiastically told me how great new mind-boggling drugs were and how the postcommunist freedom made it possible for Polish youth to open themselves to them. . . . I thought about Aldous Huxley, Timothy Leary, and finally repeated to her an acid comment of Burroughs from his visionary forerunner of a film script for Ridley Scott's film "Blade Runner": "For man has no future unless he can throw off the dead past and absorb the underground of his own being" (Burroughs, 1979, p. 6).

[5] I have mentioned the Popperian "third world" because it is the latest of European attempts to continue the Husserlian dream of philosophers serving as the functionaries of mankind's cultural treasury. It has been relatively quickly dismantled within the realm of the philosophy of science (Kuhn, Lakatos and Feyerabend precipitating this process to a point at which the very existence of a philosophy of science has been threatened). Needless to say, Popper's concept of intersubjectivity ("epistemology without a knowing subject") does not make him a philosophical follower of Husserl's phenomenology.

[6] The appearance of Rambo signifies the rebellion of the young, well-educated, academic and media-employed intellectuals against their postmodernist, avant-garde, high culture artists-priests. Rambo has been created by David Morrell, who kindly helped me in 1973 by supplying me with a then still unpublished typescript of his critical study of John Barth's fiction. John Barth went on to become a "Prince of PoMo" (PostModernism), while Morrell became a formula writer applying the Hollywood principle to produce successful pulp fiction.

[7] She is tempted by this possibility and devotes to it the following footnote: "Would it not be better to speak of the "axiologizing" of reality in lieu of its "aesthetization?" It seems to me that the term "axiologizing" is very general and abstract, does not appeal to imagination, and poses numerous philosophical problems. Selecting the structures which function in art, susceptible to aesthetic and artistic processing, we stress the fact that they have achieved the highest volume and capacity with the lowest generality and abstraction. We also think that art is not foreign to life, that in observing life from an aesthetic point of view, we do so in a commonsense meaning of the word, that it is rooted in our natural modes of relating to the world. Finally, we also think that many para-artistic structures are shared also by moral sense (e.g., harmony, which we search for in art, nature, but also in moral life, coherence – valued in art as purity of style and fidelity to a genre or convention, the equivalents of which in real life include sense of responsibility, perseverance in upholding principles, etc.)" (Golaszewska, 1984, p. 236).

[8] One can interpret some of Ingarden's later lectures as symptomatic of his general uneasiness about this rigorous exclusion of gambling from aesthetics. In the XIth Lecture (May 10, 1960), Ingarden acknowledges the necessity of human individuals' searching for a community of values, to overcome an essentially individualistic mode of experience. And he comes close to Andy Warhol's prophecy that due to TV everybody will be famous for fifteen minutes: "And another need is also involved – the need to overcome the passage of everything, the passage of time. Briefly speaking, I will not repeat after my Washington friend 'every human wants to be god,' but I do think that every human wants to overcome the transitoriness of his own life in some way. Since a man knows that he has to die, that he will cease to be, he wants to leave something which will survive him, even if only for fifteen minutes, and somehow will remain" (Ingarden, 1981, p. 184).

REFERENCES

Burroughs, William S., *Blade Runner* (a movie) (Berkeley: Blue Wind Press, 1979).

Golaszewska, Maria, *Estetyka rzeczywistosci* [Aesthetics of Reality] (Warsaw: PAX, 1984).

Ingarden, Roman, *Wstep do Fenomenologii Husserla* [Introduction to Husserl's Phenomenology] (Warsaw: PWN, 1974).

Ingarden, Roman, *Wyklady z estetyki* [Lectures on Aesthetics] (Warsaw: PWN, 1981).

Ingarden, Roman, "O odpowiedzialnosci i jej podstawach ontycznych" (a Polish translation of "*Über die Verantwortung. Ihre ontischen Fundamente*," "On Ontological Bases of Responsibility") in *Ksiazeczka o czlowieku* (Krakow: Wydawnictwo Literackie, 1987).

Ingarden, Roman, "Artistic and Aesthetic Value. The Problem of the Relativity of Values," in *Ontology of the Work of Art* (an English translation of *Untersuchungen zur Ontologie der Kunst*) (Athens, Ohio: Ohio University Press, 1989).

Popper, Karl R., *Objective Knowledge. An Evolutionary Approach* (Oxford: Clarendon Press, 1973).

Searle, John R., *The Construction of Social Reality* (London: Penguin, 1995).

Vivante, Leone, *Essays on Art and Ontology* (Salt Lake City: Utah University Press, 1980).

NANCY CAMPI DE CASTRO

THE CREATIVE VIRTUALITY OF NEGRITUDE IN BRAZILIAN LITERATURE

In order to place negritude within the sphere of Brazilian literature, the notion of black literature must be established, and its existence in Brazil debated, as negritude and black literature are complementary concepts.

From a descriptive point of view, "black literature" is a semantic composite, made up of a substantive unit (literature) and an adjective (black), which, independently of their unitary values, start to interact once they become part of the whole, demonstrating our fundamental need to define explicitly the meaning of each term, in order to be able to read the aggregate.

Setting to the side all the theoretical discussions on the concept of literature, the term "literature" will be employed here as a semeiotic system that materialises by way of the literary text. On the other hand, how can we define the term "black," when used qualitatively? If literature is a semeiotic system that materialises through the written word, black literature would materialise through the black written word. And what would be the black written word, or "black literary text?"

Descriptively, the possible existing forms of the black literary text are as follows:

(a) the literary text produced by writers with physical/racial characteristics typical of the biological group known as black;
(b) the literary text that is the vehicle for black issues, but is however produced by writers without a direct black ethnic link;
(c) the literary text produced as a vehicle of black issues and written by black writers.

These three possibilities are present in world literature. The first one is represented by black literature practised by writers from the black biological group, regardless of theme, as in the case of Wole Soyinka, the African winner of the Nobel prize; the second possibility is found in that black literature stamped with the subject of the black man, but written by whites, or half-castes, or by one of any other ethnic origin, an example of which could be Eugene O'Neill, in American literature; the third possibility is embodied in the literature that is truly black: that written by black writers, exploring themes reflecting their own ethnic

A-T. Tymieniecka (ed.), Analecta Husserliana, Vol. LIII, 49–58.

origins, such as the creative poetry of Aimé Césaire, the West Indian author.

There is no way one can address the problem of black literature just by using objective examples, or with only a descriptive focus. In truth, black literature is much more than this. From a descriptive viewpoint, as applied to black literature, we pass then to another, the normative.

In this way, from another theoretical perspective, intrinsically geared to the aspect of process, that is, facing the question "How is black literature made?," it is impossible not to recognize and to reevaluate a factor inherent in black literature, that which is represented by the word "negritude." And on speaking of negritude, it also becomes impossible not to consider the work of Leopold Sedar Senghor, whose writings gave the word "negritude" world-wide recognition.

His critics maintain basically that Senghor is the author of an "anti-racist racism," for his having resuscitated Black-African art and thinking. Analysing "negritude" from the Senghorian point of view, and going back in time, we see Senghor at age 22, in 1928, in Paris; meeting with Picasso, and the painting "Les demoiselles d'Avignon," inspire the poem "Black Mask"; later, in 1934, comes the dissident publication "L'Étudiant noir," which unites Aimé Césaire and Léon Damas and other West Indian authors, such as Étienne Léro and Jacques Roumain, and the Madagascans Jean-Joseph Rabéarivelo and Jacques Rabémananjara, all who would be reunited by Senghor in his *Anthologie de la nouvelle poésie nègre et malgache*, a book for which Sartre wrote the controversial essay "Black Orpheus." Sartre himself had said that this poetry was "the only revolutionary poetry of our times," reaffirming Senghor's contention, that it was poetry with an express functionality which brought back a real or mythical past: the flow of the past within the ever-changing present. In 1959, at the 2nd Congress of Black Writers and Artists, in Rome, Senghor presented theoretical texts such as "constructive elements of a Black-African civilization," in which one can discern the limits of the Senghorian concept of "negritude," which, according to his critics, may have been in the decolonization movement an original and effective discovery, but can not now solve the problems of the modern independent African states. This negritude of Senghor's searched for a point of confluence, an opportunity to rebuild dignity in the wake of a reevaluation of a massacred culture, giving birth to a true battle cry: "The ancient tribe declaims 'assimilation'; we reply 'resurrection.' " Senghor would have fixed the point of convergence of his theory in the nation, and not in the social classes, or in doctrines or ideologies. For Senghor, negritude would be a cultural heritage, the values and, above all, the spirit of the Black-African civilization. It was necessary to oppose the original nostalgic negritude, attempting to embody the Black-African culture within the reality of the twentieth century, as an instrument of liberation, placing it in its moment and within all its potentiality. The black intellectuals of the British sphere, however, severely criticised what they called "French theorising," a sentiment epitomized in Wole Soyinka's well-known declaration: "A tiger doesn't proclaim its tigerhood, a tiger springs."

The Senghorian concept of negritude had evolved during Senghor's mature phase, to the point of encompassing even the economic modernisation of the continent. In 1971, his participation in a colloquy on negritude was aimed at this culture, with priority status within the overall theme of the project being given to socio-economic transformation.

Senghor would, in this way, be utilizing the theoretical texts of Marx and Engels, although from the perspective of his country and his continent. The negritude practised by Senghor, however, is seen by his critics as a raw racism, because it only serves to reaffirm the assimilation and submission of the black race to the white, for his political maneuvers differ from his ideas; he has assumed the role of champion of the French presence in Africa, placing himself, along with Aimé Césaire, as the person responsible for assimilationism and for maintaining colonial bonds. There is dissent also, among the critics, regarding the theoretical links between Senghor and Marx and Engels; according to some, perhaps Senghor would have stronger links with cultural theorists such as Spengler, who spoke of superior and inferior cultures.

This much questioned negritude was defined by Sartre, in *Orphée Noir*, as "l'être-dans-le-monde du Noir," using Heideggerian terms. Sartre himself, in the same text, explains what this negritude means: "une manière définie de vivre son rapport au monde qui l'entoure, (. . .) une certaine compréhension de cet univers, (. . .) une tension de l'âme, un choix de soi-même et d'autrui, une façon de dépasser les données brutes de l'expérience, bref un projet" (p. xxix). For Senghor, this would represent the spirit of the African civilizations, or "a set of values from the black world." What would these values be? Scholars of black culture point to marked characteristics, such as the rhythm and the symbolism expressed in religious and artistic manifestations; the conjunction of black mind with the cosmic forces, without identifying it as pre-logical, or alogical, but which is not forced to run the track of traditional logic; and, above all, solidarity, the sometimes inexpressible link between black and black, as if the primitive clan was becoming gigantesque, encompassing all the blacks in the world.

In truth, negritude is not a static concept and, consequently, it's in a constant flux of debate. Other categories might be implied within its scope, principally that of culture, revealing for instance that the North-American black belongs to something that is internationalised, that is, the North-American culture, and the same applies to the West Indian black. Another category is that of religion; there is strict separation between social classes; and then there are the imposed languages, as carriers of unrenounceable cultural forms; and then there is still, and essentially, the existent tribalizations in Africa that contribute to the creation of the black community. All these categories together do not dilute negritude. Since its own category corresponds to that of "race," they actually enlarge negritude and raise questions, lifting it to a situational superexistence in which it will continue to function as negritude, not a conservative negritude, which works towards colonial distortion and paralysis, but a negritude of liberation and of the dignity of man's humanity, a negritude that reveals the black face of the mass of the world's population.

It remains for us to touch on the possibility of the existence of black literature produced in Brazil, from a normative and descriptive point of view. In order to make a proper diagnosis of this possibility, a rapid retrospective view of Brazilian culture is indispensable, as well as a survey of the trends in literature within the current Brazilian cultural scene.

So the basic questions are: What is the role of the black in the formulation of a distinctly Brazilian culture? Was the supposedly peaceful cohabitation of blacks, half-castes and whites always a truly peaceful cohabitation in Brazil?

The role of the black is fundamental to the constitution of Brazilian culture. However, it is not certain or true that this cohabitation has always been peaceful. Laying aside the whole horrific history of slavery, lived through in Brazil with the same inhuman stains that were part and parcel of it in other countries, the black had to employ all possible means of combat to gain his freedom on Brazilian soil. Ever since the seventeenth century, in Rio de Janeiro, the "slave-catchers" were responsible for repelling invasions by the inhabitants of the "quilombos" – slave refugees, who came down from these settlements in the Organ Mountains to attack the suburbs of São Cristóvão and Inhaúma. The longest attempt at self-government by blacks, known as the Palmares episode, covered a period of sixty-five years of incessant fighting. The eighteenth century will be remembered for the resistance of the "quilombos" in Carlota in Mato Grosso, São Tomé in Maranhão, Cumbe in Paraíba; 1751 saw the massacre of blacks in the "quilombo" of Rio das Mortes by the pioneer Bartolomeu Bueno do Prado, a massacre proved by 3900 pairs of black ears. The nineteenth century witnessed the peak in the number of insurrections by slaves – who had been suffering under slavery since the end of the sixteenth century when it was initiated by the Portuguese. The Dutch are responsible, just in a period of 9 years, from 1636 to 1645, for bringing a contingent of black slaves into Brazil when it became recognised that sugar production was not possible without black slave labour. It has been certified that around 1850 there were in Rio de Janeiro 660,000 blacks to 37,000 whites and that in Bahia one could count 19 blacks for every white, proof of the existence of large groups of the black race on Brazilian soil. Then in the nineteenth century, in the midst of insurrections in Maranhão under Manuel Balaio, and from the "quilombo" between the Tutóia and Praiá rivers under Cosme, the self-proclaimed "Protector and Perpetual Defender of Brazil," and with slaves escaping from the Paulista ranches with the help of Luiz Gama and Antônio Bento, it became obvious that the country was in a state of economic crisis, as the agricultural sector was buried in debt. The total abolition of slavery in 1888 was preceded by laws abolishing the slave traffic, freeing newborns and septuagenarians. The monarchy was recognizing a de facto situation in which some of the more liberal slave owners had already given freedom to their slaves, converting them into salaried workers. Historically however, it is a fact that the 1888 law freed around 720,000 slaves living on Brazilian soil, representing a minimum loss to their masters of approximately 500,000 "contos de réis," that being the purchase price of the number of slaves freed.

This immense black mass has coloured Brazilian culture, endowing it with a new tonality, an unceasing process continuing from the sixteenth century until the present day. Natives of the African west coast, the Gulf of Guinea and of the southern part of Africa, of Bantu origin, the Ardrenses came to Brazil, people known for being "lazy, stubborn and anarchist"; the Calabrenses, "bad workers"; the Angolans "argued to be good workers," as well as Congos and Sombrenses. The Angolans stood out from the start as teachers of the Portuguese language to other blacks, whom they easily manipulated. Those from Guinea were "pugnacious, hard-headed, bad workers," although remarkable for their neat appearance and physical beauty, for which reason they were used as domestic servants, which fostered an intimacy with the Brazilian family to the point of interbreeding and the production of illegitimate children, the start of a long history of miscegenation. Those from Cape Verde were also noted for physical beauty, and those from the Cape, the Congo, São Tomé, Sierra Leone and other places were of varied physical characteristics and competence in work. There was however, one common point of convergence – their culture, appearing in various stages of evolution, and coming like an impetuous torrent to enrich the waters of the developing culture in Brazil. Joining the black heritage with Brazilian culture, the process of acculturation commenced, a process of mixing and transformation by the black cultures of Brazilian culture as a whole. Religious syncretism also took place, another aspect of acculturation, manifest today in the Afro-Bahian religious beliefs. There are as well obvious cultural traces in popular festivals, in music, dance, and language, signs of the massive cultural blending that has been achieved in Brazil. Among the names of scholars of black culture, one should remember that of Brazilian Raimundo Nina Rodrigues, a doctor from Bahia, considered to be the first doctor in the Americas to dedicate himself to the study of the presence of the black person in the New World.

Occurring as it has in this immense cultural cauldron, the course of Brazil's literary development needs to be followed. A history of Brazilian literature could be synthesized into three great moments: a long start, a period of formation and then full autonomy. The long start stretches until the appearance of the Arcadianism of the Mineira School, when Portugal still was the arbiter from a literary point of view. During the period of formation, with its waves of Romanticism, Realism and Symbolism, the use of a Brazilian language and Brazilian subjects for literary expression began to develop. But full autonomy was only attained as of 1922, with the Week of Modern Art in São Paulo, a public event, when the artists united the landscape, the picturesque Brazilian scene in the first instance, and knowledge of Brazilian man, in the second instance, and further conducted research into the tools of literary expression. They were responsible for the genesis of a Brazilian literary language, one both artistic and autonomous.

How has black literature manifested itself during the course of development of Brazilian literature? The first moment in Brazilian literature, the long start, did not produce any black literary works that are worthy of mention, in any of the three modes of "black literature" presented in this work. What did appear was an enormous strength in the area of folklore, derived from an oral literature hard to recall and difficult to identify, awaiting the work of plagiarists, as it became widely disseminated even in the literary arena.

The second moment simply shaped itself as the moment of Brazilian national identity, principally through Romanticism. In literature with a black theme, a brilliant young romantic poet shines – Castro Alves, who, among other things, is a militant in the black cause, an abolitionist, taking up the "Voices of Africa," questioning even a divinity who appears to fail when faced with the horror of slavery: "God! Oh God! Where are you that you do not reply!/ In what world, in which star are you hidden, folded into the sky?/ For two thousand years I have called out your name . . ./ on what purposeless mission, since then, do you traverse the infinite . . ./ Where art thou, Lord God?" It is the black theme that will inflame the verses of the young white poet. Cruz e Sousa is another voice lifted in favour of free expression, but his is the voice of a black man. As a black Brazilian, Cruz e Sousa finds, in Symbolism, in the musical sound of verse, in its pure-music-poetry, the means of expression for the liberation of the race. Considered by many critics to be "a coloured Baudelaire," Cruz e Sousa rarely makes use of aspects of black issues in his poems, being in the words of Sílvio Romero, "The only case of a black, a pure black, truly superior, during the evolution of Brazilian culture." Could it be that Cruz e Sousa is "superior" precisely because he did not bind himself to a black theme?

The third moment, the moment of full autonomy, which lasts to the present day, is the most creative period of Brazilian literary development, but still with little significance in terms of black literature. In this third moment of Brazilian literature, there are examples of white writers covering black issues, such as Jorge Amado, who uses a black as the subject of the narrative in some of his writing; and Vinicius de Morais, a singer of black themes in the sphere of the literary, theatrical and musical text, perhaps the "blackest white man in Brazil." His *Orfeu da Conceição*, a theatrical play, which was presented in 1956, and produced as a film in France, with the title of *Orphée Noir* under the direction of Marcel Camus, won, in 1959, first place at the Cannes Festival. There are also some black poets, writing on a black theme or on general poetic themes. Would these literary expressions be enough to speak of black literature in Brazil? One believes not. And there exist factors that explain the non-existence of a black Brazilian literature.

According to some critics, there are in black literature, practised by black writers with black subjects, and which is distinguished by the mark of negritude, the following phases: an initial phase of rejection of negritude, generating literature that conceals negritude; a second phase, in which the literature turns in on itself in an examination of black identity, generating a phase of solitude; a third phase in which the literature revolves between the white and the black environments, generating the recognition of kinship; and a fourth phase, that in which one feels the harmony between the two worlds, the exterior and the interior, generating literature in close intimacy with negritude. The progression of the four phases, as observed in world literature, for example, in North America, has not occurred in Brazilian literature, however, due to factors that are of a distinctly social nature.

When the panorama of Brazilian culture was initially described, commencing with the sixteenth century, attention was called to the non-pacific relationship between blacks and the rest of the population. The situation of slavery in itself is to be characterised as a situation of oppressor and the oppressed, of master and slave. However, with the abolition of slavery, the legal validity of dominator and dominated came to an end on Brazilian soil. The cohabitation of the races that together formed the Brazilian people was already established, in part due to an irreversible mixing that was taking place in the bosom of the families that were owners of slaves. In fact, it is postulated that with the interruption of the black slave traffic, there were no longer any pure blacks on Brazilian territory, they being succeeded by mulattos, the mixed-race. We see, therefore, not only a process of acculturation in terms of a mix of various different ethnic groups, but also a true physical mixing, of white and black, of white and Indian, of Indian and black. The result of this process is a cohabitation which declares, in a loud voice, that "racial segregation does not exist in Brazil." In fact, there is no racial segregation in the true meaning of the isolation of a particular race to the detriment of another. However, discrimination does exist: subtle differences are established, creating a certain feeling of distance between humans, sometimes effected just by the manner in which one looks at a black when he is present in certain more sophisticated environments.

This black, however, has "learned" to behave like a white. He will remain in the sophisticated environment and will be "equal" to the whites. One observes, therefore, in Brazil, a "decolorization" of the black: the black becomes cultured, he mixes, but always keeps his distance from negritude, getting nearer to the white, imitating him, copying him, equalling him in his manner of dressing, of talking, and competing socially.

The Brazilian reality is expressed then in a Brazilian who is, today, original and authentic in many aspects, giving validity to a culture marked by a search for social and racial synthesis. It cannot be said that the pure black race does not exist in Brazil or that the Brazilian people are an enormous group of millions of mixed-race people. But this very mixture, be it racial, social or even cultural, can be observed in the literature. One can talk, then, of literature *of* Brazil and not of literature *in* Brazil. The first is a prototype of a cultural miscegenation, persecuted for 495 years; the second would be characterised as racially and socially distant, transforming the common cultural soil into sparse islands and making interpretation difficult.

A way was found for negritude in contemporary Brazilian black vocal music – through, in essence, poetry, as the poetic is universally lyrical: the "I" exists, and what he sings is a song of solitude or ingenuity in the face of the world. One cannot observe the phases of the literature of negritude in sequence; one sees however a going and coming of literary manifestations, in one direction or another, in isolated texts, the majority, poetic. There are no substantial works or works which stand out in contemporary narrative, just as there are no theatrical or film texts which are dedicated to the lyrics of negritude in the Brazilian literature of today.

If we wish to consider the future possibility of a literature of negritude in Brazil, what one might observe in the artist is the modelling of the human characteristics of negritude, with a consequent universalization of these models, or in other words, the mark of negritude, in the form of a model, in Brazilian literature, projecting itself, adding itself to and superimposing itself on universal models, already contributing profound vitality to the world's literature.

Today one speaks of the planetization of the arts, generating a production system of planetary proportions, one more than universal and akin to the idea of the eradication of nations in conventional terms. One should, then, speak not of black literature, but of a literature of negritude – negritude that we should all accept, not simply as a colour of skin, but as an indelible, undeniable, irresistible stamp of human characteristics such as conflict, strength, aggressiveness, but also, to a great extent and so profoundly, attributes such as subtletly, enchantment, seductiveness. How could Vinicius de Moraes, who, in 1953, already knew a homeland of negritude, divine out of space "Sweet intimacy and the desire to cry," "Great secular river, which drinks clouds, eats earth and urinates sea," the homeland of negritude, "with a drumbeat in the heart."

Universidade Federal de Juiz de Fora

REFERENCES

Sartre, Jean-Paul, "Orphée Noir." Préface, in: Senghor, L. S., *Anthologie de la nouvelle poésie nègre et malgache* (Paris: P.U.F., 1948).

Senghor, L. S., *Anthologie de la nouvelle poésie nègre et malgache* (Paris: P.U.F., 1948).

ALEXANDER PIGALEV

BEING AND THE CREATIVE POWER OF LANGUAGE (THE LATE HEIDEGGER)

Martin Heidegger is considered to be the leading thinker of the twentieth century, and his ideas changed the categorial structure of European philosophy (and philosophy elsewhere too). He directed philosophers' attention to the problem of Being in its constituting the main problem of philosophical thinking. The ideas of the late Heidegger are marked by an interest in the role of language in the history of Being. And what is more, he emphasized the specific creative power of language as the factor of the human condition. Heidegger's notion of Being is rather obscure, and his interpreters do everything in their power in order to disclose its hidden dimensions. Although the late Heidegger is bound up with the general paradigm of "man's self-interpretation-in-existence" (as Anna-Teresa Tymieniecka formulates it),[1] the emphasis falls on Being itself. Language in its turn becomes the main form of Being's revelation.

Is Being a substitute for God? Can it be identified with the God of Christianity? Or is it another form of transcendence? Irrespective of our answers, a comparison between Being and Life with due regard for the creative power of language can give us a new horizon.

Anna-Teresa Tymieniecka writes of the "three movements of the soul" as the specific aspect of the human condition. She emphasizes that

> the soul liberates the dynamism of its fundamental yearning and passes beyond the anonymity of empirical on-going life. The second movement of the soul, *exalted existence*, leads to the discovery of human finitude and the "laying bare" of all contingency. It would be necessary, finally, to bring into play all the spontaneities of our being in order to "prime" the third movement, that which proceeds *towards transcendence*. Whether this take the form of an understanding of "transnatural destiny," of the "meaning of life" or even of "the Messianic task," it is in *communication with other human beings* that the soul laboriously forges the *sacred message*.[2]

But "the communication with other human beings" and, all the more, "the sacred message" are impossible without language.

There is no doubting that there are notions which somehow imperceptibly become the symbols of a whole spiritual epoch. "Life" and "phenomenon" are such symbols. First of all, "life" is connected with

A-T. Tymieniecka (ed.), Analecta Husserliana, Vol. LIII, 59–72.

various kinds of so-called "philosophy of life." It is the intent of this philosophy to consider life as a universal flux in which the polarity of subject and object is overcome. Nevertheless, the heritage of transcendentalism was not completely eliminated in this attempt and life was usually analyzed through the states of consciousness. Thus, life could be explored as an integral process which penetrates the world, as a total and omnipresent substance like the Being of Parmenides. But naturalism of a biological type, the primitive cult of brute force, does not differ in the main from this identification of life with the flux of history. Here and there life is interpreted as a principle of the world. We may even speak of a "pantheism of life." Here and there life in its objective and individual forms reveals itself through an interior experience which is based not on reason, but on intuition. The mystical background of the "philosophy of life" comes to light in its insisting on joining the irrational flux of life and in the idea that such a fusion as the final act of the cosmic drama is possible. It is obvious that life in this paradigm signifies an Absolute of a pantheistic (i.e., immanent) type which somehow correlates with the transcendental subject.

Indeed, when we speak of the phenomenology of human life, we usually implicitly assume that the notion of life is quite manifest. But really this notion is determined by the general spiritual atmosphere. This atmosphere, in its turn, is not a constant one. That is why the vision of the phenomenon of life also changes depending on the prevailing mentality. Modern culture is staked on the universality of philosophical knowledge. In other words, the phenomenon of life is considered to be absolutely transparent in the "light" of the philosophical reason (as is, however, everything in the world). Such an approach, being the product of the Enlightenment, does not take into account that philosophy itself is not the only possible mode of thought and understanding.

And what is more, according to the paradigm of scientific philosophy, we are compelled to interpret the phenomenon of life on the basis of strictly scientific data. The horizon of the Enlightenment in its interpretation of the phenomenon of life is illuminated by the star of biology. Philosophy often uncritically agrees to consider the strictly scientific, the biological interpretation of life, to be the only possible interpretation of this phenomenon. Of course, such an interpretation contains a grain of truth, but this is the limited truth of a partial world outlook, one which is rather new besides. Expressions like the "life of the spirit" become then beautiful metaphors. The spirit immediately turns into an epiphenomenon of the biological, i.e., of the "only real" processes. To be sure, this comprehension may be complicated by the notion of society. But society within the limits of this comprehension remains a part of nature, although it follows its own rhythm. Thus, nature is regarded as the only reality (and we have to make do with a kind of naturalistic pantheism).

Briefly, from the scientific point of view the phenomenon of life may be described as a rather mysterious process which came into existence owing to chance. Further, the phenomenon of life is considered to be associated with a certain bearer (subject). That is why the scientific understanding allows us to attribute life to different subjects – to an organism and to biological cells, for instance – and to not impute life to other subjects. Thus, life is understood as a freak of nature. But the phenomenon of human being, of *human* life escapes strictly scientific interpretation. Indeed, this interpretation insists on understanding life as a biological (or, according to the mediating scheme, mechanical, physical, chemical, etc.) phenomenon. It is obvious that such a comprehension has nothing to do with words and with language in general, which is the distinguishing feature of the human being.

On the other hand, the strictly scientific, the objective understanding of life is impossible without a certain metaphysical matrix. First of all, there is the clear distinction between subject and object. Further, there is the understanding of the logical and metaphysical ideas that are the deepest foundation of all knowledge. But we know perfectly well that we cannot identify these eidetic worlds with the life of plants, insects, animals and men. The ideas really belong to the sphere of dead entities. The whole metaphysical (i.e., philosophical) era believes in a certain scale of ranks, wherein metaphysics and physics are the first stages of all knowledge.

But Being and Thought as the main metaphysical notions cannot be the starting point for the interpretation of life, for they have actually nothing to do with the multidimensional phenomenon of life. And now we understand the role of the accurate distinction between the subject and the object in the scientific (i.e., "objective") comprehension of life. This is the distinction between *I* and *It*, that is, between living and inert beings. We could say "it" of a stone or of a corpse, but if we say "it" of a man or even of a plant, we turn them into "objects," into wheels of a great machine.[3]

Thus does the problem of life reveal its humanistic aspect. The distinction between subject and object as the *leading methodological principle*, as the only approach to reality, leads us to a perversion of human life. Of course, these notions – I mean "subject" and "object" – describe a certain reality, but only that of the abstract possibility of being master or slave. The very core of life, and above all, its temporal character, escape such an analysis.

The main feature of the European world view is the postulation of a center that bears the construction of culture and even of Being as a whole. The masks and the transformations of this conception are very diverse – varying from the Platonic and Neoplatonic One to Spinoza's Substance, from Hegel's Absolute Spirit to Heidegger's onto-theo-ego-logy in the structuring of European metaphysics. This latter conception allows us to designate the implicit constitution of the metaphysical matrix which serves as the generative pattern of European culture.

The structure of metaphysics is characterized by the postulation of a specific center as the attractive point, as the origin of a field of force, the general semantic equivalent and even the source of every sense. The modern French philosopher J.-J. Goux especially emphasizes that Occidental civilization may be formally defined by pointing to its organization of social elements through their subordination to some general equivalent:

> The Occident resolves the conflict of equivalents by the prevalence and the monopolization of supremacy. Thus, the Occidental social organism proves to be the superposition, the interlacing, the confusion or, on the contrary, the radical separation, the division of the numerous planes or levels which are meanwhile quite isomorphic as a whole and in essence and which are structurally organized through the predominance of general equivalents, around which a homologous system of values is disposed. . . .[4]

Thus, the European civilization is not an abstract "civilization in general" exactly insofar as it is based on a specific "centrism." And what is more, the monopoly of the Absolute in the symbolic framework means that the classical European culture is based on the principle of monologue. Every sense originates from the Absolute which does not freely reveal Itself, but simply continually radiates sense as the sun radiates light. On the contrary, the elements are similar to the moon, since they are giving reflected light. This means that the elements of the system are considered to be dead, since their flickering surfaces do not express the thrill of their own life. The "philosophy of life" has adopted this mental pattern when speaking of the will (Schopenhauer), the Will-to-Power (Nietzsche), the *élan vital* (Bergson) etc. as being the zero point.

The advent of Husserl's phenomenology with its idea of the phenomenon as the appearance which reveals itself by itself as it is, signified a breaching of monistic metaphysics and of transcendental philosophy as well. From now on there is no difference between appearance and its essence: the appearance is the essence itself. This means that every entity (i.e., every element of the symbolic framework) acquires its own voice, one which is independent of the Absolute. Dialogue has taken the place of monologue, and this theoretical paradigm has opened a new horizon for the philosophical interpretation of life.

It should be said in all fairness that Husserl's phenomenology was not the sole indication of the changed human condition. The idea of the polyphonic structure of culture was developed first of all by Dostoyevsky, then by the outstanding Russian specialist in literature Mikhail M. Bakhtin and by the remarkable Russian psychologist Lev S. Vygotsky. And, of course, it is impossible to discuss the principle of dialogue without mentioning Martin Buber, since he was a pioneer in the dialogical approach to the sphere of culture. It is dialogue, according to Buber, that reveals the *logos* of the universe and of human freedom as well.

But neither Husserl nor Buber renounced utterly the ideas of transcendentalism in its classical form. Husserl insisted on the privileges of the transcendental *Ego*. That is why he constantly encountered the problem of intersubjectivity which consisted in the analysis of the conditions of the mental relations between individuals. Husserl discovered already in *Cartesian Meditations* that the problem was very profound and that it concerned neither more nor less than ". . . *the transcendental theory of the objective world*. . . ."[5] In his *Crisis* the problem acquires special significance. Husserl discusses the "paradox of human subjectivity": here is at the same time a subject which constitutes the world and the object which exists in that very world. Husserl asks: "How must a component of the world, a human subjectivity, constitute the whole world, namely, in the capacity of its intentional image?"[6] Then he rejects the naive answer of "positive religion," saying that "the riddle both of Creation and of God Himself is the essential component of positive religion."[7] But the philosopher deals with the problems, and he must *understand* the paradox. According to Husserl, the *Lebenswelt* becomes the foundation of every science, since it allows us to unite the subject of the scientist with the world of objects. But the transcendental subject, though in modified form, is the central point of Husserl's discourse, and the "centrism" of the European mentality remains valid. Nevertheless, Husserl anticipated a profounder notion of life.

On the other hand, in Buber's theory the second person, *Thou*, is understood as a certain addition to *I* in order that the subject be able to carry on a dialogue. So, *Thou* is considered to be the counterweight to the distinction between the subject and the object. The reason for this is the focus of *I* (empirical or transcendental), and the whole world is described as an object. *Thou* cannot be understood as an object, however. This is the peculiarity of Buber's theory. But such a presupposition is unable to demolish the metaphysical (resp. transcendental) paradigm. We simply have two worlds: the first one is based on the relation between *I* and *It*, and the second one on the relation between *I* and *Thou*. These worlds are separated from each other, and they may be understood each taken separately.

The heritage of Husserl's phenomenology is very important for Heidegger, but during his late period he tried to discover the right way of getting beyond the limits of the paradoxical problem of intersubjectivity. What is Being according to Heidegger? This question is not unnecessary in the light of our discussion, since Heidegger's conception of Being is often considered by perfunctory commentators to be similar to the traditional one. But Heidegger's conception of the *ontological difference* makes us think that we have to do with a radical new approach to Being itself which is closely connected with the linguistic aspects of the problem. But is this approach really so new?

I proceed from the assumption that the reader is acquainted with the main principles of Heidegger's conception of language. In any case, there are many works which deal with that. The task of my reflections is not to spell it out on paper once more, but to discover some of its hidden dimensions in order to include this conception in the general tradition of Western thought. Indeed, the original and refined terminology which is used by Heidegger must not push into the background the real sources of his thinking.

I mean first of all the mystical tradition of German culture (Meister Eckhart, Jacob Böhme and others), which determined Heidegger's understanding of the Absolute and its relation to man. The very discernment of the difference between Being and beings is similar to the Gnostic mystics' conception of "two Gods" or "God beyond God." According to this conception, there exist two Gods – one God is entirely spiritual, radically transcendent and quite inaccessible to man, whereas the other

God is the creator of the world. But in contrast to the Christian conception, creation is equal to the joining of inert matter with the spiritual substance of the radically transcendent God. That is why human soul within the limits of this conception must be interpreted as the "spark" of the divine substance, as "God within us." And what is more, knowledge of the divine essence of our soul (*gnosis*) is considered to be sufficient for our salvation, i.e., for the restoration of the initial unity. Thus, the Divinity loses its personal character, and the fusion of man and Divinity becomes possible. I have good cause to think that the idea of the *ontological difference* is a modification of the idea of the *difference between two Gods*.

According to Heidegger, Being can be grasped only apophatically, through the exposure of a hidden reality that would not be being. Thus, Being is not being, it is No-thing. Hence, it is radically transcendent like the first God of the Gnostic mystics. On the contrary, beings are easy of access and they form, so to speak, the background of Being. Thus, the ontological difference means a state of eclipse of Being, and this state must be overcome. Within the limits of Heidegger's approach the revelation of Being occurs in the process of historical development, but the latter is a moving away from the initial unity, i.e., a kind of degradation leading to the forgottenness of Being.

So, contrary to Christianity and in full accord with the Gnostic pattern, history is understood not as a field of the divine Revelation, but just as a process of degradation leading to the fallen state of the whole world, since the act of creation is at the same time the original fall. Salvation can be understood only as the restoration of the initial unity, i.e., as the fusion of our souls and their source – the radically transcendent God, i.e., Being. Hence, an instrument that enables us to penetrate into Being must be found. Heidegger is certain that precisely language is that which makes the mysterious sphere of Being accessible. Nevertheless, language itself is no less mysterious, for only it is able to promote the restoration of the initial unity. This general hypothesis needs verification.

The word "Being" is ambivalent, since it is not only a noun, but also a participle and an adjective with a verbal sense. Exactly this fact allows us to understand Being as the Being in general examined by onto-logy and as the ultimate ground examined by theo-logy. For in the participle coincide two characteristics, that of a noun and that of a verb. With the help of the notion of participation Plato described the connection between concrete things and the ideas, since for him Being (the sphere of ideas) was participat*ed* in, and beings (concrete things) were participat*ing*. In a way, participation is a kind of correspondence, and the forgottenness of Being means disregard of the ontological difference. That is why, according to Heidegger, the sense of Being can be revealed only on the basis of taking into account the ontological difference.

We must emphasize that Being for Heidegger is not a constant, but a process, it is constantly in the making. Thus, Being comes to be seen as similar to the phenomenon of life itself with its inexhaustible ability for change and development. Heidegger in fact draws on Nietzsche's conception of Being as Life-force. Nevertheless, his position differs from Nietzsche's, for the latter tries to distinguish two aspects of Becoming – Will-to-Power and the eternal recurrence of the selfsame state of affairs. According to Heidegger's interpretation of Nietzsche, Becoming, the Life-force and the Will-to-Power mean one and the same thing. Thus, Being is a process which is quite similar to life, since both Life and Will are characterized by a tendency towards growth, domination and expansion. The Will-to-Power differs from wishing or striving after something: the Will wills its own willing and that is why it brings itself under its own control. The Will-unto-Power becomes its own master and, so to speak, overcomes itself.

It is well known that Heidegger tried to expose the history of philosophy as the history of Being itself. Metaphysics becomes for him a certain symbol of degradation, it means a deviation from the right path of thought. Nevertheless, language even in this situation is the main form in which Being reveals itself. It goes without saying that Heidegger's interest in poetry is closely connected with his interest in language itself. Language becomes the bearer of the joint experience of Being, and to think for Heidegger means to be a poet. The problem of language in its turn leads to the main question of Heidegger's thought – that of the meaning of Being.

The polysemantic Greek term *logos* presupposes that the Being of being is disclosed as a word. And the giving of a name in full accord with the traditional theories of language and of its origin means for Heidegger the opening of Being. Hence, a name is not a conventional sign, but the very becoming of things. The emergence of Being and the origin of language are identified in Heidegger's thinking. One could even say that language is considered to be Being itself which is formed into a word. But we have here a very paradoxical situation. The implicit understanding of Being is very important for Heidegger's conception. If we suppose that Being had no meaning and we had no implicit comprehension of it, we would not have language at all. Indeed, as Heidegger writes, ". . . speaking about being as it is presupposes the preliminary comprehension of being as being, that is, of its Being. . . . For to be a human being means to be a speaking creature."[8]

But does this mean that man (There-being) simply invents language in order to give names to surrounding beings? Or, in other words, is language an arbitrary and conventional system of signs which one can use as a symbolic framework superimposed on Being? Such a conception would be similar to the famous hypothesis of "linguistic relativity" which proclaims the leading function of language to be the structural organization of reality itself. To be sure, the temptation to follow this way of thinking is very intense, but Heidegger chooses another approach. *Logos* is declared to be the source of language and *logos* is interpreted as "gathering-together" (the Greek verb *legein* means exactly "to gather-together," and every reading is in fact a gathering-together of letters which designate sounds). But in this case Heidegger also follows Heraclitus who understood *logos* as talking and listening or, more exactly, as the inseparable relation between talking and attending. "The coming-into-the-Open of being takes place in *logos* as gathering-together. Originally it occurs in language. That is why *logos* becomes the determinative criterion of the essence of speech."[9] But there exists a difference between "attending" and "listening." Everyday talking and listening are connected with the ordinary needs, whereas "attending" is the directing of attention to Being, i.e., to the condition of empirical "listening" (in particular). "Mere listening scatters and dissipates turning into the everyday opinions, into *doxa*, into semblance. Genuine attending has nothing to do with our ear and mouth, it means docility towards that which is *logos*: *the gathered-together totality of beings themselves*."[10] And Heidegger defines his position more accurately: "Talking and Listening are genuine only if they are preliminary, already directed to Being, to *logos*. Only if *logos* comes-into-the-Open, the twaddle turns into the Word."[11]

So, we must seek the locus of the phenomena of Being and Language in *Logos*. But the result depends on how *logos* itself is understood. What makes the problem the criterion of the modern situation depends, in turn, on the method of the interpretation of European intellectual history. And we anticipate the path that was chosen by Heidegger: the claim to attend to Being itself through poetical works. Indeed, language must disclose the original sense of words, and only this activity allows us to turn man's way through beings into the way to Being itself. This is an attempt to restore the original freshness of language, to save the language of everyday decadence and degradation. Only after such a freshening does language prove to be the coming-of-Being-into-words.

Heidegger supposes that a "primordial poetizing" must be the initial state of affinity with Being, and the zero point of popular creative activities must be found. This is obviously a heathen and at the same time a Gnostic motive, since the initial state of historic development is considered to be the most "pure," and as history develops there is "pollution" of the initial diffuse unity. Hence, according to Heidegger, only at the very beginning of history did Being authentically reveal itself. The following centuries mean only a moving away from the original Principle. As we know, Heidegger used to repeat a dictum to the effect that the original thinker is a man who comes nearer to the origin.

It is very significant that Heidegger calls Being the "Holy," because the Holy is for him older by far than gods. Heidegger insists upon the thesis that the Holy is that by which the gods exist. Hence, the Holy is not equal to the Divine, since it is precisely thanks to the Holy that the Divine becomes divine. Being itself, in turn, is the Holy, because it is an endless origin. Thus, the poet must penetrate into the sphere of the Holy in order to come into contact with the initial unity that was later destroyed on the guilty earth by historical development or, in Heidegger's term, by the fate of Being itself. The poet's openness to the Holy is the result of his affinity with the latter. But this affinity is yet to be reached, since the Holy, being absolutely immediate, is not directly accessible to the poet's creative activities. Thus, the Holy reveals itself mediately, i.e., through the beings surrounding the poet.

But the word "surrounding" has different meanings. First of all, the immanent beings that surround the poet in the world must be mentioned. But they are not the only beings which are able to lead the poet to the sphere of the Holy. One can find specific beings which do not belong to the "world" and which are superior to the poet. These beings are within the limits of tradition called "gods." Gods (even in their totality are not equal to the Holy, they form, so to speak, one-sided substitutions for the entire sphere of the Holy. Thus, the Holy is mediated through beings-in-the-World and through the gods, i.e., the ontological is mediated through the ontic. Nevertheless, the poet himself in the process of his creative activities forms an artificial world. The creative power of language does not use arbitrary words, but actually creates them under the guidance of the Holy. But there is a danger here that consists in the temptation to identify the Holy with one of the gods (i.e., to identify the ontological with the ontic). According to Heidegger, real poetry must avoid and really avoids this danger, which would be tantamount to forgetting the Holy to please this or that god. Although it is precisely a concrete god that sets the poet on fire, excites his imaginativeness, such an idolatry hinders him from recognizing the Holy for being what it is. So, language must be considered to be a kind of response to Being as the Holy, and the poet *names* the Holy, i.e., Being, as a polyvalent plenitude. After all, we see that the conception of the ontological difference forms the basis of Heidegger's conception of language as the pathway leading to a creative power, since it creates the worlds that have an ontological status.

We see that Being and beings are correlative and that they mediate each other through language. Heidegger expresses this understanding speaking in plain terms: "Language comes-to-presence in the capacity of the scission that takes place between the world and the things."[12] The third movement of the soul, in Professor Tymieniecka's terms, i.e., the striving for transcendence, is impossible without language. Language is used just in the moment of man's free response to the greeting originating from Being: "Man speaks as far as he cor-responds to language. The corresponding is attending. It attends as far as it is the attendant of the greeting of silence."[13] Thus, the negativity of Being, its apophatic character culminates in Heidegger's exalting of Being as No-thing that reveals itself only in silence as the sphere of nothingness. But Heidegger says nothing explicitly about the structure of Being as it reveals itself through Language in the creative activities of the poet. It is possible even to identify it with the One of a Platonic type. In other words, we could conclude that Heidegger's approach, despite his original, obscure and, nevertheless, fascinating terminology, does not overcome the traditional "centrism" of European metaphysics.

It has already been mentioned that recognition of the heathen background of Heidegger's thinking has become a platitude nowadays; likewise its connection with the Gnostic pattern of thought. His opposition to Christianity does not embarrass the modern philosophers who are themselves busy at the destruction of Christian doctrine as the domain of European metaphysics (this is a very interesting problem taking into account the modern situation in culture). But I must emphasize the fact that not Christianity itself, but just its metaphysical framework is attacked. The "overcoming of metaphysics" has become the leading theme of modern and postmodern discourse. One tries to "overcome" something only if he finds himself to be hampered by this "something." How can metaphysics be the restricting instance and how can the "freshening" of language abolish the forgottenness of Being that was engendered exactly by metaphysics? In other words, why does metaphysics corrupt our life at the end of the metaphysical epoch? Why do we need to go back to its origins?

It follows from the previous discussion that the structure of metaphysics is closely connected with the structure of the Will. Within the limits of the metaphysics the Will is shaped after a model of the metaphysical "centrism." The "totalitarian" structure of metaphysics is responsible for the totalitarian structure of social relations. Thus, metaphysics appears playing the role of a pretersensual framework of being. So long as Being is identified with being, the structure of Being is also shaped by "centrism."[14]

But such a shaping of Being is tantamount to a certain schematization of Life itself that is based on the principle of monologue. In view of this interpretation we could understand the efforts of late Heidegger as an attempt to overcome the "centered" structure of European metaphysics *without abandoning the metaphysical realm*. Thus, it is a kind of "thinking on the boundary-line." It is interesting that Theodor W. Adorno who was a severe critic of Heidegger's philosophy finished his *opus magnus*, the book *Negative Dialectics*, with an enigmatic sentence: "Such thinking is solidary with metaphysics at the moment of its fall."[15] The ideas of late Heidegger were consistently developed by the representatives of poststructuralism. In fact here is a very paradoxical paradigm intended for the salvation of European "centrism" by means of its overcoming, developed in the conceptions of grammatology, deconstruction, schizoanalysis and rhizomatics. Indeed, the combination of pantheism and transcendentalism supposes that the transcendental subject is situated within the substance, since, according to the pantheistic point of view, there is nothing outside this substance. Such a condition must remain firm and the only way out is marginalism, i.e., the "absolutism of the margins."

The modern American theologian Carl A. Raschke contemplates the final point of this paradigm:

In the era of theology the divine voice is represented as the Word of God, fossilized within its own epoch of history. But the end of theology does not presage the silencing of the *logos*. The *logos* speaks anew no longer as the Word of God (in its mythico-metaphysical mask *per se*), but as the self-presencing or fullness of meaning-presentation dwelling within language. The self-presencing of language arrives through dialogue. Theology becomes dialogy.[16]

Raschke also remarks that "dialogue does not take place between words, but between living persons. Such is the eschatological significance of the theology of the *logos* become flesh."[17]

Our discussion shows that the ideas of Heidegger do not overstep the limits of the metaphysical tradition and suppose a paradoxical model of thinking within metaphysics. Nevertheless, his approach opens a new horizon concerning the creative power of language. I think that the obvious connection of poststructuralism with the ideas of the late Heidegger must be analyzed in detail, since the question is that of whether we really have come nearer to the moment of the consummation of metaphysics. The anti-metaphysical model of thinking has become an imperative need.[18] But this is the task of another investigation.

Volgograd State University

NOTES

[1] Anna-Teresa Tymieniecka, *Logos and Life, Book 2: The Three Movements of the Soul, or the Spontaneous and the Creative in Man's Self-Interpretation-in-the-Sacred* (Dordrecht, Boston, London: Kluwer Academic Publishers, 1988), p. xxiv.

[2] *Ibid.*, pp. xxiv–xxxv.

[3] See Eugen Rosenstock-Huessy, *Das Geheimnis der Universität* (Stuttgart: Kohlhammer, 1958), pp. 149–168.

[4] Jean-Joseph Goux, *Freud, Marx: Économie et symbole* (Paris: Éd. du Seuil, 1973), p. 91. ("L'Occident résout le conflit des équivalents par la prévalence et la monovalence. L'organisme social occidental apparâit ainsi comme la superposition, l'enchevêtrement, l'intrication, ou au contraire la séparation radicale, le cloisonnement, de nombreux niveaux ou registres qui sont cependant globalement et fondamentalenent isomorphes, structurés par la prééminence d'équivalents généraux autour desquels se dispose un système homologue de valeurs. . . .")

[5] Edmund Husserl, *Cartesianische Meditationen und Pariser Vorträge*, Husserliana I (The Hague: Nijhoff, 1950), p. 124 (". . . *eine transzendentale Theorie der objektiven Welt* . . .").

[6] Edmund Husserl, *Die Krisis der europäischen Wissenschaften und die transzendentale Phänomenologie*, Husserliana VI (The Hague: Nijhoff, 1954), p. 183. ("Wie soll ein Teilbestand der Welt, ihre menschliche Subjektivität, die ganze Welt konstituieren, nämlich konstituieren als ihr intentionales Gebilde?")

[7] *Ibidem*, p. 184.
[8] Martin Heidegger, *Einführung in die Metaphysik* (Tübingen: Niemeyer, 1976), p. 62. ("Denn Seiendes als Seiendes sagen, schließt in sich ein: Seiendes als Seiendes, d. h. dessen Sein im voraus verstehen . . . Denn Menschsein heißt: ein Sagender sein.")
[9] *Ibidem*, p. 141. ("Die Eröffnung des Seienden geschieht im *Logos* als Sammlung. Diese vollzieht sich ursprünglich in der Sprache. Darum wird der *Logos* die maßgebende Wesensbestimmung der Rede.")
[10] *Ibidem*, p. 99. ("Das bloße Hören verstreut und zerstruet sich in dem, was man gemeinhin meint und sagt, im Hörensagen, in der *doxa*, im Schein. Das echte Hörigsein hat aber nichts mit Ohr und Mundwerk zu tun, sondern besagt: Folge leisten gegenüber dem, was der *Logos* ist: *die Gesammeltheit des Seienden selbst*.")
[11] *Ibidem*, p. 101. ("Das Sagen und Hören ist nur ein rechtes, wenn es in sich zwar schon auf das Sein, den *Logos* gerichtet ist. Nur wo dieser sich eröffnet, wird der Wortlaut zum Wort.")
[12] Martin Heidegger, *Unterwegs zur Sprache* (Pfullingen: Neske, 1959), p. 30. ("Die Sprache west als der sich ereignende Unter-schied für Welt und Dinge.")
[13] *Ibidem*, p. 33. ("Der Mensch spricht, insofern er der Sprache entspricht. Das Entsprechen ist Hören. Es hört, insofern es dem Geheiß der Stille gehört.")
[14] See: Alexander Pigalev, "The Problem of the Foundations of Social Being in Martin Heidegger's Philosophy," in *The Problems of Philosophy* 1 (1987). (In Russian)
[15] Theodor W. Adorno, *Negative Dialektik* (Frankfurt a. M.: Suhrkamp, 1966), p. 398. ("Solches Denken ist solidarisch mit Metaphysik im Augenblick ihres Sturzes.")
[16] Carl A. Raschke, *The Alchemy of the Word: Language and the End of Theology*, p. 1 (Missoula, MT: Scholars Press, American Academy of Religion, 1979), p. 85.
[17] *Ibidem*, p. 91.
[18] See some interesting considerations concerning this problem in Robert Magliola, *Derrida on the Mend* (West Lafayette: Purdue University Press, 1984).

PART TWO

EXPERIENCE, JUDGEMENT, IMAGINATION, EMOTION

Visiting San Pedro Tlaquepaque on the outskirts of Guadalajara: Marlies Kronegger with friends.

M. C. DILLON

BEYOND SEMIOLOGICAL REDUCTIONISM

Transcendental Philosophy and Transcendence

The problems to which philosophy addresses itself resist solution and recur for two reasons. The first is grounded in the brute fact which can be expressed by saying that the world is transcendent or by saying that human cognition is finite. The second reason for the perennial nature of philosophical problems follows from the nature of cognition, specifically from the fact that the paradigms through which cognition proceeds conceal the world in the process of revealing it. Paradigms capture generic features by obscuring specific differences. Furthermore, to the extent that we accept the paradigms which govern our thought, that acceptance conceals a closure that attenuates our openness to other paradigms; it dispels our wonder before the recalcitrant data that resist the governing paradigm.

These two facts, the transcendence of the world and the sedimentation of paradigms, are related: the magnitude of the specifics to be known, the overwhelming richness of the world as object of knowledge, demands simplification; the necessity that drives us to paradigms is the impossibility of knowing each and every part of the whole in its ipseity. God may know each entity in its suchness, but we cannot; God may be able to call all the ephemera by name, but we must refer to them generically in our attempt to grasp them all globally, and this commits us to paradigms or models or theories or metaphors. Reductionism of some sort is, therefore, a necessary feature of human cognition, a correlate of human finitude and worldly transcendence.

Transcendental philosophy came into being as a critique of naïve empiricism and a revolt against the excesses of speculative metaphysics. The naïveté of empiricism was held to consist in an unbridled optimism bound up with an incompatible paradigm of cognition: naïve empiricism accepted the paradigm of divine cognition by undertaking to know things as they are in themselves apart from the limitations of human cognition. This project embodies a contradiction: the very project of knowing presupposes a limit, but this limit must be erased for the project to fulfill its own criterion for knowledge. Again, if the thing itself is the measure of our knowledge of it, then knowing the thing cannot be limited

A-T. Tymieniecka (ed.), Analecta Husserliana, Vol. LIII, 75–88.

by time, space, perspective, or any other aspect of human finitude. For humans to know anything, they must know everything: the condition for finite knowledge is infinite knowledge.

Transcendental philosophy responded to this self-contradictory optimism with a tautological pessimism: the necessity that constrains human cognition is constitutive of human cognition, the nature of knowledge is defined by its limits, the conceivable is bounded by the inconceivable. Things in themselves fall on the far side of this limit: things as they are independently of us cannot be known by us. What we can know of things is only what we can know of things. Specify the limits of knowledge and you arrive at the way things have to be for us. If the limits of knowledge can be specified generally and unconditionally, then the generic attributes of the objects of knowledge, as objects of knowledge, can be established apodictically. Thus, for example, if it is unconditionally or unexceptionally the case that, for us, physical objects must occupy the three-dimensional space of simultaneity and the two-dimensional time of succession, then it is necessarily and universally the case for us that physical objects are extended in three-dimensional space and two-dimensional time.

As this example was contrived to show, there is a problem embedded in attempts to specify the limiting structures of human cognition which recapitulates the problems embedded in any attempt to establish limits or generate definitions in an unconditional or timeless way. Recapitulates them, but also compounds them. To think of a limiting point in space or time is to think beyond the limit to what it excludes: here is not there, what is on this side of the line excludes what is on that, today is not yesterday or tomorrow, etc. There is something arbitrary or conditional about all such limits: here can become there, today will become yesterday, etc. To think of unconditional limits, however, is different from the examples just given because it is to think of a limit beyond which one cannot think: thinking of limits *in* space or time is different from thinking of the limits *of* space or time. That is why it has been argued – paradoxically – that the limits of space and time cannot be thought by us, hence that, for us, space and time must be conceived as limitless, infinite, unbounded, or, in other terms, that the limit of human conceptions of space and time is limitlessness, that the definition of the magnitude of space and time for human cognition is the indefinite. In this manner transcendental philosophy establishes limits by means of the criterion of inconceivability.

Inconceivability is the engine of transcendental arguments designed to establish the limits of human cognition. Inconceivability drives Kant's transcendental deduction just as it drives Husserl's eidetic reduction: we establish how things must be for us by defining the limits of conceivability. For us, things must be this way because it is inconceivable for them to be otherwise. Here transcendental philosophy reaches a conceptual crisis which precipitates a historical turning point. Are the limits of conceivability, rationality, intelligibility in general subject to change in time? Earlier transcendental thinkers, such as Kant and Husserl,[1] thought they were not; later transcendental thinkers, such as Heidegger and Derrida,[2] thought they were.

The thoughts that precipitated this historical crisis are associated with such names as Darwin, Nietzsche, and Einstein. Darwin introduced the speculation that powers of intelligence were subject to evolution. In effect, Nietzsche argued that the classic transcendental argument based on the limits of conceivability, Anselm's ontological argument, had self-destructed, and the symbol of truth, once embodied in praxis, revealed its own falsity: if we are honest, as God demands, we shall have to acknowledge that we can conceive of the non-existence of this perfect being; if we are truly honest we may have to acknowledge that we conceived God, rather than the other way round, and did it out of weakness and self-deception. And Einstein demonstrated that the a priori conceptions of space and time codified in post-Newtonian modernity were inadequate to experimental findings. Space and time changed character, and so did the logic of the apriori upon which they were constructed.

As I view the matter, Nietzsche's challenge was the one most responsible for opening the abyss before transcendental philosophy. Nietzsche called the criterion of inconceivability into question by asserting that it is impossible from within a perspective to define the limits of that perspective, for that would require seeing the perspective from the vantage of a wider perspective beyond the putative limits of the first. In *The Gay Science* Nietzsche exultantly proclaims that "we are today at least far from the ludicrous immodesty of decreeing from our nook that there can only be legitimate perspectives from that nook."[3] This has the effect of denying the validity of Kant's transcendental deduction, which purported to establish the limits of human conceptuality from within its own vantage. Nietzsche argued that Kant could not establish the universality and necessity of his system of categories, and that the attempt to do so merely manifested his desire to bulwark his belief in a perspective aware of its finitude and troubled by it: the very project of *The Critique of Pure Reason* presupposes the uncertainty it sought to conceal in its claims to apodicticity.

A few decades later, Quine clarified and refined Nietzsche's thought in "Two Dogmas of Empiricism" where he argued that a priori truths are propositions to which we have committed ourselves "come what may": it is not that the truth is universal and necessary, it is rather that our commitment is adamant.[4]

Since transcendental philosophy is founded upon apriori reasoning and – for essential reasons –[5] has historically relegated empiricism to secondary status, this challenge to apriority in general brings transcendental philosophy to a crisis. The universal and necessary categories of understanding are now revealed to be paradigms of thinking which may be neither universal nor necessary, but rather historically (or socially or economically, etc.) contingent; to hold to them "come what may" begins to appear as a failure of thought, as hardheaded refusal to dislodge oneself from one's prejudices rather than as rigorous adherence to hardheaded logic. This crisis changed transcendental philosophy, and the change was ushered in by Heidegger under the heading of hermeneutic methodology.

Husserl's methodology of transcendental reduction was intended to reveal the invariant structures tacitly at work within consciousness immersed in the natural attitude. Heideggerian hermeneutics undertakes the same project, but with a much different purpose. Husserl sought to clarify, purify, codify, and consolidate these invariants in order to articulate the transcendental groundwork for a unified science. Heidegger's intent was to reveal the paradigms at work in human existence in order to make them available for critical scrutiny, historical analysis, and, ultimately, for something like sublation. One articulates the fore-structure of understanding in order to overhaul its specifics: the anticipants constitutive of fallen everyday understanding are revealed through a hermeneutical process designed to displace them with the authentic anticipants of resolute Being-unto-death. This was the program Heidegger undertook in *Being and Time*.

There is an ambivalence here. Are the anticipations of authentic resolve to be regarded as a genuine a priori to which we should adhere despite anxiety and other insidious forces? Does Heideggerian hermeneutics thematize one paradigm in order to displace it with another? If so, on what grounds is one to rest the case that one a priori is superior to the other? In my view, Heidegger is disingenuous on this point: he claims value neutrality for a distinction drawn in such value-laden terms as inauthentic fallenness and authentic resolution. Here is Nietzsche's abyss, mined so fruitfully by Derrida: either one's paradigm, one's a priori, is held to be grounded – in which case one tacitly appeals to an inaccessible absolute – or one acknowledges that all paradigms are grounded in nothing stronger than evanescent circumstance – in which case the choice among them is ultimately adventitious. Derrida's dilemma recapitulates the traditional dilemma which forces one to choose between the equally unacceptable alternatives of absolutism or relativism, with no third alternative held to be conceivable.

Six years before Heidegger published *Being and Time* (1927), Wittgenstein published his *Tractatus Logico-Philosophicus* (1921), which contains the well known assertion that "the limits of my language mean the limits of my world."[6] In the preface to this book, Wittgenstein proposes his own version of the argument about the limits of conceivability set forth above, but with a significant amendment that ties the limits of thought to the limits of language.

> . . . The aim of the [*Tractatus*] is to set a limit to thought, or rather – not to thought, but to the expression of thoughts: for in order to be able to set a limit to thought, we should have to find both sides of the limit thinkable (i.e., we should have to be able to think what cannot be thought).
>
> It will therefore only be in language that the limit can be set, and what lies on the other side of the limit will simply be non-sense.[7]

Language allows us to say what cannot be thought (for example, we can speak of round squares). If we can define the limits of meaningful language, we can define the limits of thought without having to encounter the problem of thinking the unthinkable. Thus the limits of language are the limits of thought.[8] Wittgenstein goes on to deploy a series of transcendental arguments which lead him to conclude that language defines the limits of meaning of the (human) world.

In 1935, Heidegger delivered a lecture that he subsequently revised and published (1953) in a book entitled *An Introduction to Metaphysics*. In that book he asserts that "it is in words and language that things first come into being and are."[9] In 1946, Heidegger delivered his often quoted proclamation that "language is the precinct [*templum*], that is, the house of Being."[10] The effect of the linguistic turn in transcendental philosophy is to ground the fore-structure of the hermeneutic circle in language: the paradigms through which we think reside in the structures of the language into which we are born. Language becomes the ground of this new conception of apriority. The new conception differs from the old because the historically evolving nature of language betokens a correlative historical relativity to the a priori embedded within language.[11]

The cumulative effect of these transitions within transcendental philosophy is at least to erode and possibly even to erase the claims to truth that can be made within its context. Transcendental philosophy began by relinquishing claims to deliver true statements about *the* world, but purported to make universal and necessary statements about the *human* world. Claims to necessity and universality were attenuated with the recognition of a plurality of a prioris, and the question arose as to the possibility of adjudicating among competing paradigms. When the paradigms are relativized to language or sign-systems and those structures are themselves relativized in time and space to history and culture,[12] then the strongest claim to truth would seem to be circumscribed within the paradigms of a given community during a given period of time. In Foucauldian terms, truth becomes localized.

The core argument at stake maintains that our understanding of the world is structured through paradigms or a priori forms and that these forms reside in historically situated sign-systems. Thus, how we experience the world depends on the system of signs into which we were born. This claim is radicalized by Derrida when he denies that there ever was any such thing as perception, if by perception we mean a direct experience of the world unmediated by the play of signifiers. Derrida assigns to the play of signifiers the same transcendental function that Kant assigned to the transcendental unity of apperception: they unify discrete experiences over the passage of time thereby providing the condition for the possibility of identification. Signifiers allow us to experience the thing as the same from one moment to the next or, in other words, they enable us to see things as identical to themselves in meaning while differing from themselves through temporal adumbrations. This self-relatedness of identity and difference of a given thing also provides the possibility of relating that thing to others under generic signifiers. Therefore the meanings of individual things and groups of things depends ultimately on relations made possible by the play of signifiers.[13] Thus, finally, the meaning of the worldly things we experience derives ultimately and entirely from the play of signifiers. This is the thesis I have designed as semiological reductionism: namely, that language or, more broadly, systems of signs function as the transcendental condition for the possibility of meaningful experience.

Several objections can be raised against this thesis, but the root problem concerns the theory of judgment at work in it. In classical terms, judgment is the act of predication, the subsumption of a particular thing under a generic category, the assignment of an entity to membership within a class or set. Things are known through the predicates which relate them to themselves, other things, and groups of things, and these relationships define their meanings. Thus, to understand a thing fully, to grasp its meaning, is to comprehend all its predicates, ultimately, to grasp its relations with all other things or to place it within a totally inclusive matrix of predicates, concepts, categories, sets, etc. In the classical or neo-Aristotelian view, the telos of science is to construct such a unitary and comprehensive matrix, thus to afford an understanding of all things by placing them appropriately within this matrix.

Transcendental philosophy understands these matrices as a priori systems or paradigms, and contends that the meaning of a thing is constituted by its placement within such a priori structures. As I have just attempted to show, the linguistic turn in contemporary transcendental thought consists in viewing these structures as grounded in language or semiological systems. Thus the meaning of a thing is constituted by its placement within a system of signs, by the signifiers that refer to it and define its place.

Empiricism maintains that we grasp the properties of things perceptually and, on the ground of that pre-predicative perception, assign the thing to its proper category or predicate, or place it within a conceptual matrix. Transcendentalism maintains that the perception of a thing as bearing a given predicate or having a given property depends upon a priori operation of the category or matrix of categories. Empiricism maintains that we see natural groupings, things that resemble one another in some respect at the perceptual level, then induce a signifier to name that grouping or property, and finally identify things bearing the property by calling them by the name that signifies the property: my perception of the black cat as a cat and as black allows me to refer to it as a black cat. Transcendentalism maintains that I could not see the cat as a cat or as black were it not for the a priori operation of the signifiers "cat" and "black."

The core argument of semiological reductionism, sketched above, identifies it as a radical form of transcendentalism. Signifiers must already be at work in the identification of a thing as the same over time because signifiers constitute the idealities or relations that allow temporally discrete appearances of the thing to be related or identified. Absent signifiers, we could not see the black cat as an abiding entity, that is, we could not see it as an object, hence *a fortiori* we could not see it as a black cat. The condition for perception of a thing is a priori placement of that thing within a sign system. Empiricism is, therefore, naïve and wrong in its assumption that there can be an immediate perception of the thing as bearing certain properties or having a certain perceptual meaning – because properties and meanings *are* places within sign systems. There is no immediacy to perception; perception is always already mediated by the play of signifiers.

The problem mentioned earlier which arises with the issue of judgment centers on the question of *appropriate* categorization or placement within a matrix. What are the conditions for subsumption of a particular thing under a proper universal category? Is it the signifier that allows the thing to be seen as such, or is it the perceptual suchness of the thing that allows us to see it as an instance of a given type? Can one distinguish between good and bad judgments, that is, between judgments that succeed and those that fail in subsuming a given particular under an *appropriate* universal or ideality? At what level of awareness does this judging take place? Is the notion of ante-predicative experience at all viable, even perhaps "necessary"? Finally, what are the moral and political consequences of semiological reductionism?

There are strong arguments on both sides of the debate. Empiricism maintains that some sort of inductive process is presupposed in the genesis of signifiers: if every identification presupposed the operation of signifiers, then it would be impossible to account for the genesis or evolution of signifying species from non-signifying species. Transcendentalism, recapitulating Kant's argument against Hume's attempt to explain causality in terms of habit, maintains that there is no strictly perceptual data corresponding to conceptual paradigms, hence no basis for the formation of the habit of associating things in given ways. In my view, this debate over origins is ill-conceived because it presupposes that the two positions are mutually exclusive. The task of the philosophers is not to choose sides and wage ideological warfare against the opposed position, but to find conceptual space in which an account can be given of *both* the genesis of sign-systems *and* the constitutive role they play once in place.

One key to unlocking this conceptual space lies in the notion of appropriateness.[14] When we make bad or misleading or pernicious judgments, it is because we have subsumed a given object under an inappropriate category, because we have applied the wrong paradigm or misapplied the right one. The problem is compounded when the same object lends itself to several predicates or to inclusion within a multiplicity of non-coincidental sets: objects are multi-determinable, hence ambiguous. Under one a priori, private ownership of firearms is essential to democracy; under another, it is a threat to the freedom of all: the meaning and value of handguns, hunting rifles, assault weapons, tanks, bombers, etc. are hardly clear to us, and the controversy surrounding these determinations is likely to continue into the indefinite future. Does one seek resolution, search for viable paradigms, by empirical means or transcendental means? Does one collect relevant data? If so, what paradigms govern the criterion of relevance? Does one analyze the conceptual presuppositions of the competing paradigms of value and meaning? If so, how are we to determine and assess the practical or empirical consequences of adopting one paradigm rather than another? Once again, my view is that the two methodologies must interpenetrate: one does not choose one to the exclusion of the other, one seeks to coordinate them.

The firearms dispute will not be easily resolved, I suspect, because antecedent commitments to competitive models of government and antagonistic ideologies will continue to stir up the muck of confusion. My point here, however, is that conceptual resolution would not just be difficult under the model of semiological reductionism; it would be impossible. It would be impossible because semiological reductionism holds that empirical meanings are necessarily predelineated by antecedent commitments to sign-systems: no empirical data, no event or series of events, could persuade participants in the debate to change their minds – just because the meaning of the data would be entirely constituted within the paradigm always already at work within the a prioris of the debaters. In a word, they would *have* to continue to see things differently – literally *see* things differently. Opponents would have to be seeing different things, things with different meanings, different identities. To shift from the debate over gun control to the debate over abortion: one group *sees* a human life tortured and murdered, while the other *sees* a life of human misery averted. Both sides are committed to interpret the data through the lenses of their own paradigms . . . come what may, all evidence is assimilated to the governing paradigm.

To say that conceptual resolution of such agons as those just cited would be impossible under the model of semiological reductionism is not to say that no form of resolution is possible. Resolution *is* possible under the model of semiological reductionism, but by means of another form of agon: an agon of power as opposed to an agon of debate. One forces others to adopt one's semiological paradigm by non-linguistic means. It is important to be clear on this point because the surface paradox can easily introduce confusion.

It was Edward George Bulwer-Lytton, Baron Lytton, who gave us the phrase, "The pen is mightier than the sword," but the distinction between the power of violent destruction and the force of argument predates documented history. The deployment of political power usually involves a mixture of the two, and it is not always easy or even possible to discern them in concrete praxis. If, however, one espouses a theoretical model within which language is deprived of persuasive force, then, within that model, all deployment of power is reducible to some form of violence. Semiological reductionism is an instance of this theoretical model.

The force of argumentation lies in its allusiveness, referentiality, and responsiveness. The measure of soundness lies as much in truth as in validity: to persuade you with my argument, I must show, not merely that the conclusion follows logically from the premises, but that the premises meet the worldly measure of truth, that they express what is the case in the world to which they refer.[15] The worldly measure of truth transcends expression. To borrow from Tarski, the string of signifiers "swans are white" is true if and only if swans are white. To borrow from Popper, one corroborates or falsifies the linguistic assertion by appeal to evidence drawn from the world. To borrow from Einstein, one corroborates the assertion that light bends and is subject to gravitational influence, by corroborating the prediction that Venus will be visible on earth before it has physically passed beyond the perihelion.

This process is impossible under the theoretical postulate that all perceptual experience derives its meaning from a semiologically grounded a priori: under this model, perceptual experience cannot be even a relatively independent measure of the truth of an assertion. Semiological reductionism is, thus, the handmaiden of dogmatic metaphysics. *De facto* commitment to the sedimented thesis of universal causality, for example, would preclude in principle the possibility of an uncaused event; it would be impossible to conceive, hence impossible to witness, an act of human freedom. One's paradigm could never collide with one's experience, nor could another person allude to worldly evidence that would force me to change my paradigm.[16]

To forsake the force of argumentation is to fall back on the power of violence as the only remaining means of resolving disputes. A currently controversial example – with which I shall conclude – may illustrate this point.

The thesis that gender identity is entirely reducible to linguistic structures and has no ground in the empirical data of physiology or biology, that all such data are constructed through the play of signifiers, is currently being promulgated under the rubric of "constructivism." Activists in this movement would change the way we speak and write about gender difference, hoping thereby to change the way in which we perceive gender differences and respond to them. Many of their arguments are persuasive, for it is indeed the case that gender stereotypes are in large measure grounded in the semiological substructure of culture, and this has been successfully demonstrated. My objection to constructivism is that it refuses no principle – the principle of semiological reductionism – to address the equally strong evidence from the empirical sciences that there are significant physiological differences among the genders. The reductionist thesis at work here is that physiological differences between human bodies have no significance that is not derived from cultural forms.

Accept this thesis, and empirical research has absolutely no bearing on gender studies. That is, I think, ludicrous and self-defeating for those of us who seek to promote gender equality. Beyond that, however, there is the more insidious issue of power. If language cannot change itself by modifying itself in the light of worldly evidence, then the only way to change language is by violence, that is, by fiat rather than by argument. How does this play out on the contemporary scene?

Political rectitude is working itself out in an ambivalent way because, as currently practiced, it operates through conflicting paradigms: one based on tolerance and respect for differences, the other based on intolerance and refusal of difference. In its tolerant mode, which I endorse, political rectitude seeks to let all voices be heard and to that end seeks to minimize violence or domination by brute force. In its intolerant mode, which I deplore, political rectitude attempts to force voices it perceives as voices of opposition into silence by deployments of power more basic than and different in nature from the force of argumentation.

I am willing to acknowledge that it is possible to reach a point in the agon of conflicting ideals where argument breaks down and recourse to violence is the only option left. I think that some stakes are high enough to vindicate the deployment of brute power, but it is the mark of a just warrior to unsheathe the sword only as a last resort.

Change the paradigm or sign-system through which a person or culture perceives the world and the behavior of that person or culture will change. Orwell taught us that in his prophetic *1984* where "newspeak" was the vehicle of behavior modification. When I read that book in my youth, I took Skinner's behaviorism to be the enemy and sought to engage it with arguments drawn from Sartrean existentialism and the categories of human freedom it propounded. Imagine my chagrin today to find the champions of my own causes using the tactics of my enemies, and attempting to justify the use of those tactics on the same grounds: pessimism with regard to truth and rational means of discerning it. Chagrin – because these tactics that made my enemies my enemies are now distancing me from my allies.

The real enemy here is the pessimism with regard to truth that condones the recourse to violence – and thereby transforms noble causes it, ignominious and self-defeating, attempts to dominate. No end is noble enough in itself to justify the use of any means, but it is a mark of noble ends that those genuinely in pursuit of them use noble means.

The measure of truth is transcendence, and the task of philosophy has always been to stretch our perception of truth to accommodate as much as it can of the world that continues to resist our understanding. The transcendent world is also the ultimate measure against which conflicting claims to truth are judged, hence the ultimate ground for the possibility of bringing about intersubjective accord. It is true that human cognition is finite – this paper begins by postulating that as a fact – and from that it follows that all our visions of the transcendent world reduce it to human proportions. It is also true, as I have argued here, that the only way we have to determine our limits is to attempt to stretch them – because the boundaries cannot be set a priori. Resignation to a reduced perspective is what philosophy is not.

Transcendental philosophy determines the conditions for prevalent paradigms, it searches for the a prioris latent in our seeing, saying, and thinking, and occasionally it succeeds in thematizing our hidden presuppositions. That, I think, is noble work, because to thematize a presupposition is already to see beyond it, the first step in self-transcendence. Transcendental philosophy mystifies itself and arrests its own progress when it attempts to carve out a turf bounded by limits it deems unsurpassable, thus to speak with certainty within those limits and defend its turf and its identity against all who would investigate other grounds. Semiological reductionism defends the principle that we can never see beyond the limits inscribed upon our senses by the systems of signs through which our vision of the world is filtered. It purports to liberate us from these systems by gnawing away at them from within, but it has the effect of reinscribing the dogmatism it sought to displace because it denies itself access to the source of wonder from which philosophy derives its life-breath. The limits of the conceivable are truly inconceivable. Amen.

Binghamton University

NOTES

[1] I would include Hegel in this list, but with the modification that progressive changes in conceptuality were themselves driven by a higher conceptuality that does not change.

[2] Here I take Derrida as emblematic of postmodern thought in general.

[3] Friedrich Nietzsche, *The Gay Science*, sect. 374. The sentences leading up to this assertion clarify Nietzsche's thought there:

> We cannot see round our corner: it is hopeless curiosity to want to know what other modes of intellect and perspective there might be: for example, whether any kind of being could perceive time backwards, and alternately forwards and backwards (by which another direction of life and another conception of cause and effect would be given).

His point is that Kant's inference from the necessity that delimited his perspective and that of the Aristotelian tradition to universal claims about rationality as such was unsound, that no such generalization could be substantiated. Kant's testimony should be regarded as evidence for the blinding or self-dissembling nature of paradigms.

In *Beyond Good and Evil*, Nietzsche makes a similar point, this time directed specifically at Kant:

> . . . It is high time to replace the Kantian question, "How are synthetic judgments *a priori* possible?" by another question, "Why is belief in such judgments *necessary*?" – and to comprehend that such judgments must be *believed* to be true, for the sake of the preservation of creatures like ourselves; though they might, of course, be *false* judgments for all that! . . . The belief in [the] truth [of such judgments] is necessary, as a foreground belief and visual evidence belonging to the perspective optics of life (Nietzsche, *Beyond Good and Evil*, trans. Walter Kaufmann (New York: Random House, 1966), sect. 11, p. 19).

Nietzsche is arguing that the necessity for believing in categorial paradigms is grounded in the will to preserve the paradigm or the perspective constitutive of the individual dwelling within it. The necessity in question, then, is not the logical necessity that drives *modus ponens*, it is the self-deceit of a contingent perspective willfully denying its contingency.

[4] W. V. O. Quine, "Two Dogmas of Empiricism," *The Philosophical Review*, 1951, 60.

[5] See below, pp. 12–13.

[6] "Die Grenzen meiner Sprache bedeuten die Grenzen meiner Welt." Ludwig Wittgenstein, *Tractatus Logico-Philosophicus*, trans. D. F. Pears and B. F. McGuinness (London: Routledge and Kegan Paul, 1961), prop. 5, 6, pp. 114–115.

[7] *Tractatus*, Preface, p. 3.

[8] I omit the personal pronoun in order to avoid rehearsing Wittgenstein's arguments against the possibility of private language which allow him to shirt the issue of solipsism signaled by his use of the personal pronoun in props. 5.6 *et seq.*

[9] Martin Heidegger, *An Introduction to Metaphysics*, trans. Ralph Manheim (New York: Doubleday and Company, 1961), p. 11.

[10] Heidegger, "What are Poets For?" in *Poetry, Language, Thought*, trans. Hofstader (New York: Harper and Row, 1971), p. 132.

[11] Heidegger had already argued in "The Origin of the Work of Art" (1935–1936) that truth happens historically through *poiesis*: ". . . language alone brings what is, as something that is, into the Open for the first time." Heidegger, *Poetry, Language, Thought*, p. 73.

[12] This step in the argument presupposes my acceptance of a general rejection among contemporary transcendental philosophers of theories purporting to establish atemporal grounds for language such as those proposed by Noam Chomsky and Edmund Husserl.

[13] Derrida's argument, summarized here, is delineated in detail in chapter two of my book, *Semiological Reductionism: A Critique of the Deconstructionist Movement in Postmodern Thought* (New York: SUNY Press, 1995), pp. 35–55.

[14] Heidegger takes up this issue in essays he wrote in the late fifties under the heading of *Ereignis* or appropriation. In my view, he poses the problem adroitly, but refuses to venture a solution to it. My interpretation is elaborated in *Merleau-Ponty's Ontology* (Bloomington: Indiana University Press, 1988), pp. 232–242.

[15] Thus, the argument that demonstrates that Newt Gingrich can fly because all carrots can fly and Newt Gingrich is a carrot is not persuasive, even though it is valid.

[16] One can, of course, demonstrate inconsistencies, contradictions, or "aporias" within a given standpoint, and that is the prime strategy of those who work within the model of semiological reduction. But, given the possibility of a plurality of competitive but nonetheless coherent paradigms (e.g., theism and atheism), this is an insufficient measure. The measure of internal coherence also loses force within the context of a meta-thesis that holds that logical inconsistency is inconsequential.

CALVIN O. SCHRAG

FROM EXPERIENCE TO JUDGMENT IN THE AFTERMATH OF POSTMODERN CRITIQUE

The story about the tracking of experience on the way to judgment is intrinsically complex. And this is the case because of a variegated emplotment of changing scenarios and a complementing variety in its cast of characters. In accepting the invitation to address the general topic, we offered to provide only a segment/chapter of the wider story – namely a travelogue on the road from experience to judgment in the aftermath of postmodern critique. But even a telling of this segment/chapter of the wider story places upon the narrator some quite extraordinary demands. It involves at least three distinct, if not separable, plots or accounts – namely, modernity's take on the issues; the postmodern response; and recommendations of what is to be done in the aftermath. At the most what we will be able to do – given limitations of allotted time and personal energy – is to isolate some of the pivotal themes and emergent aporias.

A full account of the travails of experience and judgment in the annals of modernity would bring to the fore a history of contention between the proponents of an empiricism that sought to found knowledge claims on percepts (construed as impressions and copies of impressions) and spokespersons for a rationalism that located the seat of knowledge in a theory of judgment (*Urteilstheorie*) in which concepts/categories/conceptual schemes took center stage. Kant's contribution in addressing the issues produced by this background of contention is sufficiently well known that we need not rehearse it here – concepts without percepts are empty; percepts without concepts are blind. That Kant's two-pronged rebuttal of the empiricistic skeptics on the one hand and the rationalistic dogmatists on the other hand was a momentous philosophical achievement surely cannot be denied. Yet, one might question whether his hurried acceptance of the empiricist (i.e. British/Humean empiricists') construal of experience as the reception of atomistic, granular, blind, non-intentional, sense impressions – read "sense-data" in later developments of empiricism – was not premature. If one buys into a perspective on experience as a serial succession of sense-data, a parade of bare isolated facts that somehow flick on and off, without connec-

A-T. Tymieniecka (ed.), Analecta Husserliana, Vol. LIII, 89–95.

tion, purpose, or sense, then one will have to scrounge around for principles of judgment, rules of inference, and conceptual schemes to supply the conditions for knowledge, to somehow bind that which is filtered through the forms of sensibility (time and space) into veritable contents of knowledge. That Kant should find such conditions for knowledge in the unifying power of the transcendental ego – marshalling the forms of sensibility, the schema of the imagination, and the categories of the understanding – may in the end not be all that much of a surprise. If there is to be knowledge – and Kant never questioned that such indeed obtained in the fields of mathematics and physics – and if the raw experience of traditional empiricism could not account for such, then one will indeed need to call upon the services of a transcendental ego to swoop down from on high (or emerge from below) and shape the sensory manifold into determinable objects of cognition.

But Kant's accepted bifurcation of experience and judgment, percepts and concepts, sensory manifold and categorical determinations, left certain crucial issues unresolved. How does a theory of judgment, designed against the backdrop of these bifurcations, provide directives for the decidability of sense and the determinability of reference? This question catapults us into an emergent subplot in the story of modernity, the philosophical preoccupations with the problem of sense and reference from Frege to Husserl to Russell to contemporary analytical epistemology and beyond. For want of a reliable theory of judgment, wherewith to pin down the specificity of sense and the object of reference, it would appear that all supports for a reliable theory of knowledge begin to creak and crumble.

The proverbial "rub" in the quest for sense and reference in the annals of modern epistemology had to do with the vagaries of representation and the continuing purchase on correspondence as a truth-criterion. The path from the presentational immediacy of the givens in experience to trustworthy judgments about them remained blocked because the immediacy of the given and theories of judgment alike became problematized. Not only did there seem to be problems with pinning down that which was directly/immediately present (Sellars on the "myth of the given"), but even greater problems accompanied projects of *re*-presenting the presented via categorical judgments (Derrida on the "silent cogito"). The problem of "presence," both with respect to the status of the knowing subject as somehow present to itself, and the presented object as the proper terminus of reference via a circuitous path of re-presentation, culminated in a veritable epistemological crisis in modernity.

For a time it was hoped that semiotics and semantics (and linguistics more generally) might be able to come to the rescue, restructure the epistemological problematic as at bottom a semiotic/semantic one, and then move out from there into the wider domain of knowledge, including the developing human sciences. At least such was the dream of structuralism, which rather quickly, like Husserl's early dream of phenomenology as a rigorous science, became a dream that ran its course. Substituting "signifier" for percepts and concepts and "the signified" for representable objects didn't help all that much, for it took very little effort to show that the rhizomatic signifiers in our everyday linguistic usage remain incurably lame in coming up with the properly signified. It was thus that the stage was set for the "deconstruction" of structuralism – or the advent of "post-structuralism," as the phenomenon was called in the United States.

Now the story of the deconstructive turn, and the continuing story of postmodernity (of which the deconstructive turn is only one aspect or feature), is admittedly enormously complex. It is indeed as complex as the story of modernity itself. Vis-a-vis our focal topic, "experience and judgment," the effects of postmodernity are particularly telling, and especially so in its delimitation of the resources of judgment, which in the end invites a frontal attack on the resources of rationality itself insofar as judgments stake out claims of reason. Clearly, if reason as logos, and more specifically in its logocentric guise, is deconstructed, the faculty of judgment will find itself to be philosophically shackled. It will remain bereft of rational criteria/principles on the basis of which to judge. Postmodernists find the principal fault to reside in the modernists' preoccupation with principles and criteria of judgment rooted in an obsession with unity, universality, necessity, and the requirement that these criteria be laid out in advance. Examples of epistemic criteria legislated beforehand that quickly come to mind are Descartes' "Rules of Method" and Kant's analytic and synthetic judgments. But if one jettisons the criteria of universality and necessity and the epistemic requirement for unity, it would appear that everything dissolves into a rhapsodic play of difference, diversity, particularity, and contingency – and imperial reason is found to be without clothes. Such is the story of the deconstructive/postmodern turn, quite abbreviated to be sure, and "simplified to the extreme," as Lyotard would be wont to say! But hopefully it will provide us with a point of departure in addressing the question "Where do we go from here?"

Might one agree with the postmodernists that logocentrism, and particularly of the modernist version, has outworn its usefulness, and then respond by suggesting that the Western metaphysical and epistemological concepts of the logos are not the only concepts available? Might one endorse the postmodernists' recommendation for an "incredulity toward metanarratives," and then reply that there may still be a story left to tell? Might one agree that the subject-centered rationality of modernity is too puffed up, but still find resources of reason in a subject duly de-centered? Might one applaud the problematization of an epistemological criteriology, obsessed with universality and necessity, and then respond by pointing out that universalizable criteria laid out in advance are not the only criteria in town? Might one be done with the requirement of unity as a metaphysical solace and an epistemological guarantee, but still find the drive for unification, in both the life of self and society, to be an indigenous feature of lived experience?

The above series of questions should provide the reader with some notion of how we propose to move on matters having to do with the postmodern challenge to modernity. Now how does it stand with the issue of experience and judgment in all this? The first point that needs to be made is that the subject of experience and judgment – or preferably the subject *in* experience and judgment – is redefined/refigured/transformed/transvalued. The idea of a subject-centered reason was not a particularly good one to begin with, and we are all better off for it having run its course. The implications of this for the fate of sense and reference are really quite profound. No longer will we be required to scrounge around in the alleged goings-on in a mental interior, a citadel of mind, for the proper concepts/propositions/assertions that will deliver both sense and reference. Instead, we will look for meanings already operative in our practical engagements, in our dealings with other subjects, and in our insertions in particular traditions (always in the plural).

Maurice Merleau-Ponty already pointed us in this direction in his observation that "we are condemned to meaning." Although this was more specifically a rejoinder to Sartre's notion of being "condemned to freedom," it supplied a rather profound and general indictment of the vagaries of modern epistemology. In a quite analogous manner, reference, which in the tradition from Frege to Husserl and beyond was defined as the fulfillment of sense (meaning), will no longer be construed as an epistemological vector proceeding from a subject in search of an object. Instead we will focus upon descriptions of human behavior and social practices in which we find ourselves "condemned to reference." Although we may still be able to retain the grammar of "constitution" in speaking of meaning and reference, this will no longer be a constitution inaugurated by a centered subject but rather a constitution solicited by the intervention of that which is "other" – the alterity of the perceived object, the alien text, the different culture, the encountered other self in one's social practices and institutional involvements.

The recognition of the play of alterity, the upsurge of otherness, is clearly a pivotal feature in our resituation of the dynamics of meaning and reference, and this play of alterity is usually considered to be the point of emphasis in the legacy of postmodernism. This may indeed be one of the things that we can learn from postmodernism. But one needs to be wary about construing this alterity as an *absolute exteriority* that would derail all responsivity of the subject to that which is other, and in doing so de-face the subject in its encounter with the face of the other.

The implications/spin-offs from this play of alterity entail a dismantling of the epistemological grid that constrained the accomplishment of meaning and reference in the narrative of modernity. The troublesome requirements of representation, correspondence, and justification on the basis of pre-established criteria are set aside. This proceeds in tandem with a shift from theory of mind as the locus of explanation of meaning and reference to supervenient/advential factors that occasion the happening of meaning and reference in our concrete and variegated world-experience. Reference is no longer construed as the terminus of a subject-centered mental vector, seeking a correspondence (or indeed identity) with that which is allegedly cognized, but is now seen as an event of disclosure, a disclosure *to* the experiencing subject – what we have come to call an *incursive* disclosure, a disclosure *to* rather than *by* an experiencing subject.

Traveling with this newly won incursive disclosure as the proper context and background for the dynamics of meaning and reference is a new topography of experience. The flip side of modernity's problematic *Urteilstheorie* is its problematic view of experience. And much of this has to do with the restrictive positing of experience as a serial succession of atomistic, granular, disconnected, and non-intentional sensory impressions/sense qualia – the peculiar legacy of British empiricism. If indeed the web of experience is pulverized into discrete and random sensory percepts, then one will need a theory of conceptualization and a theory of judgment wherewith to unify the discrete perceptions into objects of knowledge. Kant's singular achievement was to recognize this requirement, and he put into place an elaborate categorical scheme and theory of judgment to deliver the sensory manifold from ignorance and perceptual chaos. Percepts without concepts are blind. But suppose that one proceeds from a broader, richer, and more vibrant notion of experience than that delivered by British empiricism (which Kant appears to have taken for granted), a notion of experience in which an understanding of ourselves and the world is already operative, and operative in such a manner that it antedates the troublesome binaries of perception and conception, sensory receptivity and judging activity, as defined in the epistemological paradigm of modernity. Such a broadened notion of experience would already contain the resources for "seeing" – seeing the world and having insight into ourselves. Experience thusly articulated would be intentional from the bottom up, and it would be intentional in the sense of a *lived-through, functioning intentionality*. As such it would profoundly delimit the need for an elaborate theory of judgment to provide a back-up for what we already know.

There are, of course, precedents for such a refigured and revitalized notion of experience. One is reminded of William James's attack on the pulverization of experience at the hands of the British empiricists and his installation of a "radical empiricism" in which the "world-as-experienced" provides the requisite resources for discerning and evaluating that which we live through in our quotidian existence. There is also the contribution of phenomenology, and particularly Merleau-Ponty's existential phenomenology, with its refiguration of intentionality as lived performance in our learning of skills and our social practices. And there is the contribution of Heidegger in his analysis of the structure of understanding and interpretation as proceeding from a "fore-having" (*Vorhabe*) and "fore-sight" (*Vorsicht*), through which the primordial phenomenon of "Dasein-as-Being-in-the-World" first shows itself. This fore-having and fore-sight, Heidegger takes pains to emphasize, occurs in advance of the "assertion" (*Aussage*) in "judgment" (*Urteil*), which is relegated to a "derivative mode of interpretation." It is particularly from these three seminal twentieth-century philosophers, James, Merleau-Ponty, and Heidegger, that we have much to learn as we think our way through and beyond the aporias of the modern epistemological and semantic theories of meaning and reference as they bear upon the relation of experience and judgment.

Purdue University

RAMSEY ERIC RAMSEY

TRANSVERSAL RATIONALITY, RHETORIC, AND THE IMAGINATION

Probability and Contingency in Experience and Judgment

> At no time can one say: "I have done all my duty," except the hypocrite.
>
> Levinas

I. A HUMBLE INSIGHT

Nietzsche tells us that if we cannot be absolutely certain in our judgments, then perhaps we should not judge at all; he quickly follows, however, with "would that it were possible."[1] This insight highlights not only the risk and danger of post-metaphysical projects, but shows also the necessity of those risks. Rhetoric, by dealing specifically with that which must be probable and possible, rather than certain, understands the necessity of risks. Because a rhetorical thinking toward liberation understands that cynicism and quietism have not avoided risks, it therefore does not see risk as a prohibition from action. In the following reflections I would like to sketch the ground of judgment, the logos of judgment as transversal rationality, and the future's role in judgment *via* the imagination.

II. A CRITICAL RETURN TO RHETORIC

A fine point of departure for my current reflections is the work of Schrag as he has made major contributions to situating the role of rhetoric in the matrix of contemporary thought.[2] Following Heidegger's claim that we are at the end of philosophy, Schrag reminds us that this end occasions the beginning of the task of thinking. Schrag argues quite persuasively that at the end of philosophy we find rhetoric resituated. Philosophy in this sense, the philosophy claimed to be at its end, is a type of metaphysical philosophy that searches for universal truths and acultural certainties. As rhetoric seldom, if ever, is or was engaged in such searchings (though this was much to Plato's chagrin and we would do well still to take his objections seriously), I follow Schrag in holding that

A-T. Tymieniecka (ed.), Analecta Husserliana, Vol. LIII, 97–107.

rhetoric and its methods, along with its traditional concerns, has a major role to play in answering the questions that still remain – perhaps even haunt us – at the end of philosophy. These questions have to do with and bear directly on concerns of experience and judgment.

In his two most recent works, which I consider as volumes one and two of his project of transversal rationality, rhetoric and communicative praxis are placed at the center of his ethical reflections. This, of course, is not surprising, since Schrag as early as *Experience and Being* was demonstrating the fundamental role that communication plays in the intersubjective relations among persons in the life-world. Indeed, this early work was ahead of many in recognizing the fundamental importance of communication to philosophic study. In the first work, *Communicative Praxis*, Schrag makes a move to return to the inspiration of the major works of Aristotle that deal with probability, *viz.*, the *Politics*, the *Ethics*, the *Poetics*, and the *Rhetoric*. In the last of these, especially, Schrag finds what he will later call the "resources of rationality." That is, rhetoric is seen as the means to discern in situations, not *via* some universal method, but by a more situation-based-recognition of what is fitting. In the wake of his return to rhetoric, Schrag develops the first of his recent major contributions to contemporary philosophy in his concept of the fitting response. This return to rhetoric gives the theoretical project two key components necessary for contemporary thinking: first, it gives an inescapable concern for others and secondly, with equal importance, a recognition that others are always a part, participants, and products of the *polis*. The fitting response is the rhetorically constructed and philosophically informed intervention of speech and action into and from within the existential "dwellings" that were together the "timely" aspects of situations. Dwelling here is linked with the recovery of *ethos* as a constitutive charter of the fitting response just as it was in its own way central to Aristotle's position on rhetoric and to his reflections in the *Nicomachean Ethics*. Thus, experience here needs to be seen against the backdrop of habit and character as the sedimentations of what we have undergone and suffered. The classical connection among ethics, rhetoric, and politics is brought into contemporary focus. This understanding is exemplified when Schrag argues: "It is within the space of *ethos* that we meet rhetoric. The intentionality of the rhetorical event, its directness to the other as interlocutor and co-agent, discloses the space of *ethos* as the arena for moral discourse and action, as the abode or dwelling" within which the life-world articulates its varying positions on the public and political life of the *polis*.[3] A move to *ethos* as dwelling – which situates the fitting response within the concerns for the other, the polis, and the exigencies of this particular situation – confronts the often asked performative question.[4] In what follows I add to the very fine analysis undertaken by J. M. Fritzman, who addresses the possible manners of answering this question of how one performs the fitting response in his insightful reading and considerations of nostalgia and the sublime.[5] Fritzman ends, as shall I, by making an appeal to the future. But before the move to the future I wish to make a more Aristotelian response to the performative question.

I suggest that the demands for an answer to the question "How does one perform the fitting response?" are consequently displaced by the move to dwelling and as a result are placed onto the terrain of the question "where?" This move from "how?" to "where?," I suggest, allows the situated and embodied notion of communicative praxis to answer the questions concerning the fitting response. If one is forced, forced just by virtue of being-in-the-world, to respond, then we see now, against Schrag's notion of the texture of communicative praxis, that one performs the fitting response on the basis of *where* one finds oneself dwelling (here again taken as *ethos*, charter, and habit) in the varied and radically complex sense of the life-world. It follows, then, that asituational rules or universal performance dictums for conduct cannot be given. "Where" is always some particular dwelling, and consequently rules for universal conduct cannot be given for specific situations that can hold good for every "where." The best we might be said to do in this light of the demands of the "how?" question is to offer a set of constitutive features with which one must deal (i.e., the other, the *polis*, the relevant tradition(s), details of *this* situation, consideration of the future, etc.), but the exact manner of how one must deal with them is to be decided in the timely and fitting manner of the situation. The abstract question of "how?" gets grounded, as it were, in the concrete and praxial concerns of the where of *ethos*. While the fitting response can address the abstract question of how by moving to the concrete particularities of where, it still faces another charge.

The traditional concerns of ethical theory have been to give universal and trans-situational imperatives for behavior and conduct. This, as we have seen, the fitting response cannot do nor, for that matter, is it something it is willing to do. The notion of dwelling, however, reveals that situations are neither relative nor wholly unique. We do not, indeed we cannot, begin completely anew in each existential encounter and Schrag speaks to this when he talks of the fitting response's inescapable relation to its own situatedness. The fitting response is not, nor could it be, constructed, be made accessible to the other, or, in fact, be fitting or timely at all, if it where not the case that all of these factors are indebted to an embodied past and a tradition that precedes and makes possible the particular interventions of speech and action. This does not mean either, however, that we simply mimic the tradition and give our speech and action the imitative structure of what has been already said and done. The fitting response, to be liberatory, needs to "take off" from the tradition. As Schrag articulates it, the fitting response is: "enjoined to do double duty – to preserve the tradition without which communication could not proceed, and to critique the tradition in search of a pharmakon that might remedy its conceptual and existential ills."[6]

Positions such as the one I am articulating here are often suspected for being not only relative, but irrational as well. Facing this question raised by the turn to rhetoric, a counter-turn must be undertaken to a critically inspired return to earlier philosophical endeavors and a concern for *logos*. That is to say, even if we abandon many of the answers of metaphysical philosophy, there are lessons nonetheless to be learned from the heart of that intellectual tradition concerning questions of *logos*. To this point Schrag has spoken quite decidedly: "The problematization of the classical and modern claims for universality has sufficient force to awaken us from our dogmatic metaphysical and epistemological slumbers, but this problematization does not require that we simply tune out the voices of classical and modern thought. After the overlay of philosophical constructions has been pruned away, certain existential concerns about understanding of self and world in the tradition of Occidental thought continue to solicit our interest."[7] Schrag goes on to say emphatically that, "we cannot think and act except through an engagement with the tradition. The task is to stand in a *critical* relation to the tradition."[8] It is important to mark this orientation because it has become a platitude to assert that modernity has had deleterious effects on Western thinking and that its influence acts only to inhibit rather than foster answers. (For example, one could put together a multi-volume work containing the one paragraph with dismissals of Kant.) Although there is obviously a point to be made by critiquing the Enlightenment and the tradition of Western philosophy, nonetheless, we need a critical engagement with the tradition rather than its absolute abandonment.

In *Resources of Rationality* Schrag confronts this question of *logos* and develops a description of rationality that he calls "transversal" which is both indebted to and critically distanced from the tradition. This is not a simple return to *logos*. This return is undertaken to fashion a critically situated rationality that recognizes probability, contingency, and the specificity of context, and this rationality is indebeted to an Aristotelian backdrop. In the end it becomes clear that transversal rationality is set up by a recognition of the role that rhetoric and communication play, which was first worked out in *Communicative Praxis*. Transversal rationality can only seem to come after the resituation of rhetoric undertaken in that first work, which is one reason why I suggest that these works need to be taken together as volumes one and two of the project of transversal rationality of praxis.

In *Resources* Schrag extends the role of rhetoric developed in *Communicative Praxis*. The transversality of praxis has three major features that can be said to be rhetorically indebted. Schrag's three conceptual distinctions, *critique, articulation*, and *disclosure* mark the praxial moments of transversal rationality. As the metaphor of transversality suggests, the three moments of critique, articulation, and disclosure move across and between, and I would pluralize things here and suggest "among," the life-worlds ontical hierarchies and among its horizontal homogeneity of flattened taken-for-grantedness. By engaging in this moving across and among, transversal rationality is not reducible to either the plane of verticality as universal, synchronic, and transcendent nor the plane of horizontality as the particular, diachronic, and immanent. Transversal rationality is neither a universal and orthodox reason nor a wholly immanent and heterodox one. While Schrag does not say as much directly, one gets the sense that transversal rationality is the possibility for discernment and considered judgments about self and society between and among the vertical hierarchies and its sedimentations of horizontal homogeneity of the life-world. Considered in this light and under this description, transversal rationality finds itself playing a role similar to the other major "metaphors" in twentieth-century continental philosophy, i.e., *différance* in Derrida, power in Foucault, and will to power in Deleuze as he uses the concept from his reading and appropriation of Nietzsche.[9]

If we see the rationality from which we move away as being the certainty and necessity which traditionally defines the major premise of a syllogism, then I suggest that transversal rationality is a move toward the possibility and contingency which, Aristotle teaches us, is the rational character of the major premise of an enthymeme. Consequently, when it comes to the ethical and political interventions to be undertaken by the fitting response, the major premise that grounds one's speech and action is only probable and contingent and not necessary or universal.

The entire work of transversality and its three moments of practice generating the fitting response, as rigorous and detailed as it is, can still only give us interventions of speech and action based on probability. When we displace the discussion of fitting responses on to the ground of the probability of the enthymeme and the terrain of the where of ethos, we are able to see even more connections between the project and its Aristotelian/rhetorical heritage. The use of the enthymeme uncovers the debt transversal rationality also has to rhetoric in respect to its three praxial moments of critique, articulation, and disclosure. Each of these three practices, which must always appear together, set out from the variegated amalgam of the life-world and not from syllogistic universal principles. While these are certainly more sophisticated, they bear family resemblance to Aristotle's discussion of the construction of persuasive discourse in the *Rhetoric*. If we have a rich and complex version of what the life-world means and clearly communicative praxis gives us just such a description, then: "we have no need for grand theory to back up praxial critique, articulation, and narrative configuration [disclosure] all of which are quite capable of accomplishing their designs within the dynamics of experience itself."[10] The event then of critique, articulation and disclosure proceed and emerge in light of the probable and can be more clearly seen in light of a *logos* associated with the enthymeme which is, of course, not irrational because it is probable. As Aristotle tells us in the *Rhetoric*: "Most of the things about which we make decisions, and into which therefore we inquire, present us with alternative possibilities. For it is about our actions that we deliberate and inquire and all of our actions have a contingent character."[11] And thus our ethical and political worlds must find ways of adequately dealing with probability. Transversal rationality as the *logos* fitting to our situation, offers us an opportunity to face the challenge of contingency and post-metaphysical thinking. Perhaps it goes without saying, but here it goes nonetheless: contingency is not radical relativism that says anything goes, and probability is not irrationality. Using the concept of "insufficient reason" from Bloomenburg's work on rhetoric, Schrag says: "This principle of insufficient reason pruned of all *a priori* rules and antecedent criteria, subject to inescapable limitations and fractures is able to supply only *provisional* claims."[12] Yet, is something missing here, or if not missing still only latent and thus in need of further conceptual clarification? What these positions based on provisional claims often are, are plans of future action that take as their beginning a probable major premise. Very well, but are they not underwritten by an imaginative construction and envisioning of how things might be, and are not these visions often motivated by something beyond the specific proposal that is itself the fitting response? How does the fitting response envision the future? How do these enthymematic images come to the rhetor engaged in situations? Here transversality, even with my focus on the enthymeme, may need some assistance.

III. IMAGINATION, THE FUTURE, AND THE DANGER OF FASCINATION

In his essay "Imagination in Discourse and Action" Paul Ricoeur gives us the beginning of the type of thinking concerning the imagination that we need. In the remainder of this essay I shall outline these salient features of imagination and address its subsequent dangers.[13]

Ricoeur turns to the imagination for the purpose of bringing back its creative role. The imagination's creative function can get us to the future before the actual future arrives. It opens a space of anticipation that can imaginatively judge the effects of our proposed response before they are undertaken. Imagination occupies the other side of the scale of judgment, if you will. While our dwelling involvement grounded in experience as history and developed character weighs on one side of the scale whose fulcrum is the present, the imagination and the future weigh on the other. Perhaps we have here, taking a cue from Kant, involved non-involvement. The productive and creative imagination are able to temporarily transcend the exigencies of the life-world, but they cannot became wholly transcendent because they are still tethered by ethos to the situation. Yet this very tension, this high-strung vibration might shake loose a response we could judge as fitting. Indeed, as Ricoeur notes: "It is in this state of non-involvement that we try out new ideas, new values, new ways of being in the world."[14]

If the exigencies of existential situations motivate the embodied character of persons and the past as experience, we need to ask what motivates imagination? The creative imagination is activated in response, but what calls it forward? I wish to call these motivations of imagination – in a position developed with my colleague Lesley Di Mare – transforming and transformative ideals, and these ideals are regulative in the Kantian sense.[15] The ideals are transforming with respect to themselves, that is, they change or can be altered in the wake of each experience and undertaking. But they are also transformative with respect to situations because the motivation they supply to imagination can alter and transform the situations in which they are invoked. While motivations for the production of transforming and transformative ideals remains necessarily elsewhere, we bring back pliant, but sufficiently "firm" concepts that are able to effect the situations and relations in the life-world.

Lest we get caught in the mere repetition of tradition and the habitual, we need the imagination to generate the space or distance from the tradition so that we may be able to critically evaluate and furthermore so that we have some measure by which we may judge. Ricoeur argues that there are two moments of imaginative practice, which he calls ideology and utopia, each with their positive social function. However, the imagination brings its own attendant dangers as well. Ricoeur calls these the pathologies of imagination and both ideology and utopia have their dangerous sides.

Imagination gives us the representations that make communication and associations in the life-world possible. Ricoeur cites this as the positive sense of ideology, as a product of the imagination. The complex discursive forms we construct so as to be able to share the life-world are indispensable. The pathology of ideology comes when an authority captures the representation and controls any given situation with that construct. In these cases authority freezes the representation – a representation that it further attempts to naturalize and protect from alteration and challenge.

This marks the danger of ideology. We have already made note of utopia's function as the imaginative critique of the actual, above in the reference to transforming and transformative ideals. Marcuse had already recognized the positive function of utopian thinking and what transforming and transformative ideals are meant to represent, and Ricoeur's thought articulates the relation succinctly when he writes: "from nowhere comes the most formidable challenge to what is."[16] But there is another side as well. Utopia's danger can be said to lie in its failure to return to the life-world. As Ricoeur puts it: "Because utopia arises from a leap elsewhere, it develops unsettling features that are easy to spot . . . a tendency to subordinate reality to dreams, a fixation with perfectionists' designs and so on."[17] Ricoeur goes on to note that utopia also can fall into the danger of an all or nothing logic, and that it can ignore the difficult work and piecemeal strategy demanded of us by the material features of the life-world. Much work has been done in ideology critique, but are there safeguards to the pathologies of utopia that do not entail abandoning utopia altogether? This question bears heavily on transforming and transformative ideals because of their strict relation to utopian thinking.[18]

I think we can make one move of caution by examining how ideals are developed and function with respect to those who engage with them. Here again a lesson from rhetoric shall help us articulate our point.

The will, or at least what is left of it, does not attach itself to the ideal solely from the side of the will, nor does the will select or initiate the connection *in toto*; rather the will is as much captured by the ideal. In English we have the wonderfully telling phrase "to be taken by something" as in I was taken by her or him. In this we see that the transforming and transformative ideals are not simply created statically, but they choose those who might be said to be ready or responsive to them, they choose us. Like the dynamics of a good conversation, the rhetor and the ideal affect one another. The ideal one creates takes on a life of its own and from this living posture, begins to transform in response to the transformed rhetor who confronts it.

Thus, our wills are captured by transforming and transformative ideas and we are pulled toward the future by our interaction with them and yet that future, as we known painfully, recedes from our arrival. We never get *there*, because *there* never is right here; that is, where we finally are. And yet this must not deter.

The danger comes if we fall into fascination with the ideals once they have taken us. Fascination imprisons us within it if we believe in one of three errors: 1) that we resolutely chose the ideal as ours and that as a static entity it is something we possess (we are, to the contrary, as much possessed by it); 2) if we mistake the ever receding *there* quality of the ideal to be here in full or; 3) if we mistake the totality of the details of the ideal to be fullfillable (ever) in the time in which we now stand.

To be taken requires letting go – a type of release, but this can go too far and fall into being over-taken. This fall into being over-taken would be fascination. Recalling here Heidegger's notion of *Gelassenheit*, the concept can now take on a character that might keep it from pure quietism, and might keep us from being deluded into thinking that we resolutely forged the connection with the ideal.[19] From this side we guard against being over-taken by our arrogance with respect to our will, but the will that does not wholly choose the attachment to the ideal still has a role to play. We choose to stop the interpretation that leads to fascination. Once the will is chosen it must struggle against being over-taken.[20]

To link the imagination and transforming and transformative ideals to transversal rationality and the fitting response allows us to chart some type of course in the midst of post-metaphysical challenges. With an appeal to communication and the rationality of probability and an appeal to the productive imagination we see an interplay between the legacy of the Aristotle of the *Rhetoric* and the *Nicomachean Ethics* and the Kant of the second and third *Critiques*. We need to see these two positions as necessary components of thinking about experience and judgment, rather than in the unhelpful characterization of an either/or that demands that we choose between them. The choice between Aristotle and Kant on questions of experience and judgment is undecidable.

Nietzsche ends the aphorism with which I began by saying that we are unfair beings because we always and of necessity judge too soon, always with insufficient information, and never with full knowledge, and yet Nietzsche does not leave this observation as the last word. These very things we can know about ourselves, he tells us. This knowledge of our unfairness ought to give our judgments a sense of humility that can begin to protect us from the pathologies of imagination, an imagination we desperately need if we are going to have any chance at all to be fitting in our judgments and the experiences that will follow from them.

Arizona State University West

NOTES

[1] F. Nietzsche, *Human, All Too Human*, trans. Martin Farber (Lincoln: University of Nebraska, 1984), aphorism 32.

[2] Calvin O. Schrag, *Communicative Praxis, and the Space of Subjectivity* (Bloomington: Indiana University Press, 1984), see esp. pp. 179–196 and "Rhetoric Resituated at the End of Philosophy," *Quarterly Journal of Speech* **71**(1), pp. 164–174. See also, Calvin O. Schrag and David James Miller, "Philosophy and Communication: Convergence

Without Coincidence," in *The Critical Turn: Rhetoric and Philosophy in Postmodern Discourse*, eds. Ian Angus and Lenore Langsdorf (Carbondale: Southern Illinois University Press, 1993).

[3] Schrag, *Communicative Praxis*, p. 202.

[4] See David James Miller, "Immodest Interventions," *Phenomenological Inquiry* **II**, 1987, pp. 108–114.

[5] J. M. Fritzman, "The Future of Nostalgia and the Time of the Sublime," *Clio* **23**: 2, 1993, pp. 167–189.

[6] Schrag, *Communicative Praxis*, p. 207.

[7] Schrag, *The Resources of Rationality* (Bloomington Indiana University Press, 1992), p. 166.

[8] *Resources*, p. 166.

[9] Only a sketch can be provided here, though in just this brief outline, one can note the relation among the concepts cited by thinking about the way these concepts function in the following: Deleuze's *Nietzsche and Philosophy*, trans. Hugh Tomlinson (New York: Columbia, 1983), Derrida's "*Difference*," in *Margins of Philosophy*, trans. Alan Bass (Chicago: Univ. of Chicago Press, 1982), and Foucault's *History of Sexuality*, trans. Robert Hurley (New York: Vintage, 1980).

[10] *Resources*, p. 109.

[11] Aristotle, *Rhetoric*, 1357a, 23–27.

[12] *Resources*, pp. 135–136, italics mine.

[13] I am indebted in the following to the very fine analysis of Richard Kearney, *Poetics of Imagination* (New York: Harper Collins, 1991).

[14] Paul Ricoeur, *From Text to Action*, trans. Kathleen Blamey and John B. Thompson (Evanston: Northwestern University Press, 1991), p. 174.

[15] Ramsey Eric Ramsey and Lesley Di Mare, "Transforming and Transformative Ideals: A Contribution to Critical Rhetoric," Top Paper presented to the WSCA Convention, Portland, Oregon, February 1995. Schrag has taken this position and incorporated it into his own recent work: see *The Ryle Lectures*: forthcoming, Yale University Press.

[16] Ricoeur, p. 184. See Herbert Marcuse, *Eros and Civilization* (Boston: Beacon Press, 1962).

[17] Ricoeur, p. 185.

[18] I have already made some moves in this area dealing with the work of Ernst Bloch; see "A Politics of Dissatisfaction," *Rethinking Marxism* **8**: 2, Summer 1996.

[19] See Ramsey Eric Ramsey, "The Earth Might Be Round, But the World Is Flat," *Kinesis* **21**: 2, Fall 1994.

[20] There are lessons here to be learned still from Arendt. See *Kant's Political Philosophy*, edited with an interpretive essay by Ronald Beiner (Chicago: University of Chicago Press, 1982).

WILLIS SALOMON

POETRY AND EMOTION

Psychoanalysis and the Ontology of Lyric

Lyric poetry takes subjective consciousness as a condition of its content, certainly since romanticism, but arguably back to the origins of modernity. In this way, it has served the purposes of a formalist aesthetic in which poetic speech instantiates subjectivity in that larger Hegelian sense of the "ideal becoming the real." However, this intrinsic conception of lyric presents some obvious difficulties for psychoanalytic interpretation. Pointing to defenses, resistances, and discursive distortions, psychoanalysis exacerbates disparities in the relation of speech and content. It seeks to discover motivations based on unconscious ideas, emotions, and impulses, even if it grants the difficulties of distinguishing various kinds of unconscious articulations. This emphasis on contradictory motivations contrasts with perhaps the most salient generic expectation of lyric: a subjective wholeness in which a speaker reflects on an experience and transforms that reflection into insight, into self-understanding – transforms it, that is, into a lyric "subject." Put in another way, the pathos of self-presence, of self-understanding, of the subject-who-knows-herself, is central to our generic understanding of lyric, even if that presence is only implicit in the authorial judgment of an objectified speaker of intractable unreliability, as in the dramatic monologue. Hence, vigorously symptomatic readings seem to outrun the subjective closure that formal closure enforces in the lyric. As a result, even though lyric poems present in richly impacted form the deflections and delusions that are the hermeneutic plunder of psychoanalytic interpretation, the interpretation of lyric has lagged behind the kinds of readings suggested by the post-structuralist narrative of fragmentation with which psychoanalysis is so obviously consonant, especially Lacanian psychoanalysis, with its myriad versions of the letter in sufferance.

Raising the issue of the relation of lyric utterance to its implied content thus involves the ways in which the self-realizing subject of lyric can also be the displaced or "split" subject of psychoanalysis. The emotions are crucial here. They comprise in lyric what is pertinent to a formalistic reading, the content transformed into aesthetic wholeness; but they also include what is pertinent to a psychoanalytic reading, the affective

A-T. Tymieniecka (ed.), Analecta Husserliana, Vol. LIII, 109–121.

register of expressive content, taking affect in the broadest sense of emotions as they are involved in the responsive reenactment of fantasies and their accompanying symptoms. If read in a context of motivated behavior, the emotions of a self-referring lyric speaker can begin to show how the generic expectations of lyric and the explanatory claims of psychoanalysis are more similar than they at first appear. Moreover, I would suggest that this link is more rather than less the case in language-privileging, Lacanian versions of psychoanalysis, in which the split between imaginary fixation and symbolic interpellation emerges in language, in concatenations of signifiers.

I

No small part of the difficulty in placing the emotions in a psychoanalytic context is their relation to the unconscious. Freud acknowledged the ambiguity of the emotions in explaining the relation of conscious and unconscious motives in the 1915 metapsychological paper, "Unconscious Emotions":

> . . . the possibility of the attribution of unconsciousness would be completely excluded as far as emotions, feelings, and affects. . . . But in psychoanalytic practice we are accustomed to speak of unconscious love, hate, anger, etc. and find it impossible to avoid even the strange conjunction, "unconscious consciousness of guilt," or a paradoxical "unconscious anxiety." Is there more meaning in the use of these terms than there is in speaking of "unconscious instincts"? (*SE* 14: 177)

In this passage, Freud's classificatory ambitions are at odds with his recognition that the meaning of the seemingly paradoxical term "unconscious emotions" is in its explanatory use and that this use is defined relative to chains of reasoning that explain conscious behavior in terms of unconscious motivation. The impossibility, he says, of avoiding "even the strange conjunction 'unconscious consciousness of guilt'" is the impossibility of interpreting neurotic conflict without a grammar of conscious action. This does not mean, however, that unconscious emotions do not stand potentially in causal relations to conscious avowal or behavior. Freud's hermeneutics does not go so far as to say that, because we grasp action in a context of practices, conventions, and rules, unconscious emotions are limited to "reasons under a description" and not "causes" of conscious avowal or behavior. It means, rather, that the motivational disposition of unconscious emotions is complexly causal, but not in either law-like identity relations between brain and mind or in stable correspondences between manifest and latent content. As Freud recognizes, a strictly causal interpretation of the drives, even if understood as emanating directly from the mind's "bodily organ and scene of action, the brain (or central nervous system)," as Freud describes his materialism in the summative *Outline of Psychoanalysis*, would blunt the explanatory possibilities to be had from viewing the drives as concepts linked to an expressive picture of mind in which emotion, through affect, has illocutionary force (*SE* 23: 144).

Thus, the relation of affect to the drives is part of the overall psychic organization of the subject. But the structure of the psyche is not simply emotions converted through repression into symptoms. An emotion constitutes, rather, as Jonathan Lear has put it, "a framework . . . an orientation to the world" (47–49), rationalized in the process of its expression and in which process "recognition has a transformative effect" (7). In these terms, an emotion functions as *both* a symptom *and* a context for its interpretation, since it manifests itself in clusters of cathexion. So, as emotions arise in the compensatory gestures of the discoursing ego, they form a "ground," not in the sense of an accumulated pressure to be discharged, but rather as material to be experienced and transformed as a result of entry into a symbolic dialectic of desire and loss. Psychoanalysis in part depends on a "transcendental argument," in which behaviors, including the speech acts that reveal them, are explained by an appeal to the unconscious substratum of the drives. Neurotic conflict exists on observation. Hence, something like the unconscious drives must be true to explain it. But their causal force as reasons for action, including unconscious reasons, is part of the texture of their experience as emotions. Freud's question – "Is there more meaning in the use of these terms than there is in speaking of 'unconscious instincts'?" – is rhetorical: there can be psychoanalytic meaning *only* if terms like "unconscious emotions" mean more than "instinct," since, as Freud says in the outline, it is only by "associations provided by the dreamer himself [*sic*] to the manifest content" that "we can reinstate the latent content to the dream and 'interpret' it" (*SE* 23: 169). In this way, Freud's discussion of "unconscious emotions" attempts to preserve the distinction between affects and drives in a way that maintains a causal relation but one in which the associative links that comprise acts of consciousness are open to possibilities of explanation.

Freud thus suggests that we must speak of "unconscious affects" as long as they are not mistakenly believed to exist at the same level as "unconscious ideas," that is, as long as the distinction between a substratum of unconscious drives and the manifestations of them in the experience of an individual is maintained. "The whole difference," he goes on to state, "arises from the fact that ideas are cathexes – basically in the form of memory traces – whilst affects and emotions correspond to processes of discharge, the final manifestations of which are perceived as feelings" (*SE* 14: 178). It is not difficult to see the theoretical motive behind this distinction. If affects are the symptomatic manifestation of interpretable unconscious ideas, then they cannot be located on the same conceptual plane as those ideas. If they are, they threaten to relativize the connections between symptoms and repressed ideas and, in doing so, diminish the possibility of categorical symptomatic schema. But, given the obvious explanatory opportunities of moving between two contrasting levels of utterance, unconscious and conscious expressions of emotion in a developing story of cathexis, Freud allows an ordinary sense to prevail, in that saying that affects and emotions correspond to a process of discharge allows him to explain what such representations can *do* for the mental states they are of, and thus for the subject who has them (Wollheim: 66–67).

There are any number of passages in both Lacan and Freud that encourage this possibility of making affect not only discursively mediated, but also mediating of discourse. For example, in the seminar of 1953–1954, *Freud's Papers on Technique*, Lacan says that

> The affective is not like a special density which would escape an intellectual accounting. It is not to be found in a mythical beyond of the production of the symbol which would precede the discursive formulation. Only this can allow us from the start, I won't say to locate, but to apprehend what the full realization of speech consists in (1991a: 57).

Speech here is the constituting medium of subjectivity. This much is unproblematic. The question concerns rather the ontological status of "what precedes it." For Lacan, "apprehension" is preferred over "location," that is, explanation over cause, but explanation of the fullness of the relation of manifest and latent content and the underlying symptomatic structure that their relation implies (Zizek, 1989; Ch. 1). Instead of "a special density beyond intellectual accounting," there is the possibility of the "full realization" of affect (in "full speech"). It is also striking in this passage that Lacan does *not* reduce affectivity to language. It is rather *placed* in speech, or its place in speech, its being as speech, is acknowledged but not limited. That is, speech does not materialize in the *letter*al to the point of being rule-bound, or requiring an account of rule-governance (like the algorithms). Patterns of reasons for action are defining of affectivity in a holistic and potentially developmental understanding of the grounds of interpretation. Affectivity is not separate from, but located in, speech. It is interpreted with reference to its own representation.

Hence, for psychoanalysis, emotions have grammatical currency as both symptom and cause, both "outward criteria" and "inner process." For Freud, emotion is both the discharge of energy and the awareness of the discharge of such energy, caused by a wish and constituted in the process of repression. The wish is ideational; it has content in the memory traces of infantile gratification. The content is repressed because it is unacceptable to a deeply felt, socially constituted understanding of values and purposes. From these conflicts, symptoms develop. Thus emotions are "converted" into symptoms, or into anxiety, but the process of conversion begs the question of its own representation. That is, conversion assumes a language imbued with the symptomatic cluster of criteria that require but do not name an origin. We have the converted thing, but what was it before it was converted? Even though Freud generally reserves the idea of conversion for somaticization, the conversion model still loosely obtains in cases of anxiety, where failures of repression constitute symptoms, often phobias, in an *a posteriori* process. That is, the emergence of symptoms as deviations within patterns of speech and behavior reveals the failure of repression, assumed as the normative ground from which such deviations occur.

So how do such conversions occur? What can be the connection between mind and body that controls them? That such conversions are mysterious, Freud was often at pains to acknowledge. He more or less abandoned the idea of a link between mind and body in Cartesian terms by 1908, and in the *Outline of Psychoanalysis*, in a passage partly quoted above, he says:

> We know two kinds of things about what we call our psyche (or mental life): firstly, its bodily organ and scene of action, the brain (or nervous system) and, on the other hand, our acts of consciousness, which are immediate data and which cannot be further explained by any sort of description. Everything that lies between is unknown to us, and the data do not include any direct relation between these two terminal points of our knowledge. If it existed, it would at the most afford an exact localization of the processes of consciousness and would give us no help towards understanding them (*SE* 22: 144–145).

Monique David-Menard offers this explanation for why there can, in psychoanalytic terms, be no solution to the problem of conversion. There can be no solution to the problem of conversion, she argues, because its "mysteries" are precisely those that weaken the mind/body picture that would explain them. The hysterical body, for example, if readable, must be said to "think" (12–14). That is, from a crudely dualist point of view, when a thought is displayed physiognomically, it either ceases to become a thought or it must be said to think. If it is to be interpretable psychoanalytically, it cannot cease to be a thought, lest it become merely a physical symptom, simply manifest enervation and explained in the simplest "psychosomatic" terms. Characterizing symptoms in this way would reduce the relation of psyche and soma to a crude correspondence between sign and meaning. However, saying that the body thinks is to speak in the language of a post-Cartesian holism. As David-Menard puts it, in becoming a symptom, thought, "takes the place of an utterance" (13). This formulation reveals the inappropriateness of a mind/body dualism in rendering the relation of symptom to interpretation, since saying that "thought takes the place of an utterance" begs the question of the deep (or reified "unconscious") grammar of any utterance we would regard as symptomatic. But it also shows that at issue here is not the discharge of mind through body in act and utterance, but rather relations among statements as they reveal levels of motivation and degrees of understanding in a developing story of responsiveness to conditions and judgments about those responses.

It has been remarked that Freud's development of an instinct-based theory of the neuroses and the corresponding movement away from one based on affects prevented the development of a proper understanding of the relation of affect to the drives (Jacobson: 38). But Freud's remarks in the *Outline of Psychoanalysis* show how his scepticism about a correspondence theory about instinctual causality requires that the behavioral dimension of affect and the repressed ideational motives of the unconscious be seen in an objectivizing totality of behavior, one in which significance and meaning are taken to be essential properties of human behavior. Behavior, rather, like "personality," implies no binary choice as to the conscious or unconscious meanings beyond the register of intersubjectivity where dialogue takes place; behavior, taken in this way, gives us a sense of a non-dualist ontology where events seem to grow out of each other in a rather impersonal fashion as in nature. The idea of behavior here is thus certainly not behaviorist. It provides rather a once-removed perspective from the counter-transferential vicissitudes of compulsive fantasy. In this formulation of the relation of conscious/unconscious, the impersonal unfolding of affect holds unconscious and conscious meanings together in the same nebulous zone of interpretive potential. The conscious/unconscious distinction remains the basis for specific interpretive moves (even in the structural theory). Attributing emotions to the unconscious does not compromise the distinction. Contrarily, it reinforces the necessity of understanding unconscious emotions (and motives generally) on the model of conscious mind (Searle: 151–173). In assimilating archaic mind to active mind, psychoanalytic interpretation requires a context of plausible motives – "as if" possibilities – out of which can be constructed narratives of motivated action that contextualize specific constellations of affect.

Of course, involved here is the question of pre-linguistic understanding. The issue is a vexed one for a language-privileging conception of psychoanalysis, the kind associated with Lacan. It is vexed because, as Marcia Cavell points out, the claim that the accession to language reveals the pathos of a lost subjective plenitude assumes that there is something to be lost (114–117). Moreover, the master tropes of psychoanalytic theory – the drives, object formation, the oedipus complex, castration anxiety – all assume a pre-linguistic experiential narrative in the reactions of the child to a world defined by the familial drama. The belief that the subject's discourse contains the residue of archaic mind makes something of a transcendental deduction from the discourse of the subject to complexes realized in symptoms. Assumed, in other words, is a developmental conception of consciousness that, while it is subordinate to webs of alienating signifiers, also bases itself on a narrative of pre-linguistic ontological conflict. Understood in this way, the commonplace view that Lacan's linguistic preoccupations "banish the 'I,' " as Norman Holland has put it, misunderstand this crucial dependence on ontological perspective by taking at face value linguistico-epistemological claims that are actually ontological ones (210). Holland rightly argues, with reference to Lacan specifically, and to the post-structuralist denial of subjective presence generally, that it is wrong to think that "a formal description of language equals an account of the psychology of language users" (148). But we should not conclude from this that the ontological dimension, the contextualized narrative of psychic alienation is irrelevant. Rather, psychic alienation suggests the unacknowledged primacy in the linguistic picture of selves in stories. The reduction of person to signifier only has interpretive power if the signifying action, the "representation of a subject for another signifier," takes place within an intersubjective context, even if arguments privileging the signifier can easily be construed otherwise.

Given this assumption of an intersubjective context for signification, the emotions, from a psychoanalytic perspective, cannot simply be "effects of signification." The issue of linguistic materiality hinges for psychoanalysis on the status of the wish, of "desire." Desire acts theoretically for Lacan, more or less, as the instinct/drive confusion does for Freud. As Lear has suggested, a drive, as a "psychical representation of biological stimuli," realizes its meaning in "the mental representation which partially constitutes it" (124). Thus it cannot be conceived as an instinctual, causal "outside"; it must be part of a developing tissue of understanding, however perforated with misrecognition. For Lacan, unconscious desire is "part of the concrete discourse," that is, accessible only in language and not existing separately from it. As such, it is "transindividual," not in the possession of an individualized subject (1977a: 49).

But if desire is transindividual, how can it be the continuous force in an individual human life that it must be in order to have interpretive power psychoanalytically? This is where "affect" is so theoretically crucial. In Lacan's Seminar XI, *The Four Fundamental Concepts of Psycho-Analysis*, affects are the "representations of the representation of desire" (1977b: 217). As such, they are signifiers, but of a *structured* insufficiency, of the "hole" or "lack" represented by the *objet a*, the untranslatable mark of the impossible outside of consciousness. The turn to the object in Lacan is thus a necessary boundary condition for the realization of the subject's individuation in discourse. As objects reveal themselves as the metonymic signifiers of desire, they become crucial in the formation of fantasy. But the term "signifier" inadequately comprehends the complexity of an emotional matrix that "represents desire" in any bracketed action or constellation of actions. If "desire" represents itself in the ordinary language of emotion, it will involve a strong sense of the "I." In fact, its own reduction to the literal will imply the impossibility of such reductions, as it does in Wittgenstein. Russell Grigg gets at something similar, I think, when he reads Lacan's turn to the object as a turn away from the belief that, as can be seen in Freud's "The Dynamics of Transference," "the symptom is *just* a *symbolic* structure." If it were, it would "be fully displaceable from object to object" (111); it could not be said to be symptomatic of anything, and it would not imply a subject for whom the self-representations of mental life were internal to the organization of that individual mental life, that individual consciousness. If fantasy is to be analyzable, symptoms cannot be so freely displaceable. Lacan's "impossible real" thus functions allegorically as the enabling condition of narratives of loss and lack, their unassimilable, non-symbolic outside. Desire, in turn, becomes (as Cavell says of Freud's "instinct") Lacan's "weasel word," negotiating the obscure space between causality and the interpretive reasoning that constructs the analytic narrative (55–56).

II

I would suggest that, in the same way that the disposition of affect in the analytic situation emerges in a developing narrative of constellations of symptoms, so too does the emergence of stance, or self-positioning, in lyric poetry involve the production of affect and articulation of emotions in response to the fragments, to the "partial objects," of an implicit narrative. This is, of course, a rather large jump, perhaps even farcically so. The simple fact of the open-endedness of analysis in comparison to the well-wrought closure of lyric, indeed the very problem with which I began, speaks against it. But recommending it is the localizing function of the tension between conscious and unconscious emotion, as opposed to the narrative-of-life, death-driven masterplot of Freud's structural theory (though the death drive is obviously important for a broader understanding of the way affect is defined in its attempt to fill the void marked by the *objet a*) (Bracher: 169–174). The tension between conscious and unconscious is the tension between symptomatic relativism and the indestructibility of the wish in Freud, or between the metonymic chain of signifiers and the affective inertia of desire in Lacan. It arises in lyric as conflicts between tone and substance, between figure and apparent purpose. It arises, moreover, often with a figurative density that makes the formal symmetry of lyric itself a factor in the relation of utterance to meaning.

The utterance of a lyric "I" foregrounds an implied emotional drama. The fact of literariness, of the artificial or rhetorical mode and motive of lyric poetry, is ultimately answerable to this persistence of the personal (and the experience and judgment it embodies) in lyric. Indeed, most formalist understanding of lyric utterance – "fictions" composed of assorted grammatical modalities – will still focus on speaking action. That is, predicating emotions and states of mind and motive to the "I" of a poem will give it fictive life, an implied agency. Even the most imagistic poems imply perceptual states and emotional dispositions. Perhaps the only way of talking about lyrics that does not imply a subject in this way are the most atomized, categorical analyses of figures of speech. Yet again, the "schemes" and "tropes" distinction in the history of rhetoric developed to accommodate precisely this conception of the reflexive incorrigibility of the personal that lyric invites.

That the emotions in lyric poetry thus assume the concept of a sentient, expressive person as a *formal* principle is crucial to the possibility of reading lyric poetry in psychoanalytic terms. This is not to attribute to the content of a lyric poem an uncomplicated non-linguistic existence in which discursive expression is secondary to expressive origin and power. Obviously, in short poems, the emotions are concentratedly discursive and representational. But location in the natural body, or in the "body" of a self-contained poem, grants the emotions an undeniable facticity, and emotional authenticity, no matter how attenuated, remains a crucial aspect of our *conception* of lyric expression, emerging in the voices of lyric personae as speech most intended to move and thus appearing least artificial (or most sincere) and thus least subject to control. With emotions in lyric, it is no more plausible to imagine a self without a story than it would be to imagine that story without "emotional content," where we understand that content to be of a piece with, as Cavell has recently put it, "the world . . . we have been in touch with all along" (116). To quote Cavell further, who paraphrases Daniel Dennett but with reference to psychoanalysis, "a creature that can tell a story about itself, and only such a one, is a self, someone to whom we can attribute subjectivity" (117). Divorcing the self subject to emotions from the formal conception of lyric would mean imagining poems that are not, to some extent, stories. The analogy might be imagining language games without forms of life. Whatever the extent of the analysis of the grammar of emotional utterance, the assumption of depth, of conceptual bedrock, involves an embodied sense of activity. This sense, as Charles Taylor says, "is encoded in the body" and "activates" the "patterns of reasons for action." It constitutes an "other kind of understanding," one coextensive with what in ordinary discourse we call emotional and one presupposing of the narrative of the drives on which psychoanalysis is based (185).

III

In this way, lyric subjectivity does not necessarily premise a regressive self-presence that reaffirms the general importance of the category of the "autonomous subject" and denies its symptomatic delusions, as some contemporary pieties might suggest. Rather, its attempts at disclosure are concentrated exempla of unconscious or preconscious processes of concealment wherein motivated speech becomes symptomatic, and in becoming symptomatic, its unconscious motives are revealed for possible description. Symptom is thus conceived as contingent and retrogressive, as the constituting medium of the "returning" repressed, in order to prevent, as Slavoj Zizek argues

> a naive substantialist notion of the unconscious as a positive entity ontologically preceeding its "returns" . . . the repressed content doesn't return in symptoms, there is no way to conceive it in its purity undistorted by "compromises" that characterize the formation of the symptoms (1992: 14).

In this view, symptom, as the "surplus of what is effectively said over the intended meaning puts into words the repressed content" (14), escapes the substantialist error by comparing "the proposition with its own process of enunciation" (1992: 26). The "horizon of meaning," Zizek goes on to say, "is always linked . . . to a point *within* the field disclosed by it; the frame of our point of view is always framed (re-marked) by a part of its content." The moment of psychoanalytic insight, then, comes when the circumscribing horizon is inverted by the unseen detail, the point on the horizon that the subject doesn't see and that turns "context" into foregrounded detail. The frame of understanding is in this way enframed by a privileged part of its content. Thus the subject's "meaning," the organization of experience into a livable totality, becomes "symptomatic." But the moment is a moment of transformation in the realization of what Zizek calls "the symbolic debt," the seeing of the "knot that condenses all that one had to forget" in the encounter with "Thou art that!"

Thus Zizek's account of the symptom requires an emergent transformation in a developing narrative of compromises. But it does so within an assumed intersubjective network of vectored misrecognition. To see the place of the heretofore unseen symptom is to reposition speech in relation to others and thus to itself, but the repositioning is, of course, neither final nor privileged, nor can it be in any sense complete. Emotions, with this in mind, might then be said to comprise manifestation of apparent motive forces *recognizable in* compensatory gestures of a speaking subject. The grammar of expressions of emotion in lyric, expressions which reveal the direction of response to loss, become themselves the means by which, even in Lacanian terms, their transformation occurs. That is, we see emotion in lyric as part of a symptomatic cluster *already in* the process of transformation.

Hence, a poem becoming itself in the process of its movement to closure, is, from a psychoanalytic perspective, a process of both misrecognition and transformation. Misrecognition is prior, as emotion is disposed and displaced in relation to its object in complex ways. But, as various forms of affect are described, in both poem and transferential commentary, "a dialectic is introduced into the affective register . . . in the form of various ways of denying" and in which "we grasp the transformations, the very metabolism, which occur in the preconscious order, in a symbolic structuration" that "includes very elaborate grammatical mechanisms" (Lacan, 1991b: 237–238). Thus, the apparently conflicting critical narratives of self-interpreting self-formation and regressive misrecognition are only in apparent contradiction in psychoanalytic readings of lyric poems, especially when we consider the responses of a reader and the (re)constitution of a text in commentary. Misrecognition and self-formation appear contradictory only if we view the symptom in terms of a narrow etiological dualism, one in which it is a "sign" of some corresponding neurosis or signifiers understood independently of the inertia of the transference. But my point here has been that symptomatic misrecognition and a developing potential for at least contingent self-understanding are not irreconcilable in the lyric if we see them as part of a complex and evolving illocutionary grammar of expressions of denial, patterns of concealment, and moments of misrecognition. In these terms, the transformations of a subject appear only as glimmers of possibility, but these glimmers are linked to *formal* properties of poems in tangible, interpretable ways, ways that could be said to constitute the typical speaking action finally of lyric poetry, indeed, the ontology of its representation of subjectivity.

Trinity University, San Antonio

REFERENCES

Bracher, Mark, *Lacan, Discourse, and Social Change* (Ithaca: Cornell University Press, 1993).

Cavell, Marcia, *The Psychoanalytic Mind: From Freud to Philosophy* (Cambridge, MA: Harvard University Press, 1993).

David-Menard, Monique, *Hysteria From Freud to Lacan: Body and Language in Psychoanalysis*. Trans. Catherine Porter (Ithaca: Cornell University Press, 1983; 1989).

Freud, Sigmund, "Unconscious Emotions." In *Standard Edition of the Complete Psychological Works of Sigmund Freud*. Vol. 14, 1915. Trans. and ed. James Strachey *et al.* (London: Hogarth Press, 1966–1974).

Freud, Sigmund, "An Outline of Psychoanalysis." In *Standard Edition of the Complete Psychological Works of Sigmund Freud*. Vol. 23, 1940. Trans. and ed. James Strachey *et al.* (London: Hogarth Press, 1966–1974).

Grigg, Russell, "Signifier, Object, and the Transference." In *Lacan and the Subject of Language*. Ed. Ellie Raglund-Sullivan and Mark Bracher (New York and London: Routledge, 1991).

Holland, Norman, *The Critical I* (New York: Columbia University Press, 1992).

Jacobson, Edith, *Drives, Affects, Behavior*, Vol. I. Ed. Rudolph Loewenstein (New York: International Universities Press, 1953).

Lacan, Jacques, *Ecrits*. Trans. Alan Sheridan (New York: W. W. Norton, 1977a).

Lacan, Jacques, *The Four Fundamental Concepts of Psychoanalysis*. Trans. Alan Sheridan (New York: W. W. Norton, 1977b).

Lacan, Jacques, *The Seminar of Jacques Lacan, Book I: Freud's Papers on Technique, 1953–1954*. Ed. J.-A. Miller. Trans. John Forrester (New York: W. W. Norton, 1991a).

Lacan, Jacques, *The Seminar of Jacques Lacan, Book II: The Ego in Freud's Theory and in the Technique of Psychoanalysis*. Ed. J.-A. Miller. Trans. Sylvana Tomaselli. Notes, John Forrester (New York: W. W. Norton, 1991b).

Lear, Jonothan, *Love and Its Place in Nature: A Philosophical Interpretation of Psychoanalysis* (New York: Farrar, Strauss, and Giroux, 1990).

Searle, John, *The Rediscovery of the Mind* (Boston: MIT Press, 1992).

Taylor, Charles, "To Follow a Rule." In *Rules and Conventions: Literature, Philosophy, Social Theory*. Ed. Mette Hjort (Baltimore: Johns Hopkins University Press, 1992).

Wollheim, Richard, "The Bodily Ego." In *The Mind and Its Depths* (Cambridge, MA: Harvard University Press, 1993).

Zizek, Slavoj, *The Sublime Object of Ideology* (London: Verso, 1989).

Zizek, Slavoj, *Enjoy Your Symptom!: Jacques Lacan in Hollywood and Out* (New York: Routledge, 1992).

MICHELLE FACOS

LANDSCAPE AND ALIENATION IN THE LATE NINETEENTH CENTURY

At the end of the nineteenth century, artists working on the periphery of Europe began producing neoromantic landscape paintings. I call them neoromantic because their spiritual content and poignant imagery recall the static, emotive images of northern German Romantic painters like Caspar David Friedrich or Carl Gustav Carus. Like Friedrich and Carus, they subordinate mimesis to emotion – to the expression of personal feelings and the intimation of a universal cosmic order.

Unlike the proliferation of earlier artistic trends such as Neoclassicism or Realism, there was, in general, no web of interconnections or influences linking neoromantic artists. Furthermore, the cultures producing these paintings represented a diverse range of political and social conditions. For example, while Finland, Flanders, Norway, and Poland were pressing for political emancipation from foreign occupiers at the end of the nineteenth century, Germany, Russia, Sweden, and Switzerland were sovereign powers. The question I pose is this: why did a similar kind of landscape painting emerge at the same time in these particular places?

It is only in the past decade that western art historians have begun taking a serious look at cultural developments on the margins of Europe, an interesting historiographic issue in itself. Nonetheless, neoromantic landscape painting remains little known to scholars outside the nations where it was produced, and is generally viewed by them as a purely local phenomenon. This results in part from the fact that neoromantic paintings have an intracultural specificity that makes them difficult for outsiders to fully understand. And because of their intimate ties to the definition of nation identities, scholars have tended to read them as manifestations of highly exclusionary, culture-specific ideas. I, however, believe there is a relationship among these paintings. To me it seems clear that they express a yearning for wholeness and stability in a world undergoing cataclysmic social and economic changes; they are aesthetic responses to a world in which alienation was endemic.

Nationalism was a powerful motivating factor for neoromantic artists, and the definition of a generic national identity was a central concern

A-T. Tymieniecka (ed.), Analecta Husserliana, Vol. LIII, 123–129.

to many, but by no means all, of these artists. Thus, painters like the Norwegian Eilif Peterssen and the Flemish artist Valerius de Saedaleer, concentrated on the most unique or typical aspects of nature in their respective regions. Peterssen depicted the Norwegian wilderness during the blue hour of a Nordic summer evening in *Summer Night* (1886, Nasjonalgalleriet, Oslo). In the latitudes where darkness never comes for several weeks of the year, the blue hour is a silent time in the forest. The crisply delineated forms convey this static quality, as does the crystalline pool of water reflecting the evening sky. Like Peterssen, neoromantic artists favoured dawn and dusk because these are diurnal moments conducive to quiet contemplation and emotional receptivity. And the goal of these artists was to enable viewers to empathize with the feelings they experienced before these subjects.

De Saedeleer focussed on the expansive, cultivated Flemish landscape near the village of Latem, where he established an artist's colony. The wiry fruit trees in *Orchard at the End of Winter* (1908, Town Hall, Aalst) stand in neat rows, the bulges of buds just beginning to appear. In the distance is a typical thatched-roof, white-washed barn and a windmill. De Saedeleer evoked the rusticity and continuity of Flemish peasant life here not only by the subject matter, but also by the willful simplicity of his composition and technique. These timeless images reflect the imperative to define Flemish identity as rural and historically grounded, in opposition to Walloon identity, which at the time was closely associated with industrialization and cosmopolitanism.[1]

Although there were relatively few factories in Flanders, traditional life was threatened by a precipitous decrease in the number of home industries. For instance in 1876 Belgium had 150,000 lacemakers working at home and only 50,000 in 1896, and in the same twenty-year period the number of linen weavers working at home declined from 74,000 to 11,000.[2] Both de Saedaleer and Peterssen created images intended to foster in their compatriots a profound love of their native lands untainted by modernization.

Still, the fact that this kind of painting emerged in places with disparate socio-political situations suggests that forces beyond nationalism were also at work. At a historical juncture when cataclysmic social and economic changes were afoot, the landscape became for many an important locus of yearning for stability and continuity. This yearning was religious in character and represented a rejection of anthropocentric, Judeo-Christian ways of thinking, in favor of what Norwegian philosopher Arne Naess refers to as an ecosophical attitude toward nature. Thus, the conviction that nature is theologically subordinate to humankind and exists for their exclusive use and enjoyment was abandoned in favor of the belief that people are an integral, but not privileged, part of the organic world.

Indeed, a sudden concern for the preservation of nature and tracts of wilderness was widespread in European intellectual circles at the turn of the century. Distress over the rapid disappearance of natural resources and primordial habitats is evidenced in Russian painter Ivan Shiskin's *Ship Timber* (1898, Russian Museum, St. Petersburg). Southern and Western Europe, the British Isles, and even Iceland, had been largely deforested by the Middle Ages for shipbuilding and construction materials, and avaricious entrepreneurs turned their eye to Northern Europe's expanses of virgin forest at the turn of the century to fuel the engine of industrialization. In an otherwise pristine forest devoid of human activity, Shiskin inserts a simple, wooden fence separating the viewer's space from that marked for commercial logging. We can imagine how the dense, dark forest will soon be replaced by a sea of stumps stretching into the distance.

In Sweden, lumber exports rose at such a rapid pace during the mid-nineteenth century that parliamentary action was taken in 1874 to protect the northernmost forests. This did not solve the problem however, and a governmental committee formed in 1896 to study the forest issue. During that decade one concerned citizen remarked, "our forests have been felled for famine money, and only ruin, economic and moral, have been left in its wake."[3] This statement referred to the fact that many poverty-stricken peasants sold to lumber companies long-term leases on their land, usually for a period of fifty years. The claim that deforestation had a moral dimension suggested a relationship of identity between a land and its inhabitants, where the well-being of nature determines the health of all who depend on it. It also referred to the social calamity precipitated by the lumber leasing, which did not include reforestation and left many independent farmers at the brink of starvation. Sweden's Social Democratic Party rallied to the support of these victims in the 1890s. Artist Gustaf Ankarkrona designed a title page to author Valdemar Lindholm's book *When the Forest Dies* (1901) that presented the issue with graphic clarity. Etched on the gravestone are the words "here lies Norrland's independent country folk. Corporations erected the monument." In 1905, Nils Lundh published his book *Firs and Birch*, a book extolling Sweden's most beloved trees, the fir and the white birch, arguing that they were integral elements of Sweden's physical and spiritual landscape. Similar claims were made by other neoromantic writers.

Neoromantic landscape paintings are united by a pervasive emotional mood, and focus on either expansive views of countrysides untouched by incursions of modernity, as in Danish painter Laurits Ring's *Village Road in Zealand* (1898, Statens Musem for Konst, Copenhagen), or forest interiors, like Austrian artist Carl Moll's *Twilight* (1900, Österreichische Galerie, Vienna). Artists invested their images of nature with sacred power through a number of pictorial devices, including unified tonalities, which one finds in both the Moll and Ring paintings, and bilateral symmetry, found in Norwegian artist Harald Sohlberg's *Flower Meadow in the North* (1905, Nasjonalgalleriet, Oslo).

Furthermore, while they undoubtedly did not specifically refer to the writings of eighteenth-century English philosopher Edmund Burke, they did turn to subject matter identified by him as sublime in his 1746 essay *A Philosophical Enquiry into the Origin of Our Ideas of the Sublime and Beautiful.* Burke defined the sublime as "the strongest emotion the mind is capable of feeling." Among the qualities inducing this state were vastness, obscurity, and infinity – found in Swiss painter Ferdinand Hodler's *Jungfrau in the Fog* (1908, Musée des Beaux Arts, Geneva), and fear and danger – embodied in Swedish artist Karl Nordström's *Grave Mound, Stormy Weather* (1903, Norrköping Museum, Sweden).

Swiss painter Ferdinand Hodler witnessed major transformations in native life during the late nineteenth century. The traditional, decentralized, canton-based system of government which had existed for centuries was transformed in 1874 to a federalist state by a constitutional amendment forced through the legislature by Protestant, mostly German-speaking radicals, against the wishes of Catholic, mostly French-speaking conservatives. In addition, the population of the cities exploded – within a thirty-five year period, the inhabitants of Geneva more than doubled. Also in the last quarter of the century, there was a tourist invasion unparalleled in Swiss history, with hordes of scenery seekers trampling the countryside in search of glaciers, waterfalls, and other picturesque sights. This changed forever the character of the Swiss landscape, which was made increasingly accessible and exploited through the opening of roads, railways, and hotels. Hodler and others bemoaned these events. A visitor in search of the old Switzerland complained, "the Oberland is now nothing but a panorama, – an immense museum or show, got up and utilized by hotel-keepers of every nationality."[4]

Hodler routinely avoided depictions of this modern Switzerland, as Sharon Hirsh points out. Instead, he shows one of the nation's most beloved mountain peaks from a vantage point far above the bustling crowds. The peace and tranquility sought in the remote recesses of nature are restored here. While we, as viewers, have no secure vantage point in the foreground, but rather seem mysteriously suspended among the clouds, Hodler provided us with the possibility of achieving a moment of solitude in the alpine heights through contemplation of his image. Indeed, it is primarily through the power of imagination that such experiences can be recuperated, since in reality, they are lost or at least spoiled, forever.

Sublime vastness and fear are embodied in Nordström's *Grave Mound, Stormy Weather*. Such grave mounds, sometimes marked by a large stone or dolmen, are common sights along the Baltic. Until the end of the Viking age in the twelfth century, the graves of important persons were often marked in this way. Folklore is rife with superstitious tales about such configurations. Sometimes they are thought to be gravemarkers of historical figures or giants, other times they represent the home of trolls or the site of buried treasure.[5]

For Nordström and his contemporaries, dolmens evidenced continuity with the national past at a time when many links to it were being irrevocably severed. The vitality of village life in most Swedish provinces had all but vanished, the victim of rationalized land reforms and urban industrialization. And the bourgeoisie rejected residues of folk life and tradition in favor of the latest fashions from the continent. Many Swedish intellectuals perceived these changes as psychologically and socially catastrophic and worked assiduously to restore the sense of continuity and belonging threatened by an uncritical embrace of modernism, industrialization, and market economics. Indeed, this was a motivating factor for all of the artists producing neoromantic landscape at the end of the nineteenth century.

Even in Poland, where nationalistic feelings hovered near the boiling point after the 1797 partition, the lodestar of patriotic sentiment in painting shifted from historic narrative to landscape in the last decade of the century with the advent of the Young Poland group. While reunification of the lands controlled by Austria, Germany, and Russia remained the ultimate goal of these artists, I believe that the destruction of traditional, agrarian ways of life was the fundamental reason for the primacy of landscape painting at the turn of the century. Polish intellectuals trying to establish a linkage with their historical past realized that urbanization and the radical transformation of the Polish social structure threatened the social and psychological continuities they sought to reestablish.

In *The Old Manor House* (1903, National Museum, Warsaw), Young Poland artist Ferdynand Ruszczyc identified a traditional country home as the site of stability, authority, and continuity. Manor houses were emblematic of pre-partition Poland, the eight-hundred-year period when society was intimately tied to the land and neatly divided into two classes: aristocrats and peasants. During the nineteenth century the landed aristocracy all but disappeared as its properties were confiscated following nationalist insurrections and land reforms. It became the urban bourgeoisie, and many peasants became proletarians, as they exchanged plows for hourly wages in factories. Demographic shifts indicate the magnitude of transformations in traditional Polish life: during the second half of the century, the population of Warsaw (then the capital of Russian Poland) almost quadrupled, and that of Lodz, Poland's textile center, grew tenfold. The physical landscape was also irrevocably altered, as mineral- and timber-rich regions were exploited without scenic or ecological considerations.

I have tried here to sketch the contours of what is a broad and complex phenomenon, and to suggest that the emergence of neoromantic landscape painting at the end of the nineteenth century was not merely the manifestation of patriotic impulses in nations situated on the European periphery. Historically significant persons, events, and landmarks could also function as emotionally-charged indices of national identity, yet these were largely abandoned in favor of the more subtle imagery of landscape, where a particular sort of tree, bird or even meteorological conditions triggered in their audiences fervent nationalistic feelings. For neoromantic painters, cosmopolitanism, industrialization, urbanization, and market economics were enemies of both a cohesive society and national identity because they disrupted traditional social practices and values. They fostered historical amnesia and selfishness, both of which were anathema to their collectivizing goals. Above all, they lead to the loneliness, alienation, and social dysfunction that we have come to associate with modern life.

Indiana University

NOTES

[1] I am indebted to Mark Denaci for bringing de Saedeleer's work to my attention.
[2] B. Seebohm Rowntree, *Land & Labour. Lessons from Belgium* (London: Macmillan and Co., Limited, 1910), pp. 88–90.
[3] See Torgard Rentzhog, "Norrlandsfrågan," in *90tal. Visioner och vägval* (Stockholm: Nordiska Museet, 1991), pp. 169–182.
[4] Quoted in Sharon Hirsh, "Hodler as Genevois, Hodler as Swiss," in *Ferdinand Hodler. Views and Visions* (Zurich: Swiss Institute for Art Research, 1994), p. 72.
[5] See Donald Ray Floyd, *Attitudes toward Nature in Swedish Folklore* (dissertation, Berkeley: University of California, 1976).

ALIN CRISTIAN

THE UNBEARABLE LIGHTNESS OF SACRIFICE

Kundera's essay *Les testaments trahis* ends with a parable eulogizing the defence of textual literality against an unrestricted hermeneutic freedom. As a last wish, a dying peasant asks his son not to fell the old pear tree growing in front of the window and embodying for Kundera the will sealed with filial piety. As long as the son's pious memory of his father lasts, the tree will not be felled and by analogy, according to Kundera, our respectful memory of the author should keep the letter of her work intact. Yet no matter how strongly this parable may appeal to our emotions, the phenomenological abyss of literality continues to defy the possessive claim of the author. That which is worthy of respect in the will might elicit being kept alive otherwise than through blind submission, perhaps in the way of a circumspective inquiry into its essence. Let us not forget that the letter symbolized by the tree belongs in the first place to *life* with its power to challenge any would-be absolute foundations.

An ontological openness insinuates itself like a Trojan horse into the heart of the metaphor whose imprevisible life does not yield to any definitive inscription within fixed semantic horizons. This does not go without a certain irony in response, I believe, to the condescending gesture with which Kundera bracketed the question of the end of history. Toward the beginning of the same essay, he lightly dismissed the thorny issue of the *eschaton*, reducing it to a bizarre philosophical whim merely unimaginable in concreto.[1] The end of history, just like its origin, is necessarily linked with the concrete, present sign in which signification and with it history itself are inscribed. The life of this concrete immediacy is always protected from understanding through a fundamental obscurity sustaining the metaphor just as it sustains the tree from the parable. The origin and end of history direct the efforts of a metaphorology intent on describing precisely the withdrawal of these limits. Their appropriation, meant to perform what Blumenberg was calling "a rationalization of the lack," proceeds without guarantees.[2] It strictly forbids the instituting of an absolute metaphor, be it even through a moving apology of literality. Kant convincingly proved that one could

A-T. Tymieniecka (ed.), Analecta Husserliana, Vol. LIII, 131–140.

never find an example for this absolute metaphor, since their very relationship would have no other analogy on the basis of which to be understood. The auctorial will remains then intelligible only from within an *open* horizon, i.e., animated by a life which determines in a structural sense the limits of history.

If the sign and (vegetable) life have the same origin, one could ask in dialogue with Blanchot if the foundation of the auctorial will is not threatened by the very seriousness of the founding gesture, by the elimination of all risks from the life of the letter. What could it mean to open this foundation to contingency through a casual gesture which takes upon itself the unbearable lightness of our being and leaves us, as if through a will, the burdensome quest of its meaning? If the foundation is explicitly opened to the abyss of hazard through a rending and breaking – to remain within Kundera's metaphor let us say through the felling of the tree – how could this futile sacrifice not break the will but bind it and our destiny into a mutually releasing gesture? Could we, by granting this auctorial will its alterity, learn to assume our own sublime exile? The ontological exile can only be conceived of when departing from a metaphysics of presence, from absolute foundations and from a shallow version of literality. Correspondingly, only a flighty gesture can throw us in the obscure abyss of literality, at least as long as it does degenerate into exercise and end in itself, but consecrates the auctorial will by entrusting it to *its* destiny.

For the sake of the argument, how could one understand the gesture of another old peasant who, upon seeing his world of meaning threatened by violent misappropriations, sacrifices the "letter" in which this world is inscribed? In one single night he fells the cherry trees that he himself has planted, has seen grow old under his eyes and has silently felt to be sustaining the existence of his community. The sacrifice of the cherry trees is an attack against the metaphysical understanding of literality, somewhat reminding of Plutarch's Sardanapal. If for the peasant in Kundera's essay filial piety is the guarantee of respect toward the transgression of the limit which circumscribes his world, it might be the very consecration of this world and the proof of his supreme love for it. The deepest care for the essence of the auctorial will appears thus to be disputed between obedience and transgression.

The meaning of a world is determined to the extent to which it maintains itself within a specific horizon delimited, among others, by a concrete moment constituting its "letter." But understood this way, the concrete is as far as possible from something like "dead matter," the self-identical character of which is guaranteed by the metaphysics of presence. On the contrary, the secret life of this concrete moment manifests itself as capability to establish spontaneous connections that understanding presupposes without, however, being able to make totally transparent. Trying to rescue meaning from change, the inscription finds it most intimately and originally entangled in the life of the letter with its inevitable vicissitudes. For Heidegger, the ungraspable thick of these connections functions as resonant states of mind [*Stimmungen*]. The foundation as well as the violation of the letter in which a plurivocal meaning is inscribed amounts to a laceration of the feelings of those who belong, together with the author, to the threatened world. The Pauline reference to the inscription of Christ's revelation as a circumcision of the heart points out in the same direction.

If the inscription as concrete access to the meaning of a world is circumscribed by a horizon, and if the Greek *horizein* is understood as delimitation, cleavage or severance, then this horizon can be said to establish a limit, the letter most originally belongs together with its Other. Conversely, the threatening of a world of meaning necessarily goes through a disturbance of this harmony which defines the nature of the inscription. Yet the threat need not be explicit and therefore perceived as such. The world of meaning is primarily eroded by an imperceptible distortion covered up by the bright hues which, for the most part, keep us absorbed. Depending on how the essence of auctorial will is conceived, we have to decide whether the threat is directed primarily against the letter as such, or rather against the harmony of our mode of relating to it. The idea of ontological exile conveys the specific way in which this harmony is given to us mortals, namely, through contingency as a possibility of assuming it. What are then the ontological conditions for rescuing this harmony if one takes into account that its most insidious form of degradation is inconspicuous, for example by forgetting the limits and by immediately accepting the self-evident?

The problem for the author of a will entrusting a world to posterity is to find a reliable mode of fixating the letter, of extracting it from the context of aleatory links. This amounts to a decision about the essence of meaning and about what counts as merely unessential vicissitudes of its preservation. In other words, the life of the letter has to be sacrificed, buried at the foundation of an enduring meaning the ontological status of which should guarantee fixity to the inscription. When the quest for fixity exhausts the possibilities of metaphysics, it turns into an ontology which takes into question the founding gesture itself. The question now sacrifices each and every answer severing it from the roots of certainty strengthened by forgetfulness. Upon exploring the foundations, the doxic origin of certainty is brought to light, i.e., its thriving upon an uncritically reached consensus. An unfathomable richness of possibilities inherent in the beginning is consequently disclosed from underneath this arbitrary suppression.

But the sacrifice of widely accepted conclusions is existentially significant only insofar as it is repeated, turned into a mode of being, appropriated as ontological openness. The severing of the answer from its inauthentic roots denounces the futility of an *unrepeated* founding, of our metaphysical belief that repetition merely confirms an independently existing identity. Far from existentially comforting posterity, the auctorial transmission of a world through the letter already involves the receiver in a hermeneutic co-participation. A sterile deference that replaces the hermeneutical dialogue with the will turns the life of the letter into a stiff foundation. The fear that the world entrusted might be compromised through a renewed founding could only be overcome through the understanding that what is transmitted is, above all, a mode of exercising the founding gesture and not the foundation as such.

The sacrifice of the cherry trees explicitly jeopardizes the meaning of the world that they sustained, thus exposing it to contingency out of care for its destiny. This exposure, however, is as far as it can be from simple destruction or, as Bataille would put it, it is the ritual transgression that the law of preservation needs in order to assume its ambiguous origin and be authentic. It alerts the consciousness of posterity to the intrinsic vulnerability of that harmony which maintains it attuned to the unfathomable depth of the letter. The recipient's possibility to understand himself depends on the destiny of the letter. The striking contrast between the two wills allegorically represented through the conflicting attitudes toward the trees concerns two different modes of inscribing. They correspond to a certain and respectively uncertain literality founded upon obedience or the possibility of loss. The pear tree from Kundera's essay and the cherry trees from our example are the points of access to transcendence. A more detailed analysis of this phenomenon elicits a return to the Greek myths.

Orpheus' descent to the underworld in order to retrieve Eurydice takes place under an irrevocable interdiction: he is not allowed to look at her *there*, in that underworld which claims her, but can only bring her out on condition that he keeps his face turned away from her. As an attempt to secure a meaning for his existence, the retrieval is essential for Orpheus, it is indeed his supreme work. Art and the journey to the underworld are two moments of the same creative quest for meaning. To ignore the secret of the origin is the absolute condition upon which this work is founded. It can be rescued from contingency, brought to light, founded on clarity and distinctness only provided that the mystery of obscurity is respected. And yet, the hero of the Greek myth, archetype of the artist, cannot submit to this taboo, the essence of which is deferral, and does look at Eurydice while still in the underworld. Thus he loses her irretrievably, although at first sight the respect for the interdiction would have granted him a full-fledged foundation. As if through a hardly explicable lightness he betrays his own creation and, insofar as creation governs the meaning of his existence, he betrays himself as an artist.

Blanchot's interpretation of this myth considers the double betrayal as absolutely necessary for elevating the work above the mediocre foundation that blind obedience would have secured.[3] Apparently defying all seriousness, Orpheus sacrifices the work by exposing it to the hazard of human passion, which means by subordinating its future to a momentary whim. This sacrifice is not a simple annihilation, but rather a placing of what is most important for the artist outside all guarantees. Only what is mediocre remains accessible without reserve and without risks. Cutting off his access to the work, Orpheus exposes it without destroying it. Insofar as the meaning of his life is dependent on this retrieval, with the same stroke he cuts off his access to himself. By returning to the uncertainty of the origin, the destiny of the work is exposed to the risk of annihilation which sacrifice explicitly discloses as essential, infinite vulnerability. The transfer of a present element into absence calls up from under a sedimented oblivion the whole nexus of meanings which grounded that presence. Preservation or loss of the effective meaning of this world becomes now a task lying ahead of posterity as its very destiny. This more fundamental, though only implicit, last wish challenges the mere preservation of the letter without further ado. As conspicuous genesis of an absence, sacrifice denounces the inconspicuous dissolution of the world of meaning which presence always conceals. The vulnerability of the world experienced as painful rupture effectively elicits a response which remains inseparable from the problem of posterity's access to the essential.

Orpheus' gaze is the moment when the weight of the work's destiny is relinquished and entrusted back to the obscurity of hazard disimulated in the context. The umbilical cord of the still unconsecrated act of creation is thus cut off. Through the failure of his expedition in the underworld, Orpheus not only releases the work from the specific ontological nexus of its creation, but he abandons it precisely to the contingent intersubjective references. The obscurity in which the essence of the work withdraws is *inhabited*, and this inhabitancy claims the work in various ways – not necessarily artistic. By suspending the work in a *social* version of hazard, Orpheus' sacrifice opens up a new dimension of the phenomenon which Blanchot's interpretation does not thematize. This sacrifice involves posterity through the fact that Eurydice returns to the same obscurity that shrouds our possibility of access to ourselves. The quest for an authentic existence on our part and the quest for the heroine of the Greek myth irretrievably lost by Orpheus converge in an essential ambiguity. It is the moment when Orpheus' existence has to be freed from the ontological limitations of traditional subjectivity through a releasing gaze, this time on our part.

The tree from Kundera's essay is claimed by only one testamentary wish which is an oversimplified state of affairs insofar as the author's claim does not have to compete for its privileged status. But the ontological condition of the letter is quite different. Filial piety necessarily competes with a whole system of loyalties the hierarchical ranking of which, as psychoanalysis warns us, does not go without saying. For the sake of conciseness we should mention only that the son's unrestricted attachment to *this* letter-tree, his firm refusal to substitute for *this* individual object may appear suspect after a reading of Lacan. Kundera's allegory runs the danger of turning the letter into a fetish. I believe that emancipation from the metaphysical, rigid version of literality is the only chance that the son has to mature, which in a cultural sense means to assume the possibility of loss. A culture which turns the letter into a fetish and reduces its relationship with the genius to filial submission condemns with the same move its highest values to a nihilistic devaluation. The emancipation from its own phantasms might then have to envisage the preservation of tradition only as repetition of the founding gesture. Loyalty would in this case become an appropriation of both the prerogative and the utmost responsibility of reopening the fundamental issues dissimulated in the present letter. Disrupted by failure only in order to elicit a series of rearticulations, the constructive auctorial gesture would culminate in a symbolic destruction mobilizing posterity to a reconstructive work. The moment of auctorial intention to be rescued would be precisely its unintentional but inevitable openings. As possibilities of disputing the letter, they would "appear" as a promise of self-growth, the obscure essence of which would be respected in a Levinasian sense.

Violating the foundation of the work is the same as violating the principles of the world to which it belongs inasmuch as the failure to secure this work is assumed with a lightness verging on irresponsibility. The meaningfulness of a world implies concern about the work which human action accomplishes. For an existence whose fundamental structure is intersubjective, sacrifice understood as closure of one's access to oneself appears at the same time as one's definitive absorption in the others. The more complete the individual failure, the more profound its recuperation by the obscurity of a social context which tacitly retrieves the broken fragments of this individual action. But this failure becomes merely an empirically attestable fact devoid of social significance only when its retrieval by the context fixes the loss of meaning in designating it as an accident. This is the contrary of Orpheus' destiny whose failure acquires the value of archetype and myth. As an individual, Orpheus is not in the position to consecrate his sacrifice. This can be implicitly done only by the community through the recuperation of the fact from the empirical and through its elevation to the status of myth. The two moments are not points in a chronological sequence, but belong together most originally. The action of the hero is not first empirical and subsequently mythical, but is from the very beginning located in the cultural space created through the withdrawal of consciousness from the empirical.

From within the inner space of myth the hero cannot determine the destiny of his work in the consciousness of the community. Not only can't he do so, but such an intention could not even be conceptually articulated in the pre-Homeric *Weltanschauung*. The question of whether the failure of the artist is a will or an accident, if the auctorial intention is a sacrifice meant to mobilize our quest for the irretrievable or a simple idiosyncrasy, does not make sense in the economy of the myth. The pre-Homeric Greeks did not conceptualize subjectivity, hence an individual claim to authorship that would attempt to rescue the work from social contingency would remain incomprehensible to them. Oral creation presupposes from the ground up co-participation which places the "will" in the intersubjective realm. If one can talk about auctorial intention before the introduction of writing, then one should take into account that the good name, that utterly important *kleos* of the Greeks elicits a peculiar type of inscription quite different from literality. Good reputation is preserved through a repeated evocation implying most directly and originally the community.

Things stand differently with the explicitly articulated will, since the author is not in the position to guarantee that his failure will become myth and archetype of a world. But neither is the failure condemned to insignificance through its social recuperation as mere accident. Between these two extremes, there opens the possibility of sacrifice as attempt to consecrate the work without guarantees of success. The auctorial wish to create nothing less than a myth by demanding absolute guarantees that the letter of her work will be preserved might appear presumptuous from a Greek point of view. In directing its longing toward the divine, sacrifice elevates human existence above the vicissitudes of contingency without, though, completely isolating it from those. Salvation of the work from hazard through sacrifice is not a paradox for a consciousness which acknowledges its origin in *ethos*, i.e., in that specific mode of being together of a community. Neither can consecration be seen as an individual action any more. At the same time, even though essentially intersubjective, sacrifice reaches deeper than *doxa* since it cannot be reduced to a matter of common sense. As one's own existential resolve *par excellence* opened to the social context, sacrifice cannot constitute the object of the mainstream discourse in its limitation to most widely acceptable generalities. Between assuming contingency and transcending it there opens a gap the bridging of which calls for a profound modification of the way in which the relationship of the individual with the community is conceived.

Such a transformation replaces the metaphysics of lasting foundations based on the primacy of an eternal present with an ontology of finitude in which the key to human understanding is the circumstantial. But in the intersubjective structure of this changing context an infinite promise is already lurking. According to Heidegger, contingency constitutes the specific character of being [*Sein*] where the ultimate determinations of our experience reside.[4] On the one hand, we have the firmly founded, lasting creation which rests on a form of loyalty toward the past guaranteed by filial love, patriotism, racial or national identity and so on. All these attachments are meant to confer fixity upon a version of literality which claims to totalize the auctorial intention. On the other hand, we have a creation which renews its meaning through exposure to contingency. It thrives on the hermeneutic tensions that its context occasions without entrusting its semantic resources to a lasting principle. It means something only to those who are ready to bury their own existence at the foundation, and thus share in the destiny of the letter. In this infinite hermeneutic exercise, only the readiness to begin anew is lasting. The only thing that such an exercise could found is a style of existence exiled from the safe ground of metaphysical certainties and precariously installed in the fascination of the quest for meaning.

The futile sacrifice of the work for the sake of a momentary whim has the power to inspire other individual destinies to rescue that which is thus disclosed in its intrinsic vulnerability. The essence of the work is just as much lost by Orpheus as it is inscribed in the destiny of the community, more precisely bound up with its transcendental freedom. "What is no longer accessible" does not mean "presently available." On the contrary, the availability of the essential is constantly threatened by an inconspicuous, oblivious accumulation of sediments. The realization of the urgency to resume Orpheus' venture is inspiration. It orients other destinies whose efforts to retrieve the exposed work would have to be in their turn consummated in a capricious gesture able to free the work anew.

If the explicitly formulated will claims a normative status, the futile sacrifice points to the fact that the last grounds of existence elude any norms of understanding and ask to be permanently reconsidered. Its legacy is, properly speaking, the preservation only of our readiness to reconsider the essence of cultural transmission starting from the experience of a wounded literality. The mode of being of futile sacrifice does not attempt to maintain posterity in an infantile submission, but calls upon it to assume the whole weight of its originally fractured being as the only essential legacy. In order to decide whether it should be worshipped in daylight or released in an uncertain obscurity, we might first have to ask whether the last wish still belongs to oneself or already to the others.

Leuven University

NOTES

[1] Cf. *Les testaments trahis* (Paris: Gallimard, 1993), p. 29.
[2] "Ausblick auf eine Theorie der Unbegrifflichkeit," in *Schiffbruch mit Zuschauer* (Suhrkamp, 1978), p. 89.
[3] Cf. *L'Espace littéraire*, pp. 227–234.
[4] Cf. *Einführung in die phänomenologische Forschung*, *Gesamtausgabe* vol. 17 (Frankfurt/Main: Vittorio Klostermann, 1994), p. 52.

PART THREE

INFUSION OF THE AESTHETIC SENSE INTO HUMAN EXPERIENCE

The National Independence Day Celebration in Guadalajara.

LAWRENCE KIMMEL

TELLING STORIES

> Telling stories is an act by which Man strives to realize his capacity for wonder, meaning and delight. . . . The possibilities of storytelling are precisely those of understanding the human experience.[1]
>
> N. Scott Momaday

PREFATORY REMARK

In what follows I will be using Native American culture and literature as the primary focus for a discussion of storytelling. For this culture, the life of speech and the presencing of meaning through the sharing of stories is vital to the very existence and identity of a people. Momaday's remarks about the nature of the relationship between language and experience surely is not limited to the lives of Native Americans. His accompanying claim that we cannot exist apart from the moral dimension of language is no less applicable to our own culture, but showing the importance of an awareness of this condition requires a bracketing of the cultural insensibility that pervades the routine of our lives. The interest of this paper is not in Native American culture or literature as such, but what lessons it may have for a dominant culture which has forgotten the sense of, or lost the genius for, telling stories.

I

It may well be that in the beginning was the word, but the word was not *logos*; it was *chaos* . . . or *eros* (desire) or *moira* (fate) or, for our purposes, *mythos* (story).[2] Somewhere in the beginning, or soon after, was the impulse to tell a story, to fashion wonder and terror and hope and triumph into a story, to make and mark the sense of it all.

A description of human beings as tellers of stories invites a comprehensive aesthetic in which human existence is understood as a creative act of imagination and remembrance. The conception of Man as a storytelling animal[3] shifts philosophical attention away from an abstract reason

A-T. Tymieniecka (ed.), Analecta Husserliana, Vol. LIII, 143–154.

which reduces meaning to the truth value of propositions, and toward a more inclusive depth and diversity of meaning as sense. In terms suggested by Nietzsche, this shifts is from a narrowly conceived *logos* (reason) to a fully imagined *mythos* (story). Nietzsche holds myth to be the cultural imperative of a healthy and creative people. He also uses myth as a measure of what he regards as the degeneracy of modern culture:

> We are now ". . . able to approach the once-living reality of myth only by means of intellectual constructs. Yet every culture that has lost myth has lost, by the same token, its natural, healthy creativity. . . . Man today stripped of myth, stands famished among all his pasts and must dig frantically for roots, be it among the most remote antiquities."[4]

II

In Aristotle's definition of Man as a rational animal, human beings are: what we make (*poiesis*), what we do (*praxis*), and what we think (*theoria*). But perhaps more than all else, human beings are what they imagine and remember, and the creative act of telling stories gives coherence and continuity to the whole of human experience. Indeed, there is no "whole" of experience nor "human" in the absence of stories. Some sense and conception of narrative life and narrated world is as essential to the identity and memory of an individual as it is to a constitution of community and to the *ethos* and history of a people.

In Nietzsche's analysis, Western intellectual history was once secure in the archaic genius of myth which delighted in the manifold of sense and the expression of difference, but has suffered a gradual erosion of natural myth and a corresponding loss of an inclusive and diffusive sensibility. In its place a compensatory progressive discourse has developed, centered preclusively in logistical utility. Nietzsche's call for a revaluation of all values questions the faulty disjunction and problematic priority and dominance of *logos* over *mythos*, declaring that "truth is . . . illusion of which one has forgotten that they are illusions; worn out metaphors which have become powerless to affect the sense."[5] For Nietzsche, the discourse of truth has become so insistently familiar that it has degenerated into the leverage of domesticated agreement. In fact, "there are no facts, only interpretations."[6]

Nietzsche represents part of a movement in modern thought which has witnessed a devolution of the language of truth into languages of contending rhetorics, in which conflict is presumed and the contention for domination and power normalized. Marx, for example,[7] focusing on the material and social conditions under which dominant discourse and values are established and sustained, brought to light arbitrary and self-serving ideologies which structure the values, constructions, convictions and constraints of ordinary discourse. Freud developed a parallel thesis concerning the foundational psychology of language,[8] which further displaced rational, conscious deliberation and decision as dependable resources of knowledge, even in our own most intimate experience. Finally, even within the relatively stable and structured domain of the physical sciences, as Thomas Kuhn's reorienting (and revaluing) text[9] on paradigm shifts in science shows, social context and historical and conceptual discontinuities attend even formalized, exacting, and systematic discourse.

Having said all this, it would be a mistake to conclude that all discourse, whether factive or fictive, practical or theoretical, critical or creative, is to be understood as "just so many stories," if by doing so we simply reverse the present bias and reduce *logos* to *mythos*. There is no small irony in the absurdity of such a worry, however, since it is difficult to imagine conditions of misology such that we would abandon the romance and empowerment of science. There is a lesson here that cuts both ways. Freeman Dyson, in a discussion of science and technology, cites[10] Edith Nesbit's children's story about a magic city where one gets whatever thing one can imagine, but with a single and absolute codicil that one can then never give it up, or stop using it. This simple story, read as an allegory of the controlling metaphors of science and technology, has all the mythic force of the story of the Sibyl,[11] and is for that same reason a nervous portent for the history of our future.

Nietzsche's positive remarks about the cultural need for the essential energy of myth, and the revaluation of *logos* and *mythos*, finds resonance in James Joyce's reminder about stories – that *every* telling has a taling, and that's the he and the she of it. Once we have acknowledged the fundamental coherence of language in story, we can go on to distinguish and celebrate the diverse uses and kinds of tellings. The magnificent trilogy[12] which ends with *Finnegan's Wake* is Joyce's effort to find a language in which to express the mythic in an age of what Nietzsche calls the *epigoni*, those born late into a language and culture which has the dead weight of the past about its neck.

We should keep in mind that it is not the language of science which today lacks credibility and substance. If we accept that in some important sense "all telling is a taling," this is neither cause for despondency, nor an argument in favor of a culturally cheapening relativism or skepticism. Human beings tell stories – of many kinds and shapes, for as many reasons and in as many seasons as the world and the gods will allow. The great worry of our time should be not that biology, physiology, physics, and geology are all discourses which tell stories in differing languages about related overlappings of finally different things. Even if we regard these as stories, "useful fictions" in Nietzsche's words, they are of a special and important kind. Scientific "stories" describe and explain; they do not aim to release or enchant, but to fix and bind. The more important point and lesson is that there exists no one story which is privileged to preclude or replace any other. Even if and when we have answered all the questions of physics, we have yet to ask, much less answer, the questions of biology, or politics, much less those of morality or religion. We should rather worry that we tell and listen to so few stories which concern the deep and fundamental issues of empowerment in our mutual lives.

If our dominant culture has lost a creative and consolidating sense of the mythic, so too our individual lives are often lived out in a poverty of both place and time, and in the absence of a culturally integrating or individually redeeming story. In as broken tradition, stories often take on a mood of consolation. For example, Hannah Arendt, troubled by the decay of a common space and discourse which offers protection against isolation and the terror of totalitarian fiat, quotes Isek Dinesen's somber remark[13] that ". . . all sorrows can be borne if one can put them into a story, or tell a story about them." Martin Heidegger's philosophical version of The Fall – a story in which the question of Being is forgotten in the mundane chatter of a busily efficient culture which drowns out the archaic poetry of the human spirit –[14] similarly chronicles the lost voices of the gods, the remembered earth, embracing sky, and the storied language in which Being expresses itself.

In the most general terms, our collective cultural relation to stories has become problematic and indirect. We have become people of the book, or, further regressed, creatures of the tube. In the passive context of electronic media, we have become neither listeners nor tellers of stories, but viewers of serials, of "soap works." In the place of active imagination, traditionally requisite to the preservation of culturally significant stories, a television "audience" of viewers now becomes a consumer of packaged and predictable dialogue which intentionally goes nowhere.

There are still imaginative efforts of course – print stories in newspapers, journals, novels, the reflective and reflexive efforts of several industries. But even here, a story line on stories is that a once living tradition of communal stories, in which the origin and ideals, the history and prospects of a people were embodied, became in time, at best, products of marketing promotion, more typically a banal distraction of an abstract and specialized technology. To the degree that one can speak of media stories, they have become the province and providence of professionals with energies engendered and deployed for commercialized entertainment: stories to sell soap.

III

We no longer notice, much less lament, that there are no tellers of stories who sing the wonders of the privileged first ones, of the grandeur of the gods, of the origin of time, and things, and death, of the courage and heroic deeds of the warrior, the wisdom and great words of the sage. It is not news that our culture no longer produces scripture, or myth, or fables, or fairy tales, epic or tragic drama. Such stuff is still anthologized for the schools (again, a leveling network of agencies and industries); it forms blocks of requirements for the conferment of degrees for liberal education.

Two Native American writers who exemplify alternative cultural and literary possibilities drawn from the oral tradition of storytelling are N. Scott Momaday[15] and Leslie Marmon Silko.[16] Their experience and writing, however different, is grounded in the poetry and memory of voices from childhood, and they both try to write in the spirit of telling stories in an oral tradition. They grew up within a tradition not under the obsession or imperative of print – of speaking not yet reduced to writing, of talking not yet reduced to media sound bites. Their writing testifies to some degree that there is still an accessible cultural model preserved in the story telling and sand paintings, in the dances, the Kachinas, the curing and celebratory "sings" of the American Indian cultures – an essentially aesthetic culture which commends spiritual health. But the terms of accessibility in any important sense seem to be that the stories must be lived, must be experienced in the ritual setting, the shared silence, and the continuing tradition within which stories are told and heard. In contrast to anthologized collections in print, in the telling there is no point of beginning or end to the stories or the voices. In the oral telling there is only the telling of the same stories, which are never the same, which simply take up and leave off, a telling and listening not simply to pass or fill time, but to express and sustain an understanding of the way of a people. The heritage of time and place and the wisdom of stories comes to presence in the telling. What is vital in the culture may be found in its form as well as its content, in the process which brings together generations, creates a sense of sacred place, of mutual understanding and acknowledgement, enacts ritual celebration in singing and dancing.

There is an apparent loss, in print, even in the best of storytellers. A friend of Leslie Silko's, whose children had brought a library book home with Silko's poem "Laguna Coyote" in it, said to her: "We all enjoyed it so much,/but I was telling the children/the way my grandpa used to tell it/is longer." Silko responded, "Yes, that's the trouble with writing,/You can't go on and on the way we do/when we tell stories around here./People who aren't used to it get tired."[17]

An Indian Telling

Imagine not that you are reading the following story, or that I am reading it to you. Imagine that I am telling you the story, "intending it":

Long ago, out on the high desert below Turquoise Mountain, there was a meeting of witches.
And there was a contest to see which had the strongest medicine, the greater power.
One witch danced and brought forth rain from a cloudless summer sky.
Another beat an ancient sacred drum, and its rhythm slowed the current of the river and stirred the high mountain Aspen and Pine.
Another drew lines in the sand with a stick and snakes came to rest in the hollows.
"What I have is a story," one witch said, and the others laughed, at first.
"Go ahead, laugh if you want to, but as I tell the story, it will begin to happen:
Across the ocean in caves of dark hills, are white skin people, like the belly of a fish, covered with hair.
These people grow away from the earth. Then they grow away from the sun.
Then they grow away from the plants and animals.
They see no life.
When they look, they see only objects.
The world is a dead thing for them.
The trees and rivers are not alive, the mountains and stones are not alive.

The deer and the bear are objects. They see no life.
These are people of fear. They fear the world.
They destroy what they fear. They fear themselves.
. . . Set in motion now . . . set in motion . . . to destroy, to kill
. . . Whirling . . . whirling . . ."
So the other witches said: "Take it back; Call that story back!"
But the witch just shook its head at the others in their stinking animal skins, fur, and feathers.
"It's already turned loose. It's already coming. It can't be called back."
. . . after Leslie Marmon Silko

Is this prophesy or history? Premonition or admonition? An expression of anger, or despair, or recognition, or resignation? It all depends on how the story is told, and retold, and to whom, where . . . and each telling, even this one, will transform the story. There is, in any telling of this kind, an overlapping of personal narratives – of identity, of history, of social protest, of cultural difference and solidarity. As I have told it here, it is a story about the power of the story. But even in this abstract and removed context, one only understands the story if one feels its power, unless it touches and fits a story you are already living – white-bellied like a fish, or grown away from the earth.

IV

If there is a priority of sense to truth, and an interdependence of language and life, there are still limiting conditions within any culture with respect to what can be said, what stories can be told – not just what will be believed, but what can be conceived. Stories grow out of the life and memory of a people. In the oral cultures of traditional peoples, the mundane world is one in which plants, animals, rocks, streams, and clouds are alive. There is power for good or ill in everything, and both life and its expression in story must respect and make use of this power.

So far we have discussed story and myth with respect to apparent bridging of sense and intelligibility within literate cultures. But myth and ritual also involve the empowerment of language in spiritual life, and there is an important sense in which the life of a traditional people is grounded in the sacred, in magic, in the spirit of all things. Paula Gunn Allen refers to myth and ritual as "two wings of the bird of spirit."[18] The "literature" of a tribal people is "mythic" not simply in the sense of making intelligible but also of empowering, as having the capacity to generate and transform. The most familiar example is in the curing sings, the use of sand paintings in medicine stories which intend to purify and make whole. The language of the story shows the right way to proceed in walking the path of health, of power, of beauty, of wholeness.

The magic of words is hardly a mystery to those who read poetry, or enjoy a story, but there is a reverence for words which is special for a traditional people in a primarily oral culture. Momaday, speaking of the grandmother in *House Made of Dawn*, asks us to:

> Consider for a moment that old Kiowa woman, whose use of language was confined to speech, and be assured that her regard for words was always keen in proportion that she depended upon them. You see, for her, words were medicine; they were magic and invisible.[19]

There are general issues about myth and story which connect with convergences in contemporary philosophy. Wittgenstein once remarked that it requires a whole person to have a whole thought. He presumably had in mind Kierkegaard's reminder that paradox is the passion of the thinker, and that the task of a serious thinker is have the courage to think one thought through to completion. Heidegger's familiar and final appeal is both simple and fundamental in its resonance – to think Being. The traditional opening of stories can be modified in the following way to make the same point: "Once upon a time . . ./. . . and even yet. . . ." Leslie Silko telling a story, begins, as to children: "You should understand/the way it was/back then/because it is the same/even now."[20] The cultural foundation of the tradition of storytelling is a shared conviction about the continuity and wholeness of being and time. As in the "path of power" (the universe of medicine and healing), the story and the teller seek always to make whole, to bring into harmony, to connect "once upon a time" with "so it is, and so it shall be." This does not mean that all stories are positive or without conflict or pain, but stories are fashioned out of the residual memories and experiences of a people, and remain open to continuance. There is always more to a story than is voiced in a single telling (what leads up to and away from both tale and telling) which everyone knows – about old man Coyote, or Changing Woman, or the Hero Twins.

A parallel concern to Nietzsche's analysis of myth is to consider what is lost for a dominant culture in the absence of such telling, tradition, and stories. The spiritual power of language and stories of a traditional people seems no longer accessible to a rational, literate culture in which knowledge is power only to the degree that it can abstract, formalize, universalize and systematize. The facilitating logic of contemporary critical culture is a comprehensive activity of making distinctions and making connections, of breaking down and piecing together. An analytic culture centered in the *Logoi of veritas* constructs a world which is no longer whole, but atomized, partitioned, controlled and subdued – one constituted and governed by rule and misrule.

In contrast to this, what still remains in the traditional oral culture of the American Indian develops surrounded by, and grounded in, natural beauty. The environment, which figures in all storytelling, is vital for a people that measures itself by the distance it can see. There was a comparable importance for classical Greek culture of the "*acro/polis*," the "high city" from which one could see the whole of that which gave the *polis* and its people their identity – the tilled fields which sustained its life, the killing fields within which the citizens defended their gates, the burial fields which hold the ashes of their ancestors. In both the ancient and contemporary oral cultures, experience and speech are still natural and in first order – voices and sounds not abstracted into bound books or reconstituted in points of lights on a television monitor. In the continuing if marginal culture of the Amerindian there exists the possibility and so aspiration to harmony, of finding resonance in one's life and actions with all things. The telling of stories in traditional cultures is always an occasion for learning "the way": in Lakota the expression is "to walk in the sacred," in Keres "to walk in harmony," and in the Dine "to walk in beauty"; but common to all is the idea of the sacredness and beauty of life and world – the quest is for a vision of wholeness, of The Way.

Fundamental aesthetic elements come together in the *telling* of stories in an oral tradition, for example the voice of the teller replaces the reference of the text, and the event of the telling, the performance itself, blends into the event which is narrated. Most importantly the poetics and aesthetics of this process, the making of something and the sensuous apprehension of that something, is written on the wind: nothing will remain of the story when the voice fades back into the silence except for what memory sustains, and what is renewed in the story and tradition that has been drawn upon and nourished by the community of teller and listeners.

I have perhaps too closely identified oral tradition with the survival culture of native peoples. An oral tradition has certainly existed within literate cultures as well, for example in the theatre, and in this context great writers have surely had a sense for both the power and fragility of the voice. While working on this essay I attended an outdoor performance in the mountains near Santa Fe (at St. John's College) of a summer workshop performance of Shakespeare's *The Tempest*. Against the backdrop of the mountains, summer storms sweeping in the distance, wind billowing the silken flowing robes of Prospero and Ariel, I marveled, as always, at familiar storied voices sustained these four hundred years, now celebrated in a far distant place:

> . . . our revels now are ended. These our actors, as I foretold, are all spirits, and are melted into air, into thin air. And like the baseless fabric of this vision, the cloud-capped towers, the gorgeous palaces, the solemn temples, yea even the great globe itself, and all that it inherits, shall dissolve, and like some insubstantial pageant faded, leave not a rack behind. We are such stuff as dreams are made of. . . .[21]

This oral tradition of drama – albeit one supported in print – of "live theatre" is itself surely at risk, but attests to the importance, delight, and wisdom of voices, of telling which comes out of and resolves back into silence.

One thing missing even in the most remarkable of theatre experiences now, is any sense of the primal character of the language itself. In one of Gerald Vizenor's stories, a character refers to and comments on Scott Momaday's remarks that:

> . . . the "world did not come into being, but it was. It did not break upon the silence, but it was older than the silence and the silence was made of it." He was thinking about the oral tradition and the telling of stories. Silence and wind and water sounds were here before birds and fish. The birds knew how to sing from the wind that shaped their wings in flight.[22]

There is an aesthetics of oral culture in the telling of stories, a making (*poiesis*) not just of a work, but of a space of imagination and life. In the context of Native American culture, this making invites a response in the literal and original sense of *aesthesos*, a sensuous apprehension of the life and time and space of a people. In this context, Momaday can claim with conviction, and persuasive force, that "We are what we imagine. Our very existence consists in our imagination of ourselves. . . ."[23] It is, I think, doubtful whether the "traditional culture" of the Native American, sustained only by the ghost threads of beauty, can survive the devouring haste and progress of civilized technology. It is a question of more than philosophical interest or literary achievement whether the spiritual reality of this people is anything but a fragile vanishing dream, or whether we will learn or salvage anything from its passing. Telling stories is a collective enterprise of imagination which preserves a culture; it is itself worth preserving as a resource for the renewal and appropriation of what is valuable in those cultures which do not endure.

SUMMARY

This essay has focused on the sense of story, rather than truth in stories, in an attempt to offset the dominating discourse of *logos* and *veritas* – of reasoned truth – as the exclusive order deciding value. Once liberated from the controlling metaphor of *logos*, however, the truth in stories can be pursued in the manifold contexts of its appearance and use. In general, story (*mythos*), has a concern for truth on a broad scale of perception and perspective, of world and life, of a person or a people; the telling of a story forms ground and horizon for the truth it carries. It is only within this larger story that versions of truth made of fact and assertion, argument and justification, theory and verification, come to matter to us. Stories at a mythic level search for ways of being human within a universe which is finally beyond us, and serve to open up the truth of discourse (*aletheia*) into a comprehensive aesthetic of life and world. It is ironic, and a sign of our times, that the appeal of telling stories must be argued and defended. The serious fictive literature of our times is itself framed in irony to distance the writer from both story and the embarrassment of any pretension to truth, since on the prevailing view, truth can only be conceived as universal and objective. In this context it is not difficult to understand the appeal of a simpler world, where truth is found in the telling of a story, not in the means of its verification. The spiritual strength of a simple world is paradoxically grounded in the complexity of human relations brought together in telling stories. A person, self, is a sustained act of imagination. This is true also of a people. Both have an identity through a creative activity which includes remembrance. Human life at its roots is desire, in its fullness an adventure, and in the imagination of its telling constitutes a culture, a people, a history. In acting, remembering, telling, we come together as human beings, we come together into human being.

Trinity University

NOTES

[1] "The Man Made of Words," N. Scott Momaday, in *The Remembered Earth*, ed. Geary Hobson. Albuquerque: University of New Mexico Press, 1980, p. 168.

[2] Both Aristotle and St. John recognized *logos* (word) as fundamental or instrumental to understanding the nature of human being. We should record, however, that the primal character of *logos* is not limited to the agendas of either theology ("God"), or philosophy ("Reason").

[3] This idea now has a place even in analytic philosophy; see *After Virtue*, Alasdair MacIntyre. Notre Dame: University of Notre Dame Press, 1981, p. 205 ff.

[4] F. Nietzsche, *The Birth of Tragedy*, trans. W. Kaufmann. New York: Random House, 1967, Section 23.

[5] "Of Truth and Lies," F. Nietzsche, in *Philosophy and Truth*, ed. D. Breazeale. New Jersey: Humanities Press, 1979, p. 84.

[6] *The Will To Power*, F. Nietzsche, ed. W. Kaufmann. New York: Random House, 1967, p. 481.

[7] K. Marx, *The German Ideology*, Volume One, Preface (1845). See also *Communist Manifesto* (1848).

[8] Freud's most specific treatment of errors ("Fehlleistungen") in language and intention occurs in *The Psychopathology of Everyday Life*; but Freud's whole analysis of "the unconscious" is relevant to this point as it is developed in *The Interpretation of Dreams*.

[9] Thomas Kuhn, *The Structure of Scientific Revolutions*, Chicago: The University of Chicago Press, 1962.

[10] Freeman Dyson, *Disturbing the Universe*, New York: Harper and Rowe, 1979, p. 3.

[11] Recall that the wood nymph asked for and was granted immortality. This gift of the god was, however, no blessed thing for a creature subject to the ravages of time.

[12] Apparently Joyce intended to writes a fourth book in which mythic language would find a final place – following the scheme of development of *Portrait of the Artist, Ulysses, and Finnegan's Wake*. The book was never written.

[13] Hannah Arendt, *The Human Condition*, Chicago: University of Chicago Press, 1958, p. 175.

[14] Martin Heidegger, "What Are Poets For?" in *Poetry, Language, Thought*, trans. Albert Hofstadter. New York: Harper & Row, 1971.

[15] Momaday's major works relevant to this discussion are *The Names*, New York: Harper and Rowe, 1976 and *The Way to Rainy Mountain*, Albuquerque: University of New Mexico Press, 1969.

[16] Silko's major works relevant to this discussion are *Storyteller*, New York: Seaver Books, 1981 and *Ceremony*, New York: Penguin Books, 1986.

[17] L. M. Silko, *Storyteller*, p. 110.

[18] Paula Gunn Allen, *Grandmothers of the Light*, Boston: Beacon Press, 1991, p. 7.

[19] N. Scott Momaday, *House Made of Dawn*, Preface.

[20] Silko, *Storytellers*, p. 94.

[21] William Shakespeare, *The Tempest*, Act IV, Scene 1.

[22] G. Vizenor, *Wordarrows*, Minneapolis: University of Minnesota Press, 1978, p. 117.

[23] *The Remembered Earth*, p. 167.

ALBERTO CARRILLO CANÁN

CONSIDERING DIE AND DEATH

Heidegger's Reinterpretation of Husserl's Concept of Truth as the Concept of "Care"

In his lecture *Der Satz vom Grund*, Heidegger claims that, "death as the most extreme possibility of Dasein achieves the greatest clarification [*Lichtung*] of being and its truth" and adds, "Death is the as-yet-unthought-of measure of the immeasurable, that is, of the greatest game into which man is earthly brought, upon which he is put" (SG 186f.). This demonstrates that death continues to be central for Heidegger's theory of truth even after the publication of his magnum opus *Being and Time* (= BT). This centrality, *however*, is based *completely* upon that which I term Heidegger's *performative ontology*. This ontology respects Heidegger's reinterpretation of Husserl's idea of the "fullness of the originary Givenness of something" as the idea of "care" (*Sorge*). In the following hermeneutical essay I will trace out Heidegger's reconstruction within the context of phenomenological ontology.

1. THE IDEA OF "PRIMORDIALITY"

Heidegger's peculiar idea of the "primordiality" of an "interpretation" can be found at the beginning of the second division of BT. This "primordiality" requires a very particular "basic experience of the object to be disclosed," namely the experience of an object in – and herein lies the peculiar nature of Heidegger's idea – its "authenticity and totality" (232f.).[1] One may speak here of the "primordiality" of the "interpretation," or of the primordiality of the "basic experience" upon which the "interpretation" is based. This idea is problematic, *firstly*, because of the *generality* of its formulation, that is, it deals with the fully general concept of the "primordiality" of the "basic experience" of *something*. It is, *furthermore*, particularly problematic for the interpretation of BT that this idea anchors the so-called second division of BT from its very beginning to the concept of the "authenticity" of "Dasein," which easily leads to the impression of a *break* between the first and second divisions of BT. Heidegger's concept of the "primordiality" of a "basic experience" has, to my knowledge, not been examined in any interpre-

A-T. Tymieniecka (ed.), Analecta Husserliana, Vol. LIII, 155–180.

tation of BT, whereas the problem of the supposed discontinuity between the first and second division of BT has been discussed in a variety of ways – particularly in the hope of "separating the philosophical from the political within BT."

I suggest that we interpret "*authenticity*" as Heidegger's reinterpretation of Husserl's "*selfness*" or "*originarity* of the Given" and "*totality*" as Heidegger's reinterpretation of Husserl's "*fullness* of the originarily Given." I hope to show that Heidegger's concept of the "primordiality" of the "basic experience of an object" or its "interpretation" is *only* a reinterpretation of Husserl's fundamental concept of truth: truth as "originarity and fullness of the givenness of something."[2] In the development of BT Heidegger constantly holds to *this* reinterpretation of the concept of truth. Therefore, at least with regard to the idea of the "primordiality" as "authenticity and totality," there is *no* break in the development of BT. More importantly, the reinterpretation of the phenomenological concept of truth as the "primordiality" of the "basic experience" of something represents the primary aspect of the Heideggerian concept of "care" and the foundation of Heidegger's *performative* ontology.[3]

2. "TOTALITY" AND "AUTHENTICITY" OR THE "FULLNESS" AND "SELFNESS" OF A DIE

I would like to emphasize a passage in the so-called Analysis of Death which, although relatively inconspicuous, is decisive for the *whole* of BT. In his examination of the "*formal*" concept of "the End and Totality" (§48), Heidegger writes: "The last quarter is still outstanding until the moon gets full. The 'not yet' diminishes as the concealing shadow disappears. But here the moon is always present-at-hand as a whole already. Leaving aside the fact that we can never get the moon wholly in our grasp even when it is full, this 'not-yet' does not in any way signify a not-yet-Being-together of the parts which belong to the moon, but pertains only to the way we get it in our grasp perceptually [*wahrnehmendes Erfassen*]" (243.15–21, underlining mine). The context of this quotation concerns difficulties with "regional variations" in the concepts of "that which is still outstanding, the end and totality" (241f.). According to Heidegger, concepts such as "incompleteness," "ripeness," "fulfillment" and "finishedness" etc. (236.20,244f.) belong to such variations. In the quoted passage, Heidegger uses not only the *formal* descriptives "full" and "whole," but also "not-yet" in connection with "perceptually getting-into-grasp" [*Er-fassen*]. This is an obvious application of the typically Husserlian idea of the "degree of the fullness of an originary given" – a gaming die, for example. According to Husserl, there exists, with respect to a *perceived* die, to *the* die "before my eyes," a graduated "fullness of the given." I can hold the die in such a manner that I may "see" only one of its sides, and the *remaining* five sides are merely "modelled" upon the one "seen" side. But I can also turn the die so that I "have" at best three of its sides "before my eyes" at the same time, whereas I "have" the *remaining* three sides still "merely indicated" or "symbolized" by the three sides which I actually see. Husserl himself has a very formal conception of this situation: the "having of something" may generally be divided into an "intuitively had" (the "intuitive" or the "seen") and a "merely signitively had" (the "signitive"). If one represents "fully having" or the "fullness of Givenness" with the value 1, then the act of *perception* is given through Husserl's equation: i (intuitive) + s (signitive) = 1, where 1 stands for "fully having." We thus have, for the first perceptual situation in our die example: 1/6 (i) + 5/6 (s) = 1. In the second perceptual situation we have: 3/6 (i) + 3/6 (s) = 1. It is, in any case, clear that the greatest possible "fullness of Givenness" may never equal one. Not only the moon – as in Heidegger's example – but *every* object of perception is "never (. . .) wholly in our grasp even when it is full," as every object always has another, unseeable, side. "*Fully* having," that is, "perceptual *appre*-hension" not only of the moon, but of every object, is never possible. Even when "the not-yet diminishes," it will never reach the limit of 1.

As an example of a situation which is particularly inopportune for grasping the "fullness of Givenness," we may imagine my holding a thin wire "before my eyes" so that I only "see" one end of the wire. In this case, (i) is almost equal to zero, *though* still *greater* than zero – and *only* this is important, *in order* for the situation to remain a "*seeing* of something," in its precise meaning: I neither conceive of "a" wire in my imagination, nor do I *remember* "*the*" wire as I *saw* it before closing my eyes nor do I model how I would *expect* "*the*" wire to look when viewed from one end upon opening my eyes again. Rather, the wire "itself" "is there" or "*present*." Moreover, I "am alongside" *the* wire "itself," indeed, "now." Direct perception differs from *memory* and from *anticipation* in that, in the latter "*types*" of "having something," the intuitive element (i) is, properly speaking, non-existent (= 0) – although an intuitive or plastic connection still exists between memory or anticipation and the wire "itself."

The "having" of something in its "selfness," in the narrower sense of "selfness," distinguishes perception from memory and anticipation: the something "itself" "is there" in perception, whereas memory and anticipation still have a connection to "selfness," a connection which may be strengthened through repeated perceptual acts which lead the indirect types of "having something" (memory and anticipation) back to selfness in its "presence," that is, to *the* wire in its "selfness."

In contrast, imagination and mere symbolization – through, for example, the word "wire" which excites an "indeterminate" anticipation of "a" possibly visible wire – always refer to an object, as Heidegger puts it, "previously grasped" (62.27) as *a* wire, but not to *this* particular wire "itself." Mere imaginative *expectation* and the *anticipation* of the plastic Givenness of that which is meant[4] by the word "wire" represent a "having something" which lacks all pure connection to a "this" "*in itself*" or "*by itself*." To put it differently, the "thisness" and "selfness" of the "had" are *not a priori* constitutive elements of an act of the imagination or of a symbolic act,[5] and the "previously grasped" thus serves as a symbol for "an" otherwise indeterminate X (a wire, for example) which is no longer "*the* previously grasped" "by itself." The symbol status of the particular previously grasped means, in Husserl's terminology, that the symbol "points beyond itself," namely "to something else." To use a formulation of Heidegger's, this is, with respect to this "something else," a "synthesis structure" (34.3) which no longer possesses the character of an " 'originary' and 'intuitive' taking-hold" of something (37.1).

The preeminence of perception consists, then in its "presenting" the object "itself" in its "selfness": "this" object "is there" for the act of perception,[6] or, to employ another formulation from BT: the object is "bodily present" (346.31). As the "selfness" of the perceived always implies its "thisness," we may with *intentional* ambiguity sometimes simply speak of an object's "selfness."[7] Altogether, the "Dasein" or "presence-at-hand" of an object or its "originary Givenness" is an *a priori constitutive* accomplishment of perception and is coincident with an object's "selfness," in the sense of its "bodiliness" *and* in the sense of its "thisness." In contrast, the object is, in the case of pure signification – e.g., through the word "wire" – "merely intended"; which means, "symbolically," and thus merely "in-authentically" "given," namely as "an" object of a particular kind, e.g., "a" wire. Thus, "bodiliness" *and* "thisness" belong to the "authenticity" or "originarity" of the given. The "selfness" of the given is thus, for Husserl, equivalent to the "authenticity" of the given.[8] One can characterize the reality-accomplishment of the act of perception formally as possessing *something* "authentically" and with a certain "fullness." This means, when applied to Heidegger's example of the moon, that "*the*" – *not* "*a*" – moon[9] "is available" ("authenticity") although "not full" (the certain "fullness" which represents a "not-yet" with regard to "totality").

"Authenticity" as the *a priori constitutive* complex of "selfness" in the sense of the "bodiliness" and "thisness" of the given, implies, therefore, a certain "gradation of fullness" of the given, which can then experience a – for Heidegger's conception – crucial reinterpretation through the phenomenological concept of *truth*. Husserl names this "Being" the "truth"[10] particularly because the "Dasein" or "Being" had always implied a certain "fullness." The "truth" as "Being" belongs *constitutively a priori* to the sense of the perceived. This "truth" as "authenticity" is thus the "accomplishment" of the act of perception and its differentiation from other "*forms* of the givenness of something." The question of the "*gradation* of the fullness of the originary given" can be reinterpreted as the problem of truth in the sense of the "availability of being in its totality and authenticity" or as the problem of the "primordial and authentic truth" of *something* (cf. 301.21f., 316.22f.).

3. THE "TOTALITY" OF THE "POTENTIALITY-FOR-BEING"

In addition to being applied to perception and the perceived, the aforementioned formal determinations may also be applied to Heidegger's concepts of "understanding" and "possibility." To this aim, one must take Heidegger's claim seriously that "understanding of possibility lies *only* in the *seizing upon* [*Ergreifen*] it" (cf. 38.30f., emphasis mine), that is, the "*taking-hold* [*Erfassen*]" or "projection [*Entwerfen*] of the possibility does not lie in having it in view, but in *throwing oneself* into it as a possibility" (cf. 336.25f.). The expressions "understanding" or "projection" are here Heidegger's *purely formal* terms for an extremely specific type of "taking-hold" which is *not* a "perceptive taking-hold." The pretension of "taking-hold" of the whole moon is an aspect of perception, whereas the "*seizing-upon* of the possibility" should actually "*take* hold" of the possibility. We can say that *perception*, the *sensory experience* [*Vernehmen*] of *object*, on the one hand, and the *seizing upon of possibilities*, on the other, are "ways in which Dasein comports itself" (220.16)[11] and, furthermore, that sensory experience [*Vernehmen*] "presents" or "sees" objects, whereas seizing-upon "presents" or "sees" possibilities.

I must emphasize here that Heidegger expressly defines "understanding" *formally* as a "seeing": "In giving an existential signification to 'sight,' we have *merely* drawn upon the peculiar feature of seeing, that it lets entities which are *accessible* to it be encountered unconcealedly *in themselves*" (147, italics and underlining mine). *Formal* "seeing" corresponds to the "showing-itself-*in-itself* as it is in itself," that is, to the *formal* concept "phenomenon" (31.3–17,35).[12] The expression "in-itself as it is in itself," stands, here, for the *ambiguous* "selfness" of "the entity which has become accessible" (6.34): it refers to the entity "*itself,*" namely "as *this* entity" (85.21).[13] It is important to apply the formal concept of seeing to Being-possible and potentiality-for-being. Being-possible or the potentiality-for-being are particular forms of being: the *being* possible which "is" "Dasein" (143.32). The "access" to existent possibility lies not in a sensory experience [*Vernehmen*] rather "alone in seizing upon it." To put it differently, sensory experience [*Vernehmen*] "presents" objects, whereas seizing-upon presents possibilities "*originarily.*"

It is, here, necessary to clarify the problem by concentrating on the particular possibilities of "Dasein," as *opposed* to the possibilities of *other* entities. The "seizing-upon" of one's "own potentiality-for-being" is a "comportment of oneself to oneself" (4.16f.), but his self-referential comportment has itself many possibilities, as the particularly important passage about hammering demonstrates: "*the less* we just stare at the hammer-thing, and *the more* we *seize* hold of it and use it, the *more primordial* does our relationship to it become, and the *more unveiledly* is it encountered as that which it is – as equipment" (69) – and, namely, "as *this* equipment (. . .) *that* it is" (68.25f., emphasis mine). Hammering is *both* comportment to the hammer *and*, as such, a potentiality-for-being of Dasein: the potentiality-for-hammering. The *intensity* of the act of hammering which Heidegger demonstrates in the cited passage extends from complete failure to act ("just staring") to a "more seizing-hold use" to an "'*authentic*,' wholly devoted busying oneself with . . ."

(354.4f., emphasis mine). This "just staring or looking at" (74.8) the "hammer-thing" is a failure to act (to hammer) properly speaking, but not simply the disappearance of *my*[14] potentiality-for-hammering, rather a "dimming-down of the possible" (195.1). In contrast, the "wholly devoted busying oneself with . . ." represents the *greatest intensity* of action, whereas each halfhearted "busying oneself with . . ." implies a "not-yet" with respect to the *Being* of the action: the action is "not-yet" or "not wholly" what it could *be* due to such halfheartedness. The *extremes* in the intensity of action mean *either* that the potentiality-for-hammering "is" "wholly" "there," *or* that it "is" "there," but "covered-up." "Mere wishing" is one example of such a case in that it "closes off the possibilities; what is "there" [originary givenness!][15] [in wishing] turns into the "actual world" (195.21f., underlining mine).

The *greatest intensity* of action coincides with the *seizing-upon* of action as *possibility*. This seizing-upon, which does not yet represent the concept of "resoluteness" in BT, can be seen as the resolute execution of action. Already before writing BT, Heidegger made clear that "insofar as this *resoluteness* is actually carried out, insofar, then as *I* am *resolute*, *is* the action *there* in its *utmost* possibility (B19 150, emphasis mine.)" This "utmost Being-there" [*Da-sein*] of the action, that is, the "fullness" of its givenness, is no "accomplishment" of the act of perception, but is, rather, the very specific accomplishment of seizing-upon as resolute execution. Resoluteness in this colloquial usage takes over *and* outstrips the accomplishment of "givenness" of perception. Both acts present "something" "originarily" – in the case of resoluteness, even "wholly." The *bodiliness* (*praesentia*) of a percept is an achievement of *perception*, or, rather, of my perceptively "Being-alongside" the thing (*praesens sum* as *accipiens adsum*); complete presence (*plena praesentia*) of action is the accomplishment of the *resoluteness* of the actor that I am (*praesens sum* as *praesens ego*). We may say that the *intensity* of the action is nothing other than the *degree* of *fullness* in the *presence* of the possibility *itself*, where failure to act amounts to the mere "disguise" or "concealement" of the possible, *rather* than its disappearance. In contrast, the "transparence of the action" (B19 150)[16] lies in the aforementioned resoluteness.

4. THE ORIGINARITY OF GIVENNESS IN THE "MODE OF PHENOMENON" AND THE "MODE OF SEMBLANCE"

We must examine the problem of "concealment" more closely if we are to clarify the connection between the "totality" and "authenticity" of the givenness of *something*. In the following, therefore, I will be concentrating not on the possibilities of "Dasein," but on the possibilities of *other* entities.

The action itself, that is, "the hammering itself" (69.14), is the comportment to the *other* entities (in this case, the hammer) which "frees" this entity "for *its own* possibilities" (144.35f.). We may say that the action "presents" the potentiality-for-being of the *hammer* – its "service*ability*, its use*ability*" (144.37f.). Or, in Heidegger's formulation, "the hammering itself *uncovers* ['sees,' 'presents'] the specific 'manipul*ability*' of the hammer," that is, its "readiness-to-hand" (69.14–16, emphasis mine). By comparison, the hammer does *not* disappear before sensory experience [*Vernehmen*], staring-at or examination, it does not become a figment of the imagination. The hammer remains directly "before the eyes," and that means, precisely speaking, "present-at-hand." Sensory experience [*Vernehmen*], staring-at, etc., present the hammer as a "thing which is present-at-hand," that is, they lead to a "dimming-down of readiness-to-hand" (cf. 138.11–14), or, to use another of Heidegger's formulations, they are a "discovering of presence-at-hand, which is at the same time a *covering-up* of [the] readiness-to-hand" of the hammer (cf. 158.5–7). This idea of "covering-up" is particularly important, as Heidegger uses it as a foundation for denying Husserlian perception the character of a "seeing" in the precise sense. I am speaking here of the Heideggerian demotion of perception in the order of a priori constitution (the givenness of something) – which, although well-known, has, to my knowledge, not been examined in terms of its phenomenological import. Heidegger claims, namely, that perception actually allows no encounter with Being "in itself *uncovered*." The context becomes clear in the case of "looking-at" or "sensory experience" [*Vernehmen*], which, according to Heidegger, "completes [itself] as a determinative proposition" (62).

Readiness-to-hand is thus "covered-up" "through, and for, 'looking-at' [sensory experience, i.e., *Vernehmen*]." Not *every* "originarity" of the given must, however, be lost. Although Dasein in this case *no longer* acts, it does *not* have its eyes closed. It looks directly at the entities *them-*

selves. To put it differently, the *originarity* of the pure thing remains. Heidegger remarks, therefore, that the proposition "has in view the entity *itself* and not, let us say, a mere 'representation' [*Vorstellung*] of it" (154.29f., emphasis mine); thus, "asserting is a way of being towards the *existent* thing *itself*" (218.2, emphasis mine). This passage refers to the selfness of something present-at-hand in the sense of its "bodiliness" (346.31), though, according to Heidegger, the "determinative proposition" has a "synthesis-structure" *because* it "is always harking back to *something else* to which it points, so that it lets something be seen as something" (34.2f., emphasis mine). This synthesis-structure *contrasts* with the "exhibiting of an entity as it shows itself in itself" (35.14), in other words, an entity does *not* "become accessible as it is itself" through the "determinative proposition" due to the determinative proposition's *synthesis-structure* (6.34f.). But why not? The answer lies in a most peculiar concept of Heidegger's, one which has been fully ignored in the literature to this date.[17]

The proposition is the "form in which looking-at is performed," and, according to Heidegger, it is through "looking at" that readiness-to-hand "changes over" (157f.) into presence-at-hand. In other words, the a *priori constitutive* character of "this *is in order to* . . ." (Heidegger) changes over into the *equally a priori constitutive* character of "this *is before my eyes*" (Husserl). The *a priori* "Being" for action changes over into the *equally a priori* "Being" for the eye. The *importance* of this point is that the demonstrative in the expression, "*this* is before my eyes" *cannot* be *separated* from a "determinative proposition," that is, from a merely *empirical* determination. The "this" in this context *always* implies, whether spoken *or* unspoken, a classification. Spoken or unspoken, the "this" *in* the a priori "this is before my eyes" brings with it *identification*, namely, "the identification of *this* Thing as . . . ,"[18] for example, "this that *is a* hammer" or "*an* inkwell." Looking-at the thing *itself* occurs *necessarily* in such a way that "*this* that *is* (Being$_1$) *a* hammer *is* (Being$_2$) before my eyes." "Selfness" in the sense of "thisness" is absorbed into classificatory generality – "to be an X."[19] *A priori* "Being" (Being$_1$) as the selfness or originary givenness of a thing *cannot* be held *separate* from the classificatory, that is, the *empirical*, "Being" (Being$_2$) of predication.[20] To put it another way, the *a priori constitutive* selfness of the present-at-hand *necessarily* includes the *form* of *empirical* classification "(. . .) that is an X (. . .)" so that the act of sensory experience [*Vernehmen*] *necessarily* entails predication.

"It is not necessary (. . .) in doing so (. . .) [to] make an utterance" (318.17).

This demonstrates that the selfness of the present-at-hand is *insepa-rably* linked to the "sameness as" and "it looks *like* . . . ," that is, to the sameness of "aspect" (71.28) which allows the *classificatory* sub-sumption of an entity under a general term. This is nothing other than the "*synthetic* demonstration of something as something" (cf. 34). Heidegger terms "it looks like . . ." or "it shows itself as that, which it in itself is *not*" "semblance," "covering-up" or "deception" (28f., 36). The entity is had "in itself" but it does *not* "become accessible *as* it is in itself" (6.34f., italics mine). The phenomenon, that is, the "showing itself in itself," changes over into semblance: entities "look like . . ." *other* entities (28.37–29.1), namely, with respect to aspects which allow classification.[21]

With respect to the total "*deficiency*"[22] of comportment to other entities, "Being-toward-entities is not extinguished" (22.5f.), nor are entities "looked *away*" (361.17) from; they are, rather, only "looked *upon*." Thus, the *total deficiency* of the givenness of something is *not* an *empty* ideation; indeed, the *originary* givenness in the "mode of phe-nomenon" (36.36f.) "*changes over*" (361.19) into the *equally originary* givenness in the mode of semblance.

It is the connection between originarity *and* deception, that is, Heidegger's conception [*Auffassung*] of the *originary* givenness in the *mode of semblance* (!), which first makes clear that "looking-at" or "mere looking-at" – which includes the equivalent "determinative propositions" – does not fulfill Heidegger's *formal* definition of "understanding" as a "sight" or "seeing" (146.23f.), namely "that it lets entities which are accessible to it be encountered *unconcealedly* [without semblance or deception] *in themselves* [originarily]" (147.6f.). In this manner "we have deprived pure intuition of its priority" (147.18).

5. "TRUTH" MEANS "PRIMORDIALITY"

We have seen how Heidegger provides a wholly formal reinterpreta-tion of the problem of truth – of the phenomenon as the "fullness of the originary givenness of something." Truth means, accordingly, "orig-inarity of givenness" *but* without "semblance" ("deception," "disguise"). This has the far-reaching consequence that this particular form of originarity without deception can *only* be ensured by the "fullness" ("totality") of the given, as every loss of the givenness of an entity in its Being-possible "also results in a dimming-down of the possible as such" (194f.). Heidegger's *performative ontology* implies an Either-Or: either the fullness of the originary given or the danger of changing over to ordinary givenness in the mode of semblance.

Additionally, this schema implies the crucial Heideggerian redefinition of the *a priori*, formulated so the *empirical* givenness *always* contains comparisons between the given and "something else" which is "not itself," and is thus *always* semblance or deception. In contrast, the Being of an entity *as* the a priori pure and simple – "*the transcendent pure and simple*" (38) – precludes every form of deception, or, to put it more directly, this Being precludes every predicative Being (the copulative "is"). Thus, "every reference to something *else*" drops out; what remains is, formally speaking, the "*authentic*-ity" of the given as the contrastive concept to the "*synthetically* given." In contrast to the latter form of givenness, the "totality and authenticity" of givenness represent Heidegger's phenomenological concept of "*direct*" or "*immediate*" and "intuitive grasping" of an entity (36.36–37-2, 35.2–5).[23] Generally, the originarity of the given in the mode of phenomenon is nothing other than the "authenticity" of a "Being-there" [*Da-seienden*].[24] The crux of the matter is that only the "totality" of given secures its "authenticity," as only totality precludes every predicative element. For this reason, BT is always concerned with "hav[ing] the whole of Dasein completely 'given' " (309) where "Dasein," formally speaking, may be understood as the Being or truth of "something."

For example, "the hammering itself" is the seizing upon of the potentiality-for-being of the hammer which provides – or "presents *unconcealedly* in its givenness" – the potentiality-for-being, that is, the "readiness-to-hand," of the hammer. But this is only the most opportune example, as hammering is capable of different intensities. Thus "the more we seize hold of the hammer and use it, the more unveiledly is it encountered" (69). The greatest presence of the hammer *in its readiness-to-hand* clearly coincides with the most intensive action: this action is the comportment to the Being-possible of the *hammer* which "presents" readiness-to-hand "fully," *that is*, unconcealedly. The complete failure to act implies, on the other hand, the complete covering-up of the readiness-to-hand, although "selfness" remains – namely, as the selfness of the present-at-hand which determines [*ausmacht*] the "bodiliness" of the presence-at-hand as the originarity of its givenness in the mode of semblance.

6. THE CONCEPT OF "CARE"

Beyond the employment of the potential-for-being of *other* entities, the employment of the aforementioned conception is decisive for the potentiality-for-being of "Dasein." Both extremes in the intensity of action are only expressions of the alternatives "seizing-upon or neglecting" (12.29) of one's *own* potentiality for hammering, and these alternatives are themselves expressions of the extreme possibilities for comporting oneself *to oneself*. This leads us to the concept of "care." Halfhearted action is – like the waxing moon! – marked by a "not-yet" in the sense of an "only-just," and "all modes of the 'only-just' with regard to the possibilities of concern" are seen by Heidegger as "deficient modes" (57), whereas "concern" is a "basic form of care" (57, 176.30f.). Each *half-hearted* grasping toward one's own possibility is not a *seizing* upon of possibility and thus is responsible for an "only-just" in the givenness of its own possibility: its own *Being* as *potentiality-for-Being*, so important for Dasein according to the concept of care, is *not* brought to perfection – "Man's *perfectio* – his transformation into that which he can be (. . .) is 'accomplished' by 'care'" (199.15f., emphasis mine). It thus becomes apparent that the seizing-upon of one's *own* possibility is the specific form of comporting oneself with oneself, that is, the quite specific mode of care which brings one's own potentiality-for-being to "perfection"; in seizing-upon possibility becomes "wholly that, which it can be" (cf. 309). Only seizing-upon "presents" the potentiality-for-being "wholly." Clearly, the concept of *perfectio* employed here is a version of an *actus purus* or entelechy.[25]

Even in the "neglecting" of possibility "is" possibility "there." What Heidegger says about "fear" applies equally well to neglecting: it is the "*forgetful* backing-away in the face of the potentiality-of-being" (342, emphasis mine). Backing-away means here that "concern, fearful of itself, *seize hold of* [*ergreift*] no particular possibility" (cf. 342), and is similar to wishing, in which the "possibilities have *not* been *seized hold* of [*unergriffen*]" (195.12, emphasis mine). One's own possibilities, as well as the possibilities of other entities, may be neglected. Heidegger uses the term "forgetting" or, more properly, "forgetting *oneself*" (342.12,

339) to represent the neglecting of one's own possibilities as the opposite mode of care to seizing-upon [*Ergreifen*].[26] Wishing is *only* a particular case of *theoretical* or *practical* "*comparing oneself with* . . ." (178.10f., emphasis mine) which "takes an interest" (139.16) and in which the individual "has always [as in the case of fear] stolen away whenever Dasein presses for a decision" (127.30f.). This is a version of semblance, of covering-up, which Heidegger calls "self-deception" (146). In such a situation, Dasein possesses itself "originarily," but nevertheless "disguised." The specific formulation for this "semblance" refers to Dasein which "*is* not itself" (126.19), but rather "*like* the others" (126.34, emphasis mine), and the "harking-back to something else" (34.2) here simply means "comparing oneself with (. . .)" (178.10f.). *Every* grasping toward one's own possibility which is *not* a seizing-upon [*Ergreifen*] of that possibility implies "always having stolen away" (127.30),[27] that is, it is a certain "self-deception," and thus falls short of the "transparency of existence [of the potentiality-for-being]" (146). In contrast, the seizing-upon [*Ergreifen*] of its own potentiality-for-being is the kind of "sight" or "seeing" which "sees" this potentiality-for-being "in itself unconcealed," or which makes it "available."

7. "DEATH" AND THE "TOTALITY OF THE POTENTIALITY-FOR-BEING"

I hope now to suggest the manner in which this problem relates to the concept of death and will briefly touch upon "authenticity" in the sense of "resoluteness." For this purpose it is important to keep in mind that "death" has *no* necessary connection, *immanent* within the system, to Heidegger's express and highly formal discussion of the problem of *what* "end and totality" (241.13) as *a priori constitutive* determinations of an entity could mean "regionally" (241.5–246.14, 303.16–20). Exactly for this reason, because he orientates himself according to the Husserlian conception of the "*fullness* of originary givenness," Heidegger reinterprets the *typically* theoretical Husserlian example of the fullness of a perceived die in his own example of the never "*wholly perceptible* moon" (243.15–21, emphasis mine) – an example which better matches Heidegger's "existential" framework. The reader of BT is *generally* so spellbound by the existential significance of death that she recognizes the formality of Heidegger's construction only with difficulty. Heidegger conducts an examination of the concepts of incompletion and completion respectively, ripeness, wholeness, etc. (§48), and each case concerns the *purely formal* problem of *what* exactly totality as an *a priori constitutive* determination, that is, as a *form* of the a priori givenness of *something*, ought to mean, in that this *something* always stands for a regional, *originary* given: *something* either *as* pure presence-at-hand or *as* living thing or *as* readiness-to-hand or *as* existent Being. The "as" represents each time *not* the generality of a theoretical or practical classification, but – to use a traditional phenomenological expression – the *a priori* form of the *regionally* given,[28] and perfection, or totality, then, possesses the corresponding "regional variations" (241f.).

Heidegger addresses the totality of the present-at-hand which is given in selfness with the example of the moon as presence-at-hand: for the moon there is *no* form of the given which presents the moon as a *totality* in that the moon is only originarily given in perception, and the moon *cannot*, of course, be wholly perceived – just as the die before my eyes cannot be wholly perceived. The totality of living things lies in their ripeness; the totality of readiness-at-hand lies in its finishedness (244.30–245.11). *Pure* presence-at-hand is given only as a "dimming-down of the possible," whereas living things and readiness-to-hand are only originarily given, in each case, in one *regionally* different Being-possible. "Ripeness" is, therefore, regionally pure *performative* determination of Being through which living things "become wholly that [in their Being-possible] which they can be." The situation in the case of the aforementioned fullness of the givenness of the readiness-to-hand is exactly the *same*. The readiness-to-hand which is only given in "wholly devoted busying oneself with . . ." is the *readiness-to-hand* as the specific regional *totality* of an entity in the sense of its "finishedness" – of its "in order to." In the case of "existent" Being, that is, of the "*potentiality-for-Being* which in each case I as Dasein already am" (287.29f., emphasis mine), its Being lies in its comporting itself *to itself*, not, namely, as comportment to a presence-at-hand, but as "care," that is, as modal – seizing upon [*ergreifendes*] *or* neglecting – "Being toward one's ownmost *potentiality for Being*" (193.16, emphasis mine). This structure is intended to bring with it a certain totality, which is itself to be seen as "*perfectio*," as the " 'accomplishment' of 'care' " (199.15–17).

The following observations are therefore worthy of emphasis:

a. The concept of readiness-to-hand, so highly praised in typical interpretations of BT, connotes no cardinal status for the practical as opposed to the theoretical; furthermore, this concept represents no *objective*, phenomenological description in the traditional sense, but fits *only* into Heidegger's purely performative ontology, which itself relies on a number of unspoken *presuppositions*,[29] insofar as it carries forth the reinterpretation of the phenomenological concept of truth into the concept of care.

b. The problem of "totality" as "fullness" of "originary givenness" had, at its outset, *nothing* to do with death. So central is the concept of the "potentiality-for-being-whole" for Heidegger's performative ontology, that he *already* makes full use of this conception, even before he thematizes it, in the first division of BT. This use runs through the conception of seizing-upon [*Ergreifden*] in action as a Being-possible: it presents the action "wholly" and makes it thus a concern which is *not* "deficient" (solicitude) – Heidegger would otherwise have no criterion, *immanent within the system*, capable of differentiating between deficient concern and concern which is not deficient. Praxis is *not* simply pre-eminent, rather it is, *purely phenomenologically*, the *form of the givenness* of "Dasein's" own potentiality for being, that is, his *seizing-hold of* [*Ergreifen*] his own potentiality for being, *because* this seizing-upon [*Ergreifen*] presents in its fullness or "*totality*" that which, purely definitionally, determines "*truth.*"[30]

c. In the second division of BT, Heidegger guides the reader through a *formally fully predetermined* enquiry, and the incautious reader fails *again* to notice it; not, in *this* instance, because Heidegger makes the problem *unrecognizable* by not thematizing it, but because the reader becomes blinded by the far-reaching existential consequences of death and fails to see the purely *phenomenological* formality of the "*unwavering discipline* of the existential way of putting the question" (323.26).

The *formal turn* with which Heidegger crystallizes the problem of the "fullness" or "totality of the originarily given" into the problem of death lies simply in the definition of death as the possibility "which lies ahead of every factical potentiality for Being of Dasein, and, as such, *enters* more or less undisguisedly into every potentiality for Being which Dasein factically *seizes upon*" (302.21–24, 258.36, emphasis mine). That is, death, as possibility, is "always present already" (254.11) in *every* other possibility. For the trivial potentiality-for-hammering, for example, is that the *authentic*, the *resolute* seizing-upon [*Ergreifen*][31] the potentiality-for-hammering presents hammering as its own potentiality for being "authentically" but "not yet" wholly. Why not? Because only death ensures that "*resoluteness*" itself,[32] the aspirant of the "authenticity of *Dasein*," "wholly becomes that which it can be" (145.41, 309.5). This formal turn thus concerns the "totality (. . .) of *resoluteness*" (383.8, emphasis mine).

More concretely, this means that death belongs as a *possibility* to the *a priori constitution* of hammering understood as the *fullness* of its originary givenness in the mode of phenomenon, that is, of *authenticity*. Thus, this possibility must also be presented as "originarily given" in the mode of *phenomenon*, or of *authenticity*. The sought-after "*form* of givenness" implies the "comportment toward death" (to the possibility of which, see p. 261f.) as "forerunning[33] to [death] [*Vorlaufen zum Tode*]" (262). As long as the possibility of death is *not* presented, indeed, "authentically" to the given in the *resolute* having of one's own potentiality for being, the *authenticity* in the givenness of the *potentiality for hammering*, that is, its givenness in *resoluteness* – is marked by a "not-yet" (244.8), or, to employ the Heideggerian formulation, the particular factual potential for Being is *as* "authenticity" distinguished by "a not-yet-having-become-*accessible*" (244.8, emphasis mine). If *resoluteness*, however, is "*modalized*" into specifically "forerunning[34] resoluteness" (309.3–6, emphasis mine), then every "not-yet" is wiped out of the authentic *potentiality for hammering*; the hammering becomes "authentically and wholly" given, that is "primordially" had. This reflects Heidegger's *standard* formulation: the hammering "becomes authentically *and* wholly what it can be" (cf. 309.5f., emphasis mine; 145.41).

Generally, BT is concerned with Heidegger's version of the phenomenological concept of the presence of "something." Heidegger assumes that I am always "present" or "originarily given" to myself, which he calls the "disclosedness of [one]self" (165.7). But *how* I am present (*praesens*) to myself, whether in the mode of semblance or the phenomenon, depends upon the comportment of oneself to oneself which represents the concept of "care." Heidegger develops my being present to myself or the "*disclosedness* of myself" to: 1. "*resoluteness*," that is, one's own "authentic" presence, or presence which has been purified through "angst" of all affective or comparative reference to other entities; and 2. "*total resoluteness*," that is, one's own total and authentic presence ensured by the "authentic" companion presence of death.[35] Each case concerns comporting oneself to oneself *as* a basic structural *plane* of "being towards . . ." *as* "being alongside . . ." *other* entities. Thus,

Heidegger writes that "care (. . .) lies 'before' every factical 'attitude' (. . .) and it does so existentially a priori; this means that it always lies *in* them."[36] In hammering as a *form* of my being alongside the hammer, I certainly comport myself with *this* hammer, but I *first* comport myself to *myself*, in that I seize – or seize upon – *my* potentiality-for-hammering. The "primordiality" of "care" is, at the same time, the "primordiality" of the "basic experience" of *myself*. In this experience I have myself "wholly and authentically," and I have this hammer additionally in its "Being-possible" – indeed, "wholly and authentically" – upon the *basis* of my having myself. In general, *what*, in each case, can *be present*, and *how* it can *be present*,[37] is an accomplishment of *how* I *am present* alongside myself (being toward . . . as "comporting oneself to oneself") and alongside the object (being toward . . . as being alongside . . .), so that the *whole* and *authentic* presence of *my own* potentiality-for-being as well as of the Being-possible of the *other* entity are accomplishments of *authentically seizing-upon* as a mode of care. *What* and *how it is* given to me depends upon (phenomenologically: "correlates with . . .") *how* I comport myself to myself. Heidegger writes, therefore, that "the fact that such entities ['innerworldly' entities] are discovered [given] along with Dasein's own 'there' [its own resoluteness] of existence, is not left to Dasein's discretion. Only *what* it discovers and discloses on occasion, in *what* direction it does so [the phenomenon or self-deception in the sense of *its own* authenticity or inauthenticity], how [the mode of phenomenon or of semblance with respect to *other* entities] and *how far* [the degree of fullness] it does so – only these are matters for Dasein's freedom (. . .)" (366.7–10).

The further consequences of these ideas cannot be discussed here.[38] Perhaps an example will suffice. The "world" according to Heidegger is primarily nothing other than the "<u>whole</u> of a *possible* interconnection of the ready-to-hand" (144, underlining mine), and *in this manner*, as "equipmental *totality*," that *whole* of a *possibility* "belongs" to the "being of the equipment as that which it is [*als dieses, das es ist*]" (68). On the other hand, "potentiality-for-being [is] always potentiality-for-being-in-the world" (144, italics mine), and, thus, the seizing-upon of one's own potentiality-for-Being represents *eo ipso* exactly that seizing-upon of the "world" which *presents* or "discloses" *this* world as the "*whole* of a possibility" – *so that* the world belongs "to the Being of the equipment," so that the equipment "can show itself unconcealedly in its readiness-to-hand" (69). The "disclosedness" (having) of the potentiality for being which lies in seizing-upon [*Ergreifen*] *converges* with the "disclosedness [having] of the world"; the "*disclosedness [having] of the world*" implies the "disclosedness" [having] of the ready-to-hand. The seizing upon [*Ergreifen*] of one's own potentiality for being consists of the "understanding of the world (. . .) in the mode of its genuineness" (148). If seizing upon is "authentic," that is, purified through "angst"-catharsis in "resoluteness," then the world is not only "genuine," but also "authentic," "disclosed" "out of one's own self" (146). *Additionally*, the having of the world and the *how* of this having are a priori constitutive accomplishments of the seizing upon [*Ergreifen*] of *oneself* and its *how* – in the trivial act of hammering, for example, all halfheartedness in hammering suffices to dim down the "world (. . .) to the uniformity of what is purely present-at-hand" (138) and thus leads to the "covering-up of readiness-to-hand." This dimming-down of the world extends to the "world" in its connotation of the "totality of those entities which can be present-at-hand" (64), as soon as I "allow" *my* possibility to act to "pass by" (144) in favor of pure contemplation. This means that not only the concepts of equipment and readiness-at-hand, but also the concept of the world and its "variations" (366.36) stand and fall with Heidegger's questionable *performative ontology*, which, through comporting oneself to oneself, changes *every* "busying oneself with . . ." into a reduplicated structure of action, according to which the ("authentic") *seizing-hold of oneself* [*Sich-ergreifen*] is crucial in order that the actor may have himself "*wholly* [and authentically]."[39]

8. CONCLUSION

The main thread running through the formal *construction* which is BT is the central *phenomenological* concept of truth as the "fullness of the originary givenness of something," a concept which serves *concurrently* as the basic phenomenological concept of *being*. Insofar as the concept of truth as defined here is reinterpreted by Heidegger as the primordiality of "care," it should be understood that the primary concern *throughout* BT is the "primordiality" of "truth" as the principal "*form* of the givenness of something" which Heidegger finds in his conception of the *actus purus* – namely, in *each* case (eventually authentic) wholly having a Being-possible through the (eventually authentic) seizing-upon [*Ergreifen*] of this Being-possible.[40] If the seizing-upon [*Ergreifen*] is "authentic," Heidegger's "*phenomenological construction*" (375.37,

302.11) demands the addition of "being towards death" so that the *authenticity* may *itself* become "*whole*." In this case, the *actus purus* is the "existential" *perfectio*, the whole – and, additionally, *wholly* – authentic having of, in each case, a Being-possible which is the accomplishment – phenomenologically: the correlate – of care. Through the formal adoption of "being towards death," "authentic care" becomes the "truth" as the absolutely primary "*form* of the givenness of something."[41]

Berlin/Mexico City

NOTES

[1] Numbers alone in parentheses refer to page numbers in the German edition of BT. Thus (234.12) refers to p. 234, line 12 of *Sein und Zeit*. I have used John Macquarrie's and Edward Robinson's translations whenever possible. References without the name of an author refer to other works of Heidegger's. Thus (B19 206) refers to p. 206 in Volume 19 of the *Gesamtausgabe*. Additional abbreviations are listed in the "List of Abbreviations" at the end of this text.

[2] Husserl, E., VI. *Log. Unt.* §§22–24. As most of my references are based on well-known Husserlian concepts, I only cite Husserl's text for the most important points of my argument.

[3] I have limited myself here to a presentation of the most basic aspects of Heidegger's performative ontology and have refrained from demonstrating the important results of this performative ontology for the concepts of the action-and-limit-situations with regard to the concept of death, results which yield the political consequences of this ontology. For these topics, see my forthcoming dissertation, *Heideggers "Sein und Zeit" oder die ontologische Fundierung des politischen Partikularismus*.

[4] This assumes, of course, that it is possible to differentiate between the two.

[5] Husserl therefore repudiates the givenness of "individuals" in imagination. See EU §40. The expression "thisness" could be replaced in the following argument by "singularity" or "individuality."

[6] Husserl typically puts the idea thus: experience, precisely speaking, is always the experience or originary givenness of individuals which takes place in various *types* of perception (Husserl, EU §6; *Ideen 3*, pp. 1–20).

[7] One must, in this case, keep in mind that the element of "thisness" – also in its connotation of "that in Dasein which is, in each case, mine" – is decisive for Heidegger. He therefore refers, for example, to "the general determination of each and every thing as an 'individual this': the individual-thisness [*Jediesheit*] (. . .)" (FD 12).

[8] Husserl, *Log. Unt.* II/1, pp. 168.8–11, 44.28–33.

[9] Compare: "The table there is in the room there (not 'a' table . . .)" (B61 90).

[10] See Husserl's third concept of truth in *Log. Unt.* II/2, p. 123.

[11] "Dasein" is meant here in Heidegger's sense: "Dasein is (. . .), in each case, Being that I myself am" (53.2f).

[12] See: "entities themselves turn out to be (. . .) questioned as regards their Being. But if the characteristics of their Being can be yielded without falsification, then these entities must, on their part, have become *accessible* as they are *in themselves*" (6, emphasis mine), and "the signification of phenomenon, as conceived both formally and in the ordinary manner, is such that any *exhibiting* of an entity *as* it shows itself *in itself*, may be called 'phenomenology' with formal justification" (35.12–15, emphasis mine).

[13] I refrain here from discussing the demonstrative formulation, "as that, which it is" (84.2f., 68.25 and 354.20).

[14] Recall that "Dasein is (. . .), in each case, Being that I myself am" (53.2f.).

[15] Bracketed matter within quotations are always my additions.

[16] Volume 19 of the *Gesamtausgabe* is the publication of Heidegger's lecture in WS 1924/25. See also: note 40.

[17] This answer is central to the complex of themes relating to Heidegger's conception of language, a subject which cannot be treated here.

[18] See: Husserl, *Log. Unt.* II/2, p. 25, emphasis mine.

[19] This represents the structure of that which Heidegger calls the "apophantic as" (the copulative "to be").

[20] Heidegger had worked through this concept a number of times long before BT; it appears in a number of passages in BT, but is not developed thematically. The problem is stated more clearly in a number of lectures – for example, in the lectures of SS 1925: "We do not really *see* objects and things primarily and primordially, rather we first speak about them, more precisely, we do not verbalize that which we see, on the contrary, we see that which *one says* about the subject in question. This unusual determination of the world [is caused] by its expressedness, by the already-said-and-already-spoken-about (. . .) (B20 75, emphasis mine). In BT this is the "sensory experience [*Vernehmen*] of the already-known" (62.19, 62.27). In the lectures from WS 1923/24 this is stated even more clearly: "Language does not only have a say in sensory experience [*Vernehmen*], rather it *leads* sensory experience [*Vernehmen*]; we *see* through language. Insofar as language has not been primordially learned, it *covers* things *up* (. . .)" (B17 30, underlining mine).

[21] Heidegger provides a number of presentations of this idea in his lectures. In most of these it is quite clear that he is polemicizing against Husserl's conception of truth as identification in recognition [*Erkennen*] (*Log. Unt.* II/2, particularly 6). Husserl argues that "*the thought is based on intuition and thus directed toward its object*. For example, I speak about *my inkwell*, and the inkwell *itself* stands before me" (*Log. Unt.* II/2, p. 24, underlining mine) and further, "this relationship, as namely, is carried not only by acts of signification, but also by acts of identification (. . .). The perceived object is recognized as an inkwell (. . .). (. . .) The experience [is] the identification of this thing as *my inkwell*" (*Log. Unt.* II/2, p. 25). Clearly, there is an inconsistency in Husserl's presentation: the classification of the thing "itself" as *an* inkwell is completely different from the identification of something as *my* inkwell. Nevertheless, Husserl generally holds to the idea of classification, and it is exactly this which Heidegger will not accept. In the lectures from SS 1923, for example, Heidegger begins by presenting a typically Husserlian "description," having left no doubt that he sees such description as "a false description of the everyday environment" (B63 88), and proceeds by presenting his own description: "*The* table there is in *the* room there (not 'a' table among many others in other

rooms and houses) (. . .)" (B63 90). Thus rejecting the idea of classification, Heidegger proceeds to introduce the situational "in order to" from BT, namely, through the intentional overuse of demonstratives: "This side [of the table] is not that side which points east, and not this narrow side which is so much shorter than that other side, rather it is that side at which the woman sits, evenings, when she wishes to read; at this table there we had this or that discussion on that occasion; we reached that decision here with a *friend*, we wrote that paper, celebrated that party, there" (B63 90, underlining mine). Heidegger introduces his demotion of predication in an *aside*: "the characteristic of 'to something' will not be pushed upon it due to a *comparative relationship to something else that it is not*" (B63 90, italics mine). Heidegger is also very frank in the lectures from WS 1919/20, to which he *explicitly* refers with regard to the *continuity* of the analyses of environment and equipment in BT (72n.): "What is the meaning of *this* 'existence'? [That is, Dasein in its connotation of the originarily given.] (. . .) While drinking tea *I* take *my* teacup in *my* hand [not: I perceive my teacup]; during a conversation *I* have *my* teacup before *me* there [before . . . there represents here *Da-stehen* as originarity]. It is not as though I conceive of something colored as *a thing* [a priori 'as' of the region of material objects] and *this* thing as [a] teacup [empirical, that is, predicative, 'as' of a classification] which is determined in space and time (. . .). 'My teacup from which I am drinking' – its reality [originarity], which it itself is, *completes* itself in meaningfulness [earlier term for the in-order-to or readiness-to hand]" (B58 104, italics mine). In the previously mentioned lectures from WS 1923/24, Heidegger claims (see: Footnote 21) that language does not only have a say in sensory experience [*Vernehmen*], rather it leads sensory experience [*Vernehmen*] in *such a way* as to *cover* things *up*. More precisely, this means that "for every Being-there [*Da-seiende*] [originarily given] as such, there is exactly one lógos with which I can address the things themselves in their being (. . .). Additionally, there are lógoi with which I can address things from many different perspectives. (. . .) Now such lógoi with which I address things in such a way that I have not obtained that which I say to them purely from them themselves, but, rather, with respect to *something else*, which I already know – such perspectives themselves, from which I can address things, lie outside. (. . .) Such a lógos is not simple [compare BT 34.14; synthetic pointing-out]. The simple lógos is that with which I determine the things themselves, with which I do not move away from them. (. . .) Aristotle argues that there is exactly one way of addressing each entity (. . .). Apart from this single lógos which is designated for entities, there are a number of lógoi which are free-floating. This leads to (. . .) deception. (. . .) In every instance it is possible to address a Dasein [phenomenologically speaking: originarily given] not only in the way in which it itself is available, but also in another way from any other perspective. One can say that the many lógoi deceive in every situation (. . .)" (B17 34, underlining mine). Heidegger is thinking here of the correctness of the proposition as opposed to purely "allowing sight" in "circumspect speech," as may be seen in the following passage: "It is assumed that truth is validity (. . .). A phenomenon of decay was made absolute up through Plato's 'ideas.' Validity is a characteristic of spoken sentences, of finished knowledge, as soon as it becomes *public*. Validity is the way in which truth is publicly there. (. . .) The possibility of preserving the primordiality of encounters with entities (. . .) belongs to truth (. . .). The conflation of truth and validity is not that readily apparent" (B17 98, underlining mine). Language, Heidegger claims, exerts an unusual influence, "insofar as it drives us away from the relationship to objects [*Umgang mit den Sachen*]. A *multifaceted interconnection of potentialities for deception, as of a potentiality for Being with the Dasein of speaking*, thus demonstrates itself (. . .). We notice that our very language hides objects. Insofar, then, as our access to that which is initially there is, as a matter of fact, led through speaking [sensory experience (*Vernehmen*) as recognition] (. . .) insofar is the existent world hidden and approached in a certain presenting-oneself (. . .): presenting oneself as something. That which appears, in the primordial sense of showing-oneself, becomes a *semblance*" (B1940, underlining mine). Bergson's influence on Heidegger, namely, in Bergson's view of "concepts" as "perceptives" rather than "simple intuition," is unmistakable here: "It thus follows, that an absolute can only be given in an intuition, whereas everything else belongs to the realm of *analysis*. (. . .) Analysis [is] the operation which reduces objects to elements which are already known, that is, to elements which these objects share with other objects. Analysis relies, then, on expressing an object as a function of something which is not the object itself. Every analysis is thus a translation (. . .) an image which one can obtain from many perspectives, perspectives from which one attempts to find as many correspondences between the new object which one is examining and the other objects which one believes that one already knows" (H. Bergson, DW 183f., underlining mine). "*Thinking* consists of moving from concepts to things (. . .). Knowing [*Erkennen*] a reality means (. . .) employing concepts which are already established (. . .). What we are seeking to establish insofar as the object to be known is a *this* or *that*, in which *well-known* type it fits into (. . .)" (Bergson, DW 199, underlining mine). "These different concepts were, then, only so many external perspectives (. . .) according to which I judge" (Bergson, DW 191, underlining mine). "The concept (. . .) alters it [the object], then, through the expansiveness which the concept gives it [the object]" (Bergson, DW 189).

22 See below, the concept of "care."

23 For Bergson's influence on this Heideggerian thought see above (Note 22): the contrast between intuition and analysis.

24 Heidegger uses this term, before BT, in his lectures. In SS 1923, for example, he refers to an "object (. . .) as an everyday Being-there (. . .)" (B63 95).

25 See below.

26 Heidegger claims in his lectures from SS 1925 that "Neglect only exists because Dasein is determined as care" (B20 185). In his lectures from WS 1925/26 he says, "If I neglect something, it is not though I do nothing, rather, I do something in the mode of Nothingness (. . .). Neglect only exists there, where care is" (B21 225).

27 Heidegger derives this structure from *Kierkegaard*, who claims that every argument, every thought is merely an excuse for putting-off the decision, a contenting-oneself by comparing oneself to others rather than seizing hold of oneself (K30 120; K13 118; K14 76n., 246n.; K4 174).

28 This is nothing other than the established differentiation between the "apophantic" and the "hermeneutic as" (158).

29 One may see this in a number of Heidegger's lectures, beginning at the latest in the lectures from WS 1923/24, and this remains a recurrent theme. In Heidegger's text *Vom Wesen und Begriff der Physis* (WS 1930/31), for example, in the collection *Wegmarken*: "We already translate dynamis as unity and suitability for . . . ; alone in this respect lies the danger that we are not Greek enough in our thinking and that we are freeing

ourselves of the responsibility of making the suitability for . . . clear to us as the way of the yet restrained and controlled coming-out into the appearance in which suitability is fulfilled. Dynamis is a way of had-presence [originarily given]; enérgeia (entelécheia), however, is próteron, "earlier," than dynamis according to Aristotle; "earlier," namely, with respect to ousía (. . .). Enérgeia fills the essence of purely had-presence [the fullness of the originarily given] *more primordially*, insofar as it means the having-oneself-in-a-work-and-an-end which has left every 'not-yet' of the suitability for . . . behind, or, better, which has taken the suitability for . . . with it toward the completion of full-'filled' appearance. (. . .) Entelécheia is 'more' ousía [originarily given] than dynamis (. . .)" (WM 284., emphasis mine). Similarly: "Enérgeia means, when thinking in Greek: standing-in-a-work: the work as that which stands wholly at an 'end' (. . .)(WM 282). Compare BT: "Hence ending, as stopping, can signify either passing over into no-presence-at-hand or else Being-present-at-hand precisely in the moment in which the end comes. The latter kind of ending, in turn, may (. . .) constitute the 'finishedness' of something present-at-hand, as the painting is finished with the last stroke of the brush" (244f.). For further discussion of "finishedness as a particular type of potentiality" for living as well as in its connotation as "power (. . .) entelechy" see the lectures from WS 1929/30 (B29/30 322ff.).

[30] The practical is, without a doubt, pre-eminent in BT, but the reason for this is not self-explanatory. It is, *logically* speaking, a side-effect of Heidegger's conception of "care." Indeed, Heidegger himself claims *repeatedly* that "care by no means claims a privileged status for practical, as opposed to theoretical, comportment" (193, 57, 300). The concern here is not praxis, but "care" as the archetype of the forms of the givenness of oneself ("disclosedness of oneself," see below).

[31] I cannot deal here with the problem of "resoluteness" as "wishing-to-have-a-consciousness." The following may, perhaps, be of use as a guideline for further investigation. Seizing-upon determines that which Heidegger terms " 'genuine' understanding," but "genuineness" has no influence over whether this understanding is "authentic" or "inauthentic" (146.12f., 148.30). In the case of "inauthentic" concern, for example, the actor "possesses himself in that which he undertakes" (cf. 126, 221), or, to put it another way, the actor experiences himself "in relationship to that *which* he undertakes" (cf. 388). This holds, for example, for the setting of a goal, as in hammering "*for the sake of* protection against bad weather" (cf. 84). Every action which is determined by one's own setting a goal falls under a purely formal concept of heteronymity: the actor "did not himself, alone and in solitude, make" the particular action "possible for himself" to perform (cf. 263.29–37; 188). Heteronymity, however, also means falsehood in a strict *phenomenological* sense. The "covering-up" character of sensory experience [*Vernehmen*] and its related theoretical concepts (the terms of types) becomes extended to cover action which relates, even very loosely, to *practical* criteria. The covering-up character of the determinative proposition becomes, very *loosely* speaking, taken over in practical contexts by *explicated* action. This relates to merely reactive action which is *affectively* motivated "by something else" as well as – and, in particular – all *moral* criteria. When an individual does this or that because he believes his action to be good, or correct, he puts himself in a relationship to criteria – however questionable they may be – which always imply a "comparing oneself" to others and leads, simply put, to "alienation" (178). In contrast, resoluteness – as *groundless*, fully indeterminate readiness for action (298)

– is the criterion for the Being-"authentic" of seizing-upon, for example, of the potentiality-for-hammering: "angst" is a part of "resoluteness," and *angst* ensures the "insignificance" of "readiness-to-hand" and of the "Dasein-with of others" (187.32–35, 343). This "insignificance" means, finally, that resoluteness is fully empty, as the resolute actor does not need any reason or motive for acting; indeed, a reason or motive would necessarily refer to "something else" which the actor *himself* is not (the concept of "individualization"). Heidegger defines a higher plane of his unusual "phenomenon"-concept through the use of this "individualization"-accomplishment: Dasein in the angst of resoluteness "becomes accessible to itself as it is in itself," that is, without any affective or comparative reference "to something else," "which it itself is not." It may be shown on the basis of this, that "resoluteness" thus becomes an equivalent of the seizing upon of *seizing-upon itself* – that is, leaving aside the concrete potentiality-for-Being which is to be seized hold of, as it has become "a matter of indifference" (352 11f.). Hence the term "resolute seizing-upon" (of a concrete potentiality-for-Being, as, for example, the potentiality-for-hammering).

32 Resoluteness is used here in the manner of its narrower connotation in BT.

33 I make use here of Hubert Dreyfus' suggested translation of *Vorlaufen zum Tode* as "forerunning to death," as opposed to Macquarrie's and Robinson's "anticipation of death."

34 See Note 33.

35 Heidegger's word-play "disclosedness"-"resoluteness" (*Erschlossenheir-Entschlossenheit*) has a supposed basis in "the things themselves" through the phenomenological idea of being as presence – only, of course, insofar as one thinks of BT *not* in terms of the Greek tradition (much to the dismay of its author!), but in terms of the *Latin* tradition. *Praesens* means bodily, present, contemporary (Heidegger: disclosedness as Being-there [Da-sein]), as well as resolute (Heidegger: Being-there as resoluteness).

36 In the same passage, Heidegger claims that "the phenomenon by no means claims a privileged status for 'practical,' as opposed to theoretical, comportment" (193.30–34).

37 See also: "Equipment is 'in order to.' This sentence has an ontological meaning (. . .), that is, entities are not <u>what</u> and <u>how</u> they are, a hammer, for example, and, further, also something 'with which to hammer,' rather, that <u>what</u> and <u>how</u> it as <u>this</u> entity is, its *Being-what* and *-how*, is constituted (. . .) through this in order to as such" (B24 415, underlining mine).

38 For these consequences, see my aforementioned dissertation.

39 I can only agree completely with K. Löwith's rejection (1929) of a division of BT into a purely phenomenological and a purely constructive section: "(. . .) in Heidegger's 'Analysis of Being,' for example, the systematic concepts of the 'world' or the 'They,' or even that of the 'totality of equipment' are, philosophically speaking, not a whit less reliant upon a wide range of presuppositions as the concept of 'authentic existence,' 'freedom,' 'death,' 'guilt' and 'conscience.' Not only the 'additional construction' (. . .) stands in Heidegger's system of Dasein (. . .) in the shadow of the contrast between either authentic or inauthentic existence, rather, through this basic and questionable differentiation between authentic and inauthentic Dasein (a differentiation which is no less reliant on presupposition as Kierkegaard's "Either-Or"), the positive, central concept of existence, namely in the sense of 'authenticity,' is, from the start, formally indicated, and runs through the *whole* construction from its very beginning (. . .). (. . .) The construction with regard to a particular existential ideal necessarily determines consistently

[Heidegger's phenomenological vision] from its very beginning (Löwith, SS3 p. 4). The individual determinations of Heidegger's systematic phenomenological *philosophy* [are] without exception *equally* reliant upon presupposition (. . .)."

40 This is the situation of the runner and the potentiality-for-running which Heidegger describes in his lectures from SS 1931. "He [the runner] is in a position to start running; only the cry 'go!' is lacking. It only takes this cry and then he is already off running, in a run, that is, in performance. (. . .) Then all that is present of which he is capability; he runs, and nothing remains behind of which he is not capable; running, he exploits [*ausführt*] capability [*Vermögen*]. This exploitation of capability, however, is not the exhaustion of capability (. . .), rather the bringing forth of capability itself into that to which it pushes itself, as capability. The performer is simply he who, with regard to capability, does not leave anything unexploited (. . .). To be sure, this is only the case when he, as the capable one, comes forward in the complete readiness to run, when he fully extends himself within this readiness (. . .). *In a position to* . . . really-Being-capable is the Being-in-a-position-to which is filled with readiness, for whom only the *uninhibitedness* in the performance is lacking, so that, when uninhibitedness (. . .) sets in, that is, when the capable starts running, the performance is truly an *ef*fectuation [*Ausübung*] – and only this. The performance is nothing other than (. . .) enérgeia" (B33 218f., underlining mine). Nothing has yet been established about the "authenticity" or "inauthenticity" of this seizing-upon of one's own possibility for running, unless the "cry" is interpreted as the "cry of consciousness" – or "resoluteness" in the narrower connotation from BT.

41 Heidegger takes the structure of one's-comportment-to-oneself with the primacy of seizing-hold-of-oneself from Kierkegaard. He reinterprets this structure in terms of the Aristotelian-Thomistic *actus purus* and, finally, formalizes the "existentialized" *actus purus* into the phenomenological schema of truth.

BIBLIOGRAPHY AND ABBREVIATIONS

BT = Heidegger, M. *Sein und Zeit*,[16] Tübingen: 1986.
SG = Heidegger, M. *Der Satz vom Grund*,[6] Pfullingen: 1986.
FD = Heidegger, M. *Die Frage nach dem Ding*,[1] Tübingen: 1962.
WM = Heidegger, M. *Wegmarken*,[2] Frankfurt am Main: 1978.
B17 = Heidegger, M. Band 17 der *Gesamtausgabe*,[1] Frankfurt am Main: 1992.
B19 = Heidegger, M. Band 19 der *Gesamtausgabe*,[1] Frankfurt am Main: 1994.
B20 = Heidegger, M. Band 20 der *Gesamtausgabe*,[1] Frankfurt am Main: 1979.
B21 = Heidegger, M. Band 21 der *Gesamtausgabe*,[1] Frankfurt am Main: 1976.
B24 = Heidegger, M. Band 24 der *Gesamtausgabe*,[1] Frankfurt am Main: 1975.
B29/30 = Heidegger, M. Band 29/30 der *Gesamtausgabe*,[1] Frankfurt am Main: 1983.
B33 = Heidegger, M. Band 33 der *Gesamtausgabe*,[1] Frankfurt am Main: 1981.
B58 = Heidegger, M. Band 58 der *Gesamtausgabe*,[1] Frankfurt am Main: 1993.
B61 = Heidegger, M. Band 61 der *Gesamtausgabe*,[1] Frankfurt am Main: 1985.
B63 = Heidegger, M. Band 63 der *Gesamtausgabe*,[1] Frankfurt am Main: 1988.
DW = Bergson, H. *Denken und schöpferisches Werden*, Frankfurt am Main: 1985.
Log. Unt. = Husserl, E. *Logische Untersuchungen*,[5] Tübingen: 1980.

EU =	Husserl, E., *Erfahrung und Urteil*,[6] Hamburg: 1985.
Ideen 3 =	Husserl, E., *Ideen zu einer reinen Phänomenologie und phänominologischem Philosophie*,[1] Band 3 in Husserliana Bd. IX.
K4 =	Kierkegaard, S. *Gesammelte Werke* (GTB 603),[2] Gütersloh: 1987.
K13 =	Kierkegaard, S. *Gesammelte Werke* (GTB 612),[2] Gütersloh: 1988.
K14 =	Kierkegaard, S. *Gesammelte Werke* (GTB 613),[1] Gütersloh: 1982.
K30 =	Kierkegaard, S. *Gesammelte Werke* (GTB 629),[1] Gütersloh: 1986.
SS3 =	Löwith, K., *Sämtliche Schriften*,[1] Vol. 3, Stuttgart: 1985.

SS =	Summer Term
WS =	Winter Term

SAĪD TAWFĪK

THE PHENOMENOLOGICAL MOTIVES OF HEIDEGGER'S AND GADAMER'S HERMENEUTICS OF THE LITERARY TEXT

In this essay, I attempt to show that both Heidegger and his follower Gadamer were trying, through their hermeneutics of the literary text, to disclose and transcend the enclosed situation of modernist literary theory and criticism. In doing so, they were initially and originally motivated by phenomenological motives. The first part of this essay attempts to show that the crisis of modernism, which here serves as an introduction for explicating and justifying the Heideggerian/Gadamerian approach, lies in its formalist tendency and its conception of method. In its second and third parts successively, I attempt to show how both Heidegger and Gadamer transcend phenomenologically these twofold roots of the crisis.

I. THE CRISIS OF MODERNISM AS A CRISIS OF FORM AND METHOD: AN INTRODUCTION TO THE HEIDEGGERIAN/GADAMERIAN APPROACH

Since the beginnings of the twentieth century there has been an increasing interest in the conception of form in art and literature, which has become dominant in the middle of the century. The moderns, both on the creative and theoretical level in art and literature, have asserted the primacy of form and its adjunct concepts, such as style and perspective, as decisive and essential factors which give the work its name, i.e., make it a work of art. However, it seems that most of the disputes and debates about the role and properties of form in art and literature, which the moderns often undertake against those they describe as traditionalists, is like an involvement in an illusory quarrel, for most of these disputes neglect the fact that form in art and literature is, in the end, nothing but the way an artist or writer sees his life-world. The laws of artistic form or perspective, as Merleau-Ponty taught us,[1] are ways of seeing the world which itself does not insist upon any one of them and is not of the order of laws. In this sense, which holds good for the artist as well as for the writer, artistic form is not an end in itself; it is the means or

A-T. Tymieniecka (ed.), Analecta Husserliana, Vol. LIII, 181–207.

the style which the artist or the writer practices and acquires through his endeavours to discover the world in which he lives.

It follows from this that the main and real predicament of modernism in literary theory and criticism, as well as in art in general, lies in its conception of form as an end and absolute reference for the critic or the interpreter of the literary work. Thus, modernism remains caught in the trap of the concept of aesthetic or artistic form as a criterion for evaluating the literary work as well as the work of art in general. This can obviously be seen in the attitudes which have dominated literary theory and criticism in modern times, especially structuralism and semiotics. In respect to such attitudes, the concept of artistic form has come to represent a major part of its interest in the question of structure or abstract form which is governed by a network of relations. We often find this direction of research in studies which deal with the structural analysis of the text or what is called *discourse analysis*, and which understands its task to be a description of the internal relations of the structures of the text on its various levels. Similarly, semiotics understands the text through the study of the general structures of the literary text which is considered as an enclosed system of signs and which does not refer to anything outside itself whether it is social reality or human existence in the broad sense. In this enclosed system, the verbal sign is to be interpreted by other signs; and the main purpose, in the end, is to obtain the general laws that govern the relations of signs to each other. Thus both structuralism and semiotics are concerned with the conception of form and abstract relations as a purpose for their linguistic researches. This situation, in the end, leads to a loss of the essence of language and, henceforth, to a loss of the identity of the literary text itself as it is reduced to a system of symbolic relations in which it loses its individuality and the power expressive of the reality of human life and existence, that is, loses its historical role and ontological dimension. In short, we can say that the attitudes of modernism have led literary theory and art criticism to a situation which we may describe as "*an alienation of the text in abstract form*."

The character of generalization on which the attitudes of modernism insist, and through which they ascribe to themselves the scientific character of science with the *proviso* that they are concerned with general theoretical laws and abstract formal relations, is a character which brings us to another aspect of the crisis of modernism, that is, *the crisis of method*. By the crisis of method I mean that process through which a text is "en-framed," a process which, in turn, means subjecting the text to general theoretical laws that govern its potential structures and according to which the text is to be located, inspected and applied. This procedure which treats the text as an "empirical phenomenon" that is subjected to general laws, a procedure which is common in structural and semiotic studies, not only leads to an overcoming of the individuality of the text to be interpreted, but also allows any text to be an object for evaluation even if it were of no value, in so far as the purpose of such studies is no longer the text itself, but the en-framing of it. Thus the attitudes of modernism on the level of theorizing share in the aggravation of the crisis of modernism on the level of creation itself. The crisis of method which en-frames the text is a crisis of a method which is motivated by the illusion of objectivity that conceals the personality of the text by forcing it to enter into ready-made moulds or frames by which it is measured. The illusion of objectivity, in turn, is an illusion of scientism in human studies which attempt to follow the model of the natural sciences, even as the natural sciences themselves have completely abandoned their traditional model aiming at absolute objectivity and certainty.

It may seem from what is said above that we are quite apart from the subject of our essay and that we are concerned not with the hermeneutics of the literary text, but with a criticism of structuralism and semiotics. But, in fact, we are concerned with the crisis of modernism itself, which has culminated in these two attitudes, and which has come to a dead end. Because phenomenological hermeneutics takes its point of departure from a pre-consciousness of this crisis (which is epistemological in its origin), the above introductory discussion has been necessary for perceiving the background against which phenomenological hermeneutics in literary theory and criticism has emerged and has taken as its starting point: For its original starting point was oriented in a quite different and more open direction. However, the voices of a phenomenological hermeneutics of the text, whose underpinnings were laid down by Heidegger, did not resonate in the Anglo-Saxon world until the seventies of our century when Heidegger's gleams began to be seen in the horizon as a shining sign for the surpassing of the age of modernism which was moving toward its closure. That is why Heidegger's thought, especially on human understanding and linguistic interpretation, was a source of inspiration for postmodern and post-structural literary studies which were aware of the crisis of Western

Literature and the New Criticism, a crisis which, to use Spanos' words, is "grounded in a strategy that spatializes the temporal process of existence. It is, in other words, a strategy that is subject to a vicious circularity that closes off the phenomenological/existential understanding of the temporal being of existence, and, analogously, the temporal being – the sequence of words – of the literary text."[2] It is worth noting here that what we often describe as modernism, Heidegger and Gadamer would consider as a continuation of the tradition of Western metaphysical thought, in so far as the formalist tendency of modern criticism and literary theory originated in the spatial interpretative methodology of metaphysics which was used to look beyond things and thus from above – a methodology from which Heidegger deflected Gadamer. Thus, Heidegger's criticism of what we may call "the formalist interpretative methodology of the text," which serves as part of his criticism and transvaluation of aesthetics and is closely connected to his destruction of traditional Western metaphysics, taking its origin in the phenomenological approach which proceeds from the Husserlian slogan: "Back to the things themselves" (*Zu den Sachen selbst*). It is also worth noting, however, that both Heideggerian and Gadamerian hermeneutics of the literary text aimed to transcend not only modernist literary studies but also the classical hermeneutics of the text.

In its origin, *hermeneutics* is a scholastic theological term which designated the methodological science of interpreting the religious texts whose meanings were obscure and could not be harmonized with, or assimilated into, the tenets of faith unless they had become understandable so that we were no longer alienated from their meanings. It is true that the signification of the term was extended through the classical hermeneutics of Schleiermacher and Dilthey, so that it was no longer restricted to religious texts or linguistic texts in general, but has come to designate a universal science of understanding and a method for interpreting the phenomena of the human sciences. It is also true that this classical hermeneutics has attempted since its beginning to penetrate into the heart of existence and the human soul, and, thus, has not been entangled in the concept of the self-enclosed linguistic form or structure. Nevertheless, classical hermeneutics remained captive to method, and, therefore, could not release itself from the objective tendency in interpreting texts, being committed to the model of method which was dominant in modern natural science. Thus, Schleiermacher considered the text as an objective linguistic medium through which the thought of the author is transmitted to the reader or the interpreter. According to him, this linguistic medium is objective because it represents the matter held in common which makes possible the process of understanding. Similarly, Dilthey considered linguistic signs as a true expression or an externalization of the internal and spiritual life of the creative author or poet. Thus, Heidegger's and Gadamer's criticism of this objective classical hermeneutics of the phenomena of existence (such as art and language, and, consequently, the literary text itself) is, in fact, a criticism of a way of interpreting and understanding that dominated Western thought and was extended to literary studies.

Heideggerian hermeneutics, which is phenomenologically motivated, is concerned with the interpretation of essences or meanings of the phenomena of existence in their intimacy with Being itself. Among the phenomena with which Heidegger was concerned were art and poetry; and he found that the way to understand their meanings is through the understanding of the essence of language itself. On the other hand, Gadamer's hermeneutics, which is known as *philosophical hermeneutics*, is concerned with the universality of the hermeneutical problem; that is, it has to do, as Gadamer puts it, "with the universe of the reasonable, that is, with anything and everything about which human beings can seek to reach agreement."[3] But since this agreement or understanding is only possible through "finding a common language,"[4] the hermeneutical task begins with language. Although the undertaking of this task in Gadamer's hermeneutics is not a repetition of that in Heidegger's, it is an extension of Heidegger's departures and a widening of his hermeneutical horizon; and in this sense it has a phenomenological origin.

There are two main peculiarities which distinguish Heidegger's hermeneutics of the literary text, and which have reverberated in Gadamer's. The first is that the text unconceals Being when it reveals a truth or an ontological meaning that transcends the frame of its formal structure. The second is that the interpretation of the text, hence the understanding of it, requires surpassing the frame of both subjectivity and objectivity. According to Heidegger, these two peculiarities are at the same time peculiarities of language itself.

It is time now to show how the understanding of these two peculiarities can lead us to the understanding of how both Heideggerian and Gadamerian hermeneutics of the literary text transcend both form and method.

II. HERMENEUTICS OF THE LITERARY TEXT BEYOND FORM

The overcoming of the formalist attitude which has dominated aesthetics since the beginning of the century, was one of the most important purposes of phenomenological aesthetics, a purpose which was motivated by the phenomenological concept of experience itself. For in so far as every experience or consciousness is a consciousness *of something*, and in so far as this something (as an object of our experience) always implies a meaning, phenomenological aesthetics challenges the formalist attitude which deprives our aesthetic consciousness of its content, i.e., of the truth or the meaning which the work of art implies and conveys to us. This holds good for the Heideggerian/Gadamerian hermeneutics of the literary text, which depend on their understanding of language.

Heidegger's concern for language is obvious throughout the period of his early thinking, when he created *Sein und Zeit*, his *magnum opus*. This concern, however, became dominant in his later thinking. Generally speaking, Heidegger's writings on language take us away from the conception of language which is predominant in linguistics, semiotics, structuralism and analytic philosophy, and which depends on the study of the structures of sentences through some system of formal rules and logical relations which control language. As Gerald Bruns noticed,[5] Heidegger's calling for the emancipation of language from this tradition which belongs to what he called in *Sein und Zeit* "representational-calculative thinking," is very like his appeal in his last major essay on language titled "The Way to Language" (*Der Weg zur Sprache*, 1959). When Heidegger speaks of the way to language, he does not intend the naive approach which designates some way of representing language through formal rules, structural analysis and logical formations; that is, he does not designate another strategy for gaining mastery over, or taking possession of, language. Rather, the way to language means that process which Heidegger designates as "letting language be experienced as a language";[6] that is, bringing language to its origin, to the predetermined realm in which all language, and in fact any attempt to reflect on language, must live, and in which we may let ourselves to be entrapped instead of laying a trap for it.[7] Thus, the way to language means understanding language through language, that is, through what belongs to language or, to put it in one word, through its essence.

According to Heidegger, understanding the essence of language is the basis for understanding the literary text, especially poetry; and it is even the basis for understanding the essence of art itself. When Heidegger says that "the nature of art is poetry,"[8] he means by this that all art in its essence shares in the essence of poetry which, in turn, "is the founding of truth."[9] Thus, every art is poetizing, that is practising poetry in the essential meaning of poetry as a projection of the light of truth in some form. If the projection of the light of truth occurs in words, then the artistic form in this case is called "poesy" which is poetry in the strict or narrower sense. The primacy of this artistic form does not lie in the fact that the essence of poetry is manifested in it (since all other artistic forms share with it the same property), rather it lies in the fact that it preserves the essence of poetry and allows us to have acquaintance with it through language. Thus, ". . . poesy takes place in language, because language preserves the original nature of poetry."[10] If all art is poetry in the essential meaning of poetry, and if the essence of poetry, in turn, is revealed through language where the founding of truth and revealing Being takes place, then every art is a kind of language in the revealing sense of language. For the art of any people reveals the "language" of that people in expressing their "world." All this leads us to raise once again the question of the essence of language.

According to Heidegger, the essence of language lies in the fact that it reveals Being when it dis-closes the mode of Being which belongs to the human being and to individual existents (or the things-in-being) as well. But how does this revealing or dis-closure take place in language? Heidegger says:

> In the current view, language is held to be a kind of communication. It serves for verbal exchange and agreement, and in general for communicating. But language is not only and not primarily an audible and written expression of what is to be communicated; language brings what is, as something that is, into the Open for the first time. Where there is no language, as in the being of stone, plant and animal, there is also no openness of what is, and consequently no openness either of that which is not and is empty. Language by naming beings for the first time, first brings beings to word and to appearance.[11]

Heidegger frequently repeats that language utters or speaks Being. He means by this that language denominates things and individual beings and bestows upon them their essences and modes of being. In this sense, the world reveals itself through language; and only wherever a language is, is there a world. The world in Heidegger's thought is the open horizon in which truth happens when something is dis-closed or its mode of being is revealed to the human being. Language, then, is the property of man;

it belongs to his world, and it is the blessed vehicle which he was endowed with, and it is even, as Hölderlin says, the "most dangerous of possessions."[12] Language, then, is not a mere vehicle or instrument which man uses in the same way he uses other instruments, nor is it what man utters, for man is he who utters and speaks through it.

It is obvious that the language which Heidegger means is that in which the essence of language as such is achieved and its createdness is manifested, as is the case with the literary text, especially the poetic text. The logical, symbolic and lexical language, on the other hand, is language which is used in a way which conceals the essence of language; that is, it conceals the "creativity" of language in revealing a world or a mode of being. The creativity of language, however, does not mean a "creation" of language, but rather it means allowing language to express its creativity. Analogously, the interpreter approaching the text would understand the creativity of its language in revealing and bringing forth Being. And even the symbolic function of language should not be reduced to a self-enclosed and a self-sufficient world of signs which might, as it were, constitute in itself another world independent of Being and the world in which we live. This fact has been assimilated by the French philosopher Paul Ricoeur who adheres to the German traditions of phenomenological hermeneutics, and who, therefore, has criticized structural linguistics since it proceeds from an epistemology which encloses language in a universe of signs, in a system which has no outside. Like Heidegger, he considers this epistemological approach to language as an approach which misses the experience of language; thus, in one of his confrontations with this approach he maintains that "The phonological level, the lexical level, the syntactic level . . . these levels constitute the dead part of language."[13]

In fact, Heidegger's approach to language, as Robert Magliola carefully describes it, "strikes the middle ground between the structuralist position and that of the more conventional linguistic schools," that is, between a self-enclosed system of language in which words refer only to each other, and, on the other hand, a referential system of language in which words point to the real world:

> Martin Heidegger denies that poetic language is referential in the ordinary sense, that of pointing towards the outside. But (unlike the Parisian Structuralists) he does not exclude referentiality in a broader sense, that of language which refers to Being, but not by pointing at it. Rather, language refers to Being by making it "present" within words. Heidegger takes the "outside" and brings it "inside" the house of language.[14]

Language in Heidegger's thought is "the house of Being," or as Father Richardson puts it in his explanation of this expression: "The sense is that since Being makes being accessible, we cannot gain access to being except by passing through the house of language. Being 'dwells' in the words by which beings are named."[15]

Thus, the literary text in Heidegger no longer has a self-enclosed structure which reveals nothing more than its verbal combinations, and at the same time it is no longer directed to the outside in such a way that words become signs which lose their identity in what they indicate. Words, then, are not conventional tools or symbols; they indicate by revealing and they reveal by hinting. Revealing by hinting means that there is always something which remains dark, concealed or unsaid. This, in turn, is the meaning of Heidegger's sentence: "Everything spoken stems in a variety of ways from the unspoken, whether this be something not yet spoken, or whether it be what must remain unspoken in the sense that it is beyond the reach of speaking."[16]

The unsayable is not something that needs sound, but it is what dwells in concealment and remains as a secret. In this sense, the language of the literary text dis-closes Being and reveals a world from within the concealment which dwells in this language itself.

* * *

These Heideggerian summons motivated Gadamer's hermeneutics of the literary text. Like Heidegger, Gadamer has the insight that understanding and interpreting the literary text relies on the understanding of language itself being considered not only as a written text, but also as spoken words. Understanding the essence of language presupposes that hermeneutics, unlike semantics which describes the range of linguistic facts externally, "focuses upon the internal side of our use of . . . signs, or better said, on the internal process of speaking."[17] According to Gadamer, language is not a manifold of linguistic signs or verbal expressions in which a word or an expression can substitute for another on account of their equivalence, and which is used for the sake of communication. Rather, language is an individualized entity of syntax which is characterized by expressionism, the power of discourse and evocation. This character is obvious when we notice the difference between terminology or technical expressions and living language. This character, however, is more apparent in the poetic use of language and reaches its culmination in the lyric; for "the fact characterizing the extreme case of the lyric poem – that it is untranslatable to the point that it can no longer be rendered in another language at all without losing its entire poetic expressiveness – plainly demonstrates the failure of the idea of substitution, of replacing one expression with another."[18] Nevertheless, the failure of the idea of substitution even holds good for language generally and independently of the special phenomenon of a highly individualized poetic language, for that idea is ". . . contradicted by the moment of individualization in the speaking of language as such."[19]

Just as he rejects the idea of linguistic substitution, Gadamer rejects the related conception of language as a tool or an instrument. According to this conception, language is put to use just like other tools which we use for human purposes, then laid aside when it has done its service. But this is not true, for one does not master his language in the same way one masters his tools, but in such a way whereby one lives in it and makes it present in its medium as is the case with free speaking. Language, according to Gadamer, always implies a concealed meaning; and unlike the semantic research which is concerned with analysing the linguistic form of a text and laying bare its semantic structures, "Hermeneutical inquiry is based on the fact that language always leads behind itself and behind the façade of overt verbal expression that it first presents."[20] In explaining this point Gadamer differentiates between two forms in which speaking extends behind itself: first, in that which is unsaid and nevertheless made present by speaking, and second, in that which for all practical purposes is concealed by speaking.

We can understand from Gadamer's point here that the instrumentalistic use of language is one form in which the meaning of speech is not equivalent to linguistic expressions or verbal signs. But more important, in Gadamer's approach to language, is the point that the essence of language is concealed in the strategic instrumentalistic use of language: in such use, language either does not have any revealing meaning and depends on the idea of the substitution of words which presupposes an equivalence of meanings as is the case with linguistic communication for the purpose of exchanging information, or language, spoken or written, conceals the meaning which is unsaid or needing to be said. On the other hand, understanding the essence of language according to Gadamer, and I try to read Gadamer here, lies in *revealing something*: revealing the meaning which is concealed behind the formal structure of expressions and verbal signs. However, we should notice here that the meaning which is revealed there is not a logical one, but an understanding of the world and things-in-being that is present and manifested in our use of language. This way of understanding the world and gaining an acquaintance with it through language is absent in the instrumentalistic use of language, which thereby neglects its essence. In this respect, Gadamer says:

> In all our knowledge of ourselves and in all our knowledge of the world, we are always already encompassed by the language that is our own. We grow up, and we become acquainted with men and in the last analysis with ourselves when we learn to speak. Learning to speak does not mean learning to use a preexistent tool for designating a world already somehow familiar to us; it means acquiring a familiarity and acquaintance with the world itself and how it confronts us.[21]

Now, the characteristics which distinguish our understanding of the essence of language can be applied to our understanding of the literary text. It is true that the literary text opens up another level of the problem of meaning: for "the 'meaning' of such a text is not motivated by an occasion, but, on the contrary, claims to be understandable at any time, that is, to be an answer always, and that means inevitably also to raise the question to which the text is an answer."[22] However, the main character which is still valid and which we can draw from what is mentioned above is that the literary text, in which the essence of language is manifested, *always implies a meaning and says something*, and it is a way of our acquaintance with the world, beings and human beings: that is a way of revealing truth and Being.

Gadamer understands the literary text in the broad sense which is not restricted to the literary text as a work of art. It is common for all literary texts, as linguistic texts, that they speak to us in terms of their contents. Thus, Gadamer's understanding of the literary text which is, at the same time, a work of art, is not specifically concerned with the achievement of form that belongs to it as a work of art, but with what it says to us.[23] "It is true that," Gadamer asserts:

> there is a difference between the language of poetry and the language of prose, and again between the language of poetic prose and of "scientific" prose. These differences can certainly also be considered from the point of view of literary form. But the essential difference of these various "languages" obviously lies elsewhere: namely, in the distinction between the claims to truth that each makes. All literary works have a profound community in that the linguistic form makes effective the significance of the contents to be expressed.[24]

Obviously, we should not understand from this that Gadamer drops the differences between literary genera. In fact he devoted separate studies to the hermeneutics of literary genera such as drama, poetry, . . . etc. It is enough to recall here, for example, his essay "On the Contribution of Poetry to the Search for Truth" in which he explains how linguistic form confirms and preserves the poetic content which is said: for if the essence of language as a language, that is, as a genuine process of communication which relies on dialogue and not on transmitting information, lies in the fact that it says something to us through dialogue in which word binds one with another, this character then reaches its highest intensive form in poetry which says to us what is to be said in words which "stands written" and cannot be replaced with any other words.[25] Gadamer, then, does not drop the literary genera; for the actual dropping of them and even of the individualization of the literary works themselves rather occurs when they are alienated in the abstract form and when the truth each says to us is forgotten.

The problem of aesthetic form in Gadamer is the problem of the modern aesthetic consciousness which has missed the truth which both art and literature convey to us, to the extent that literature (as well as art in general) has become alienated from our world; that is, it has become no longer understandable by us, and we, in turn, have become unable to enter into a genuine dialogue with it. This holds good for the literature of the past, the historical truth of which has become alienated from us, as well as for the literature of the present when it becomes involved in a subjective experience of the aesthetic form. Like the art, the literature of antiquity revealed their way of expressing their mythical, religious and social world. Is it not true that Herodotus tells us that Homer and Hesiod gave the Greeks their gods? But this historical role that literature and art played in the world of antiquity is no longer understandable for us, exactly because in our own time we are no longer able to assign a historical role for literature and art in general, especially when "emancipation from an objectively interpreted experience of the world appears to be a basic principle of contemporary art."[26] But "the poet," says Gadamer, "cannot participate in this process."[27] Therefore, when poetry becomes involved in such a process, that is, when it becomes non-objective and alienated in an abstract linguistic form, it will then drop the essence of language which lies in the fact that it says something to us, or at least misses the experience of language, which can never fully emancipate itself from meaning, for "a genuinely nonobjective poetry would simply be gibberish."[28]

Consequently, letting the literary text be absorbed in form will necessarily mean eliminating the historicity of the text, as if the text were speaking to an atemporal reader through the literary form. But the literary text is not an abstract form which speaks to an abstract reader. This "fallacy of the absolute text," which is common in the New Criticism and which confers an absolute meaning upon the text, has neglected the fact that the text has a temporal character which speaks to a temporal being, that is, to a temporal reader who lives in a historical background which may be different from that of the text; and that fact allows the possibility of a plurality of interpretation. Thus, the historicity of the text means, as Mario Valdes puts it, that "The text was written by someone, about something, for someone to read. . . . It is impossible to cancel out the historicity of a text without reducing it to a physical phenomenon such as waves in the ocean, for even rocks have a geological history."[29]

At this crossroads, we find ourselves challenged by many questions. How can we retain phenomenologically the concept of identity within, or alongside with, the concept of difference which may lead us to a historical relativism? Again, if the identity of the literary work of art, or the literary text in general, is determined by its historical being, i.e., takes its shape within its historical context, how then can we understand and interpret the text from the perspective of another context in which the reader or the interpreter may live? Again, is it possible then for this understanding and interpretation to retain the identity or the truth of the text? Must the truth which the work or the text communicates be an objective truth so that it may achieve historical continuation?

It seems that these questions and the like, which are concerned with the process of understanding and interpretation, appear to be drawing us to the idea of method. But in fact the hermeneutical process goes beyond the idea of method; it is a cognitive approach which takes the shape of an experience and occurs in it. But, again, how is it possible for this hermeneutics to be phenomenologically motivated apart from the methodology by which phenomenology was originally motivated? Heidegger's and Gadamer's descriptions of the hermeneutical experience of the literary text give a view integrative enough to answer our previous questions.

III. HERMENEUTICS OF THE LITERARY TEXT BEYOND METHOD

The main starting point of the hermeneutical experience of the text is that famous phenomenological principle which claims an understanding of experience on the basis of transcending the subject/object traditional dichotomy. As Husserl showed us, understanding experience brings to light a very simple idea, that is, that every experience is a consciousness and "every consciousness is a consciousness *of* something." This simple idea was neglected in traditional epistemology which involved itself in internecine skirmishes for the sake of either the subject or the object. This idea means simply that neither the world of things and objects is created by our consciousness or conceptions, nor is our consciousness created by that world: both exist at the same time. Consciousness itself is nothing but a directedness which aims at gaining access to its object, attempting to gain acquaintance with and an understanding of it, instead of attempting to gain possession of it and to submit it to our conceptions. Although this task of understanding is imposed on each of us, it is of the nature of true philosophizing to help us to achieve such understanding.

It seems at first sight that this idea has no intimate relation with the literary text and the process of its interpretation, but we shall gradually see that it brings us to the heart of the question which shall be our concern through two main ideas: the first idea is that of how hermeneutics transcends the objective methodological tendency in interpreting the literary text; the second is that of how hermeneutics explicates the role of the subject in the process of interpretation, creating a dialogue between the subject and the text.

1. *The Illusion of Objectivity . . . The Illusion of Method*

The literary text and even any work of art, like any other object of the external world, is an objective entity which is not created by our conceptions of it. Nevertheless, it does not represent an objective truth whose meaning could be constituted independently of us as if there were one objective meaning which could be measured and calculated leaving us nothing to do but to agree on it and acknowledge it. Both Heidegger and Gadamer have denied such alleged objective and atemporal truth. For truth belongs to a human world in which Being is revealed in a certain historical moment of the temporality of the *Da-Sein.* Art in general presents a good example of this sense of truth. And because the truth which the work reveals, or the summons of Being which resonates in the work, is not an abstract objective meaning, we cannot understand it by making use of some abstract categories and moulds which the subject tries to impose upon the work. Instead, truth is revealed only when the subject enters into a genuine dialogue with the summons whose echo resonates in the work. In the same manner we can consider the status of the literary text.

Because the literary text has no one atemporal meaning, it could not be interpreted objectively by submitting it to normative rules and methodological procedures. Such a conception, which understands the process of understanding and interpreting the text as a photographic reproduction of the meaning of the text, is one of the illusions of objectivism from which classical hermeneutics could not release itself, to such an extent that Schleiermacher conceived the process of understanding the text as a process of understanding "the other" in the same way that or even "better than the other understands himself."[30] This is the same illusion which in some sense has dominated modern criticism, which adopts methods seeking objectivity through description, measurement and analysis. Such approaches have adopted objectivism in the name of scientism, with the aim of following the model of the natural sciences even though this model has been abandoned by the contemporary natural sciences themselves which now rely more on making more room for meaning and interpretation than on describing and analysing facts objectively, i.e., in that manner in which a fact or a given is described through some objective and preconceived principles or rules; for such describing does not in itself produce a science, nor does it reveal a meaning.

This is the conception of method which both Heidegger and Gadamer reject. Consequently, it is of a paramount importance to note here that both of them have no antagonism against method in itself: both of them are aware of their phenomenological origin, and both of them are aware that phenomenology in its origin is methodological. Heidegger writes, "the expression 'phenomenological' signifies primarily a *methodological conception*,"[31] and Gadamer acknowledges that his own method is phenomenological,[32] for, as he says, elsewhere, ". . . what one calls *hermeneutical philosophy* is based to a large extent on a phenomenology."[33] And what is phenomenology? It is, Gadamer replies, ". . . a methodological manner of describing phenomena without biases, one which was a methodological renunciation of all explanation of physiological-psychological origins and of all attempts at derivations from preconceived principles."[34] This emancipatory character, which has its origin in the phenomenological method itself, has influenced Heideggerian/Gadamerian hermeneutics. The conception of method which they reject and from which they call us to release ourselves, is what we may call "the strategic instrumentalistic method" which depends on rules and preconceived principles with the aim of controlling phenomena, and which is conceived as the sole means for gaining access to truth. This illusion of method which has dominated the "Geisteswissenschaften" was rejected by Heidegger and Gadamer in their dealing with the text as well as other phenomena.

Accordingly, we can say that the important lesson one can learn from the phenomenological hermeneutics of which Heidegger laid the foundations, is the endeavour to release ourselves from the patterns of the analytical methodological disciplines through which we used to interpret texts. In fact, Heidegger's hermeneutics of the text does not attempt to present an alternative discipline to challenge or compete with the methods and theories which have dominated literary studies, for Heidegger's own manner is release from all manner of -isms, or as Gerald Bruns puts it:

> . . . Heidegger undermines the whole idea of literary study as the application or response of techniques of reading to texts conceived as structures or systems, whether purely formal, linguistic, or textual systems, or whether as forms and configurations that are continuous with the social and ideological systems in which human life is framed. Heidegger takes you out of the vocabulary of theory and method, form and configuration, structure and system.[35]

In fact, this emancipatory character not only distinguishes Heidegger's hermeneutics of the text, but also his thought as a whole which is a call for an emancipation of thought itself so that it may experience an intimate encounter with things, beings and, in the end, with Being itself. Thus, we can here conclude that just as the hermeneutics of Heidegger is a call for the emancipation of language from rules, form and logic, it is also a call for the emancipation of our understanding and interpreting of the text from the theoretical and calculative methodological thinking which adopts the analytical and objective tendency in the name of science or, better said, for the sake of scientism.

* * *

This emancipatory character of the process of understanding and interpretation, is what Gadamer considers to be the necessary negative character of the hermeneutical experience. This means that the understanding which happens in the hermeneutical experience as an openness to the other (or the text) is an understanding which cannot take place unless we become aware that the other cannot be contained within our ideological and methodological conceptions, and that our traditional tools of interpreting are broken down and no longer capable of gaining access to the other. This hermeneutical destruction means shaking the ground on which our conceptions rely in order that we can stay in the openness in which we become caught up, apart from the region of conceptual, representative and calculative thinking.

This hermeneutical experience is the appropriate way for knowing much about the truth which happens in art and texts as well as in history and tradition. This means that the method through which we gain conceptual and objective knowledge is not the sole way for gaining truth. For Gadamer, method in this sense is a kind of strategic thinking which has deepened the alienation of the contemporary human being. For, in spite of the fact that the idea of method, which was developed alongside modern science and was adopted by the human sciences, was an attempt to overcome the alienation which man feels in his confrontation with the world, method has replied to this alienation with another alienation. For, method, as strategic rules and tools which are to be imposed on the object, does not let the subject understand its object as it is given in a direct experience, but compels it to interpret the object according to conceptions and methodological tools.

How are we to understand this theoretical generalization by applying and specifying it to the literary text per se?

The approaches of modern literary theory and criticism became involved in the same crisis when they shared in the deepening of alienation rather than understanding in the process of interpretation. Even classical hermeneutics, which raised for the first time the question of understanding as a universal problem, could not, due to its objective tendency, release itself from this crisis of alienation.

According to Gadamer, hermeneutics in terms of interpretation which relies on dialogue or conversation, is completely different from two other main forms of interpretation: one is interpretation which *speaks about* a text, the other is interpretation which *speaks for* a text: in the first case the interpreter attempts to objectify the text by submitting it to general rules and making general truth-claims about it. In this case interpretation is a *monologue* in which the text is not allowed to speak about itself or for itself. In the second case the interpreter, in his speaking for the text by allowing it to claim its own truth, asserts that he understands the claims of the *other* [the text] and "claims to understand the other even better than the other understands it." Even though interpretation in this case attempts to let the text claim its own truth, it does not allow the give and take of a genuine dialogue. In both cases, though in two different forms, an authority of the interpreter and the *text of interpretation* is maintained over the *text to be interpreted.*[36]

Thus, we can conclude here that the objective methodological tendency, whatever its form may be, leads us to the illusion of believing in the possibility of taking hold of the text and controlling it as an object for interpretation. But in doing so, we find that the meaning of the text always slips away from us, and we come to a situation of alienation as a result of being detached from the text itself, so that we find ourselves, in the end, confronting the text of interpretation instead of the text to be interpreted. This issue leads us to the further question which is most significant in terms of the hermeneutical process, i.e., understanding the role of the subject in the process of interpretation as partaker in a genuine dialogue or two-sided conversation between a subject and a text.

2. *The Role of the Subject in the Process of Interpretation (The Meaning of Interpretation as a Dialogue with the Text)*

What is described in phenomenological hermeneutics as "conversational interpretation" is an interpretation which aims at surpassing the subject/object dichotomy in which one side is maintained at the cost of the other. This conversational interpretation as it is understood by Gadamer is distinguished by, to use Dostal's expression, "faithful productivity" of the original text to be interpreted; for ". . . a requirement of his hermeneutics is the interpreter's modesty before the text. Humility and trust are the leading notions of what could be called, with some caution, the ethics of Gadamerian hermeneutics."[37] This does not mean, however, that the interpreter should be receptive before the text, in so far as interpretation for Gadamer should be productive, not reproductive, even if the productivity of the interpretation of the text should not imply creativity or creation, words which Gadamer avoids, as Dostal observes, "because they exaggerate the power and authority of the interpreting subject over the interpreted text and the interpretation provided."[38] It is obvious that Gadamer here is influenced by Heidegger who also avoids the word "creation" not only in his hermeneutics of the literary text, but also in his description of artistic experience itself, so that he speaks instead of the "createdness" of the work.[39]

Interpretation, then, should be modest before the text in the sense that it should attempt neither to create a new text, nor to reproduce the original text, for in both cases interpretation does not set up a dialogue with the text. A dialogue with a text should be understood as a mutual relation between a subject and an object or, better said, as an I-other relation, since for Gadamer, as well as for Heidegger, hermeneutical experience as a dialogue or conversation is a pre-objective experience, that is, an experience which is prior to the subject/object dichotomy.

According to Gadamer, conversational interpretation as an I-other relation, is a relation in which the I stands for the interpreter, and the other stands for the text and not for the author of the text. In this sense, Kathleen Wright observes, "the text is, accordingly, more than just the *subject matter* of the interpretive conversation; it is a *subject matter* within the interpretive conversation. For this reason, Gadamer claims that the text expresses itself like a 'Thou.' "[40]

It follows clearly from this that the role of the subject in the process of conversation with the text is distinguished by two main characteristics which are closely connected: the first is the mutual relation between the I and the other; the second is that the other is the text itself, not the author. Below we shall try to throw light on these two characteristics successively.

A. *Dialogue as a Mutual Relation between the I and the Other*

It is obvious enough that a dialogue as a dialogue requires a mutual relation between two parties: an I and an other (with the proviso that the other should be understood here as a "Thou," as an addressor who is able by himself to speak and to share in conversation). For the mutual relation here means that the process of understanding and interpretation cannot occur in one direction that proceeds from the I to the other, that is to say, this process should also proceed from the other to the I. It is the I here who holds the power of undertaking this job properly when the I, i.e., the interpreter, enters into the openness of the other and lets the other speak to him just as much as he lets himself listen to it, so that it seems, as it were, that they are exchanging positions with each other. That is to say, that the I or the interpreter's subject deals neither with the other as a mere object to be controlled, nor does it consider itself to be, as it were, a neutral subject or a "Cartesian cogito," i.e., a cogitative subject which in its dealing with the other considers itself to be a detached entity in which the echo of the other does not resonate. For all these reasons, the modern methodological tendency to en-frame the text, controlling it and submitting it to objective thinking which relies on rules and analysis, is a tendency in the wrong direction, that is, towards a kind of strategic thinking in which not only openness to the text disappears, but also the true dialogue or conversation like that in which friends are involved and through which, as Gadamer frequently asserts, a true kind of familiarity and living in solidarity is achieved as a result of having something in common that they love and honor.[41]

Therefore, we can conclude here that the subject, when adopting an objective methodological approach towards the text, is seeking to take on a neutral form in which it can no longer be a true subject, but a subject that is dressed in masks which are those strategic tools through which it deals with the other, i.e., the other which the subject attempts to transform into an object for use among the tools which it uses in its contemporary technological world or, strictly speaking, the alienated world in which it no longer listens to the summons of the things and beings among which it lives.

In fact, this idea, i.e., the idea of the reciprocity of dialogue which hardly exists in our contemporary world, is not only present in Gadamerian hermeneutics, but we find its deep roots in Heidegger:

In Heidegger's essay on "Hölderlin and the Essence of Poetry," he picks a few lines of Hölderlin, his beloved poet of poets, in which Hölderlin speaks of the dialogue that has become missed in our world. Heidegger takes a stand at those lines to enter into a dialogue with Hölderlin's text; that is, he attempts to interpret it conversationally.

Let us listen first to Hölderlin:

> Much has man learnt,
> many of the heavenly ones has he named,
> since we have been a conversation
> and have been able to hear from one another.[42]

And let us listen then to Heidegger:

> "Since we have been a conversation. . . ." We – mankind – are a conversation. The being of men is founded in language. But this only becomes actual in *conversation*. Nevertheless the latter is not merely a matter in which language is put into effect, rather it is only as conversation that language is essential. What we usually mean by language, namely, a stock of words and syntactical rules, is only a threshold of language. But now what is meant by "a conversation"? Plainly, the act of speaking with others about something. Then speaking also brings about the process of coming together. But Hölderlin says: "Since we have been a conversation and have been able to hear from one another." Being able to hear is not a mere consequence of speaking with one another, on the contrary it is rather pre-supposed in the latter process. But even the ability to hear is itself also adapted to the possibility of the word and makes use of it. The ability to speak and the ability to hear are equally fundamental. We are a conversation – and that means we can hear from one another. . . . But Hölderlin does not say simply: we are a conversation – but "Since we have been a conversation. . . ." . . . Since when have we been a conversation?[43]

It is not our concern here to continue Heidegger's dialogue with Hölderlin on the meaning implied in this part of the stanza in which Hölderlin says "since we were a dialogue" which means, among all that it means, for Heidegger: since we were a historical temporalized existence where man learned much and named much, etc. What concerns us here is to take a stand at this deep characteristic which distinguishes the meaning of dialogue as understood by Heidegger. Dialogue cannot take place except through a reciprocity of speaking and listening: speaking alone does not constitute a genuine dialogue, since the word to be a word requires the capability of being heard. Yet, listening to language or what is said is, in turn, a manner of speaking, for speaking is presupposed by it and implied in it, or as Heidegger puts it in his *The Way to Language*: ". . . whenever we are listening to something *we are letting something be said to us*."[44] At this point we come to understand that what we call "the reciprocity of speaking and listening" does not mean that speaking and listening are two separate acts; nor does it only mean that listening accompanies and surrounds speaking such as takes place in conversation when someone speaks and the other listens. In fact speaking and listening, according to Heidegger, are one act or, better said, one process in which we experience something that is said to us. For speaking *as saying* means "to show, to let appear, to let be seen and heard."[45] Reciprocally, speaking *as listening*, that is, as presupposed by listening, means to let ourselves hear and enter into Saying.

It seems that listening, Bruns observes, is Heidegger's way of going beyond structuralism. The way of listening is different from the way of seeing, which is objectifying and possessive, for listening means the loss of subjectivity and self-possession because of being overtaken, involved and entangled in something; and whereas the eye keeps itself at a distance from what it sees, the ear gives the other access to us, allows it to enter us, occupy and obsess us.[46]

Applying this to the conversational interpretation of the text means, then, that listening is a prior requirement in that process; for listening herein means to let something be said to us in our dialogue with the text. Yet, it is worth noting here that the alternation or, better said, the mutual implication of speaking and listening, on which saying something is constituted in dialogues, is not a process which is restricted to the subject or the I, but it is also a characteristic of the text when the text is understood as being the other or the "Thou" in the process of dialogue. How can we to apply this characteristic to the text? For, while it is obvious that the text or language is what speaks to us when something is said to us, it is not obvious how the text may listen to us. It is at this point that "The Origin of the Work of Art" is helpful. In fact, and here we are trying to read Heidegger, not to interpret him literarily, the text practices listening when it is silent and seems as if it were waiting for someone to let it articulate something, that is to say its truth. The text, then, sometimes speaks and sometimes stays silent, reveals itself and conceals itself. And it is the task of understanding and interpretation to uncover the concealed and the hidden through the unconcealed and the manifested, that is, to uncover what is unsaid by the text through what the text actually says. It is through this tension between unconcealedness and concealment, between speaking and silence, in terms of the text as well as the interpreter, that the truth which the text utters is revealed. Phenomenologically, this means to reveal the ungiven through what is given.

We can turn now to the second related characteristic which distinguishes the I-other relation in the process of dialogue, that is, that the other in dialogue is the text itself.

B. *Dialogue with the Other as a Dialogue with the Text*

If the concept of true dialogue which relies on the reprocity of an I-other relation is closely connected, as was said above, to the concept of dialogue as a dialogue with the text, then how can we understand this connection? Here we should primarily observe the self-evident fact that what is important in any true dialogue is that there is a common subject matter which concerns those who are involved in the dialogue; hence, it unites them. For, as Gadamer puts it, "the only way in which we perceive the possibility of speaking with one another is what we have something to say to each other."[47] According to Gadamer, this is a way of communicating which is completely different from that of the mere exchange of information (through signs, for example), that is, from the way in which it is enough that there should be a recipient to receive the information, "For over and above that, there must be a readiness to allow something to be said to us. It is only in this way that the word becomes binding, as it were: it binds one human being with another."[48] Thus, the meaning of dialogue or discourse in the end is, as Heidegger explains, ". . . to make manifest what one is 'talking about,' in one's discourse," for the word as discourse "lets something be seen . . . , namely, what the discourse is about. . . ."[49] What can we draw from this?

There is a main idea which was previously revealed for us, that is: language, or speaking in dialogue, is not a tool for transmitting or communicating information, for what is important in dialogue, as always a dialogue about something, is what is said. And if we also know already that the text as a language always says something to us, this consequently means that our dialogue with the text, as a dialogue with the other, should not overcome or go beyond what the text says, i.e., should not use the text as a tool for perceiving something which lies outside what it says. Here is one of the main points of difference between Heideggerian/Gadamerian hermeneutics and the classical hermeneutics of Schleiermacher and Dilthey who considered the language of the text as a cipher of something which lies outside the text, such as: the creative personality of the author and his world view, etc.

The traditional model of the direction of communication has it moving from the author to the text and then to the reader and sees the text as an objective medium which transmits the message of the author to the reader. This is a model which does not deal with the text as the other or the "Thou" which is the chief point in the process of dialogue; it rather deals with the text as a medium through which to have dialogue with the author. It is not only Gadamer who rejects this model, but this rejection, as Mario Valdes observes, is a common principle in phenomenological hermeneutics: the traditional author-text-reader relationship is now resolved into two separate parts: first, the writer-text relationship, and second, the text-reader relationship. The first relationship is a matter of a creative process in which a human endeavour produces an individual and distinct work or text. But once the (literary) work is created, this relationship is ended; there is no retroactive control over it.[50] For, once the literary work is completed and given to the reader, textual meaning and author intentions have separated to find their own way apart from each other. That is why Valdes holds to Gadamer's idea that the textual meaning does not coincide with what the author meant.[51] This phenomenologically means that the psychological intentions of the author (as well as of the artist) belong to himself alone, whereas textual intentions should be considered as a part of the reader's or critic's experience.

It is worth noting here, however, that the above analysis, which is phenomenologically constituted, does not conflict with the phenomenological idea that the literary work (or the work of art generally) is itself a process of communication through which the work says something to the percipient about something. What the work communicates, however, is not the empirical ego of the writer or the artist, but the artistic and intentional meaning which has become the work itself and nothing more. That is why it is common among the phenomenologists that they agree on the idea that the experience of the creator and his own conceptions are not a part of the work itself and do not, or should not, enter into the context of the percipient's experience of the work.

It follows from this that what must concern the interpreter in his dialogue with the literary text (or a work of art) is the meaning of the text itself apart from the historicity of its author. We should always distinguish the historicity of the author from the *historicity of the text*: the historicity of the author is the mental, psychological and sociological experiences which belong to himself alone and which determine the conditions of his production; the historicity of the text, on the other hand, is the historical truth which the text says, as long as the text is always directed toward people who live in a certain age with all its cultural, sociological and religious forms. It is through the text-reader relationship that the text may enter into an alien socio-cultural context: the context of a reader who lives in another place and time. It is the task of hermeneutics here to transcend this historical alienation when the text enters into an alien context and cannot be assimilated in it. Transcending this alienation, however, requires a process of *appropria-*

tion (*Aneigung*). This term means to make one's own what was initially alien, and thus it designates that process in which the interpreter has to apply the meaning and the horizon of the text to the context or the horizon of the reader. This phenomenologically means to let the text enter in and belong to what Husserl called "the life-world" (*das Lebenswelt*) and to what Heidegger called "being-in-the-world" (*in-der-welt-Sein*). However, this process of appropriation cannot be accomplished unless it is done through a dialogue which seeks understanding, understanding the historicity of the text and what it says for us.

According to Gadamer, understanding the meaning of the text is to know how to apply it to our situation. However, this application, as Grondin observes, is one event in Gadamerian hermeneutics: for there is not first understanding and only later an application of what is understood to the present. Understanding a text of the past means to let it challenge us today and failing to apply it to our situation means to understand nothing of it.[52]

Moreover, through the process of that appropriation we not only understand the text or the other, but we also understand and get acquainted with ourselves. For, according to Gadamer, we know ourselves properly in the mirror of the other, that is, through the human traits that have been frequently manifested in the cultural products of tradition. As Ricoeur puts it ". . . what would we know of love and hate, of moral feelings and, in general, of all that we call the *self*, if these had not been brought to language and articulated by literature."[53]

Is it not true then that the hermeneutics of Heidegger and Gadamer seeks to rectify the rift which has been produced by our modern aesthetic consciousness as a result of its neglect of what literature and art in general say to us, to the extent that what is said to us no longer echoes in us? Is it not also true that in such seeking both Heidegger and Gadamer share in the original project of phenomenology which seeks to transcend the crisis of the human sciences in general? It is true that both of them have concentrated hard on the ontological dimension of the literary text, to the extent that their treatment seems to have a one-sidedness. However, this concentration itself has created a weight parallel to that of literary theory and art criticism, giving these a density which they very much needed.

Cairo University

NOTES

[1] For a detailed explanation of Merleau-Ponty's point here, see his essay on "Indirect Language and the Voice of Silence," in his *Signs*, trans. with an introd. by Richard McCleary (Evanston: Northwestern University Press, 1964), pp. 49ff.

[2] William V. Spanos, "Heidegger, Kierkegaard, and the Hermeneutical Circle: Toward a Post-modern Theory of Interpretation as Disclosure," in *Martin Heidegger and the Question of Literature: Toward a Postmodern Literary Hermeneutics*, ed. William V. Spanos (Bloomington: Indiana University Press, 1976), p. 116.

[3] Hans-Georg Gadamer, "On the Origin of Philosophical Hermeneutics," in *Philosophical Apprenticeships*, trans. Robert Sullivan (Cambridge, Massachusetts and London: MIT Press, 1985), p. 180.

[4] *Ibid.*, loc. cit.

[5] Gerald Bruns, "Disappeared: Heidegger and the Emancipation of Language," in *Languages of the Unsayable, the Play of Negativity in Literature and Literary Theory*, ed. Sanford Budick and Wolfgang Iser (New York: Columbia University Press, 1989), p. 117.

[6] Martin Heidegger, *On the Way to Language*, trans. Peter D. Hertz (New York: Harper & Row Publishers, 1971), p. 112.

[7] For a detailed discussion of this point, see: Bruns, *op. cit.*, passim.

[8] Martin Heidegger, "The Origin of the Work of Art," in *Poetry, Language and Thought*, trans. Albert Hofstadter (New York: Harper & Row Publishers, 1975), p. 75.

[9] *Ibid.*, loc. cit.

[10] *Ibid.*, p. 74.

[11] *Ibid.*, p. 73.

[12] Quoted in Martin Heidegger, "Hölderlin and the Essence of Poetry," in *Existence and Being*, trans. Werner Brock (Chicago: Regnery, 1949), p. 293. For a Heideggerian reading of the phrase see pp. 299–300.

[13] From a lecture delivered by Ricoeur in English at Wheaton College, 1969. Quoted in Robert Magliola, *Phenomenology and Literature* (West Lafayette: Indiana, Purdue University Press, 1987), p. 81.

[14] *Ibid.*, p. 69.

[15] William J. Richardson, *Heidegger through Phenomenology to Thought* (The Hague: Martinus Nijhoff, 1976), p. 528.

[16] Heidegger, *On the Way to Language*, *op. cit.*, p. 120.

[17] Hans-Georg Gadamer, "Semantics and Hermeneutics," in *Philosophical Hermeneutics*, trans. and ed. David E. Linge (Berkeley: University of California Press, 1976), p. 82.

[18] *Ibid.*, p. 87.

[19] *Ibid.*, loc. cit.

[20] *Ibid.*, p. 88.

[21] Gadamer, "Man and Language," in *ibid.*, p. 63.

[22] Gadamer, "Semantics and Hermeneutics," in *ibid.*, p. 90.

[23] H.-G. Gadamer, *Truth and Method* (New York: Seabury Press, 1975), pp. 144–145.

[24] *Ibid.*, p. 145.

[25] H.-G. Gadamer, "On the Contribution of Poetry to the Search for Truth," in *The Relevance of the Beautiful and Other Essays*, trans. Nicholas Walker, ed. Robert Bernasconi (Cambridge: Cambridge University Press, pp. 106ff.).

[26] H.-G. Gadamer, "Composition and Interpretation," in *ibid.*, p. 69. It is worth noting here that the concept of "objectively interpreted experience" in this context does not mean "objective interpretation," rather it means an experience which has an object other than our subjective representation of the aesthetic form.
[27] *Ibid.*, loc. cit.
[28] *Ibid.*, loc. cit.
[29] Mario J. Valdes, *Phenomenological Hermeneutics and the Study of Literature* (Toronto: University of Toronto Press, 1987), p. 33.
[30] Gadamer, *Truth and Method, op. cit.*, p. 322.
[31] Martin Heidegger, *Being and Time*, trans. John Maquarrie and Edward Robinson (New York: Harper & Row Publishers, 1976), p. 50.
[32] Gadamer, *Truth and Method, op. cit.*, p. xxiv.
[33] H.-G. Gadamer, *Heidegger's Ways*, trans. John W. Stanley (Albany: State University of New York Press, 1994), p. 51.
[34] *Ibid.*, loc. cit.
[35] Gerald Bruns, *Heidegger's Estrangements: Language, Truth and Poetry in Later Writings* (New Haven: Yale University Press, 1989), *op. cit.*, p. 6.
[36] Gadamer, *Truth and Method, op. cit.*, pp. 322ff; see also Kathleen Wright, "Literature and Philosophy at the Crossroads," in *Festivals of Interpretation: Essays on Hans-Georg Gadamer's Work*, ed. Kathleen Wright (Albany: State University of New York Press, 1990), pp. 237–238.
[37] Robert J. Dostal, "Philosophical Discourse and the Ethics of Hermeneutics," in Wright (ed.), *Festivals of Interpretation . . . , op. cit.*, p. 65.
[38] *Ibid.*, loc. cit.
[39] Heidegger, "The Origin of the Work of Art," *op. cit.*, p. 65.
[40] Wright, *op. cit.*, p. 237.
[41] For a detailed analysis of this point, see Dieter Misgeld, "Poetry, Dialogue and Negotiation: Liberal Culture and Conservative Politics in Hans-Georg Gadamer's Thought," in Wright (ed.), *op. cit.*, pp. 165–169.
[42] Quoted in Heidegger, "Hölderlin and the Essence of Poetry," *op. cit.*, p. 293.
[43] *Ibid.*, pp. 301–302.
[44] Heidegger, *On the Way to Language, op. cit.*, p. 124.
[45] *Ibid.*, p. 125.
[46] Gerald Bruns, "Disappeared: Heidegger and the Emancipation of Language," *op. cit.*, pp. 127–128.
[47] Gadamer, "On the Contribution of Poetry to the Search for Truth," *op. cit.*, p. 106.
[48] *Ibid.*, loc. cit.
[49] Heidegger, *Being and Time, op. cit.*, p. 56.
[50] Valdes, *op. cit.*, pp. 60f.
[51] *Ibid.*, p. 38.
[52] Jean Grondin, "Hermeneutics and Relativism," in Wright (ed.), *Festivals of Interpretation, op. cit.*, pp. 50–51.
[53] Paul Ricoeur, *Hermeneutics and the Human Sciences: Essay on Language, Action and Interpretation*, ed. and trans. John Thompson (Cambridge: Cambridge University Press, 1981), p. 143.

DANUTA ULICKA

LANGUAGE AND EXPERIENCE

On the Subject and Method of Roman Ingarden's Philosophy of Literature

Among Ingarden's writings on the philosophy of literature there is a special group of treatises which are rarely commented upon by those who examine his ideas. The group includes articles on Aristotle's *Poetics* and Lessing's *Laocoön*, on "various understandings of truthfulness in the work of art," on the form and content of the work of art, and a number of studies concerned with the aesthetics of empathy. Indeed, if one reads these studies concentrating on the theses explicitly put forward, especially if one confronts them with the object they deal with, they may not arouse any special interest. These treatises, however, turn out to be extremely valuable material, shedding light on a number of statements the philosopher made about literature and the study of it that he initiated, provided they are looked at from the perspective of the method he applied (which he termed "style") to his analyses of the tradition of aesthetics and literary theory.

From this point of view the above-mentioned studies show, above all, that Ingarden's poetics, which specifies and concretizes his phenomenological philosophy (this is why we term it "philosophy of literature," an expression that covers all the aspects of Ingarden's reflection on the art of writing literature), can be placed in the context of twentieth-century analytic philosophy or, more precisely, within a linguistic approach to it, within the analytic philosophy of language. The facts that justify this classification are the semantic angle of Ingarden's analyses, their order and type of reasoning as well as the priority he gave to so-called ordinary language as the criterion for the choice and initial treatment of an object, on the one hand, and at last resort, for determining whether the reasoning is correct, on the other. This is also why Ingarden's concepts seem to be akin to the research into ordinary language as carried out by English-speaking philosophers and to similar inquiries made by some of the scholars of the Lvov-Warsaw school.

In Ingarden's philosophy of literature, however, so-called ordinary language functions not only as the criterion delimiting the scope of investigations or the frame of reference for the results obtained later, through

A-T. Tymieniecka (ed.), Analecta Husserliana, Vol. LIII, 209–212.

"direct" phenomenological experience, but also as a tool with which the results are conveyed and which makes it possible to get through to the reader.

The fact that the philosopher chose to use this particular type of language in his discourse on the philosophy of literature is also one of the reasons why his thought can be placed in yet another context, namely, that of the neo-idealistic approach. Placing Ingarden's poetics in this realm of thought is fairly reasonable, especially if we consider the functions ("the mission," as he said) which his reflections were meant to fulfill.

The above-mentioned schools of thought played a significant role in Polish philosophy in the twenty-year period between the two world wars. It followed two directions. Still, the philosopher laid particular stress on the influence the neo-idealists had on his research whereas he completely disregarded the inspiration he drew from the philosophers of analysis. The latter trend (which he termed, with a certain tinge of irony, "so-called analytical philosophy") was one he never spoke favourably of, at least not in the studies he published. This was probably because of the fact that Ingarden was not well-versed in the different varieties of analytic philosophy, and he treated them all as one and the same thing, namely, he identified them with the neo-positivism of the Vienna Circle, or with formal logic, and he totally ignored the analytic philosophy of language – which was the nearest thing to his own approach.

Thus, on reading Ingarden's philosophy of literature from the point of view of its links with its native historical background, one can detect the specific configuration of the trends on the map during the twenty interwar years. At that time, unlike anywhere else, Poland witnessed an unprecedented interaction of various intellectual tendencies. This was evident in the careers of the scholars of the Lvov-Warsaw school (Kazimierz Twardowski, its founder, and his followers – Kazimierz Ajdukiewicz, Jan Łukasiewicz, Leopold Blaustein, Izydora Dąmbska, Władysław Tatarkiewicz, Stefan Szuman, Władysław Witwicki) and the neo-idealistic literary historians (Zygmunt Łempicki, Juliusz Kleiner, Wacław Borowy). To give just a few examples: Twardowski came from Brentano's Austrian school, just like Husserl, Ingarden's mentor; Ingarden himself studied at Göttingen under Husserl and in Lvov under Twardowski, and, in a manner, also under Bergson; Blaustein was probably more strongly influenced by Husserl and Ingarden than by Twardowski; Ajdukiewicz studied in Germany under the same professors as Ingarden; Łempicki and Kleiner came from the Lvov-Warsaw school, but then they popularized the neo-idealistic philosophers, Dilthey and Bergson; the earliest article on Husserl's antipsychologism was written by Twardowski's pupil, Łukasiewicz.

The development of the analytic movement in the sixties, and of late post-Heideggerian hermeneutics, which was, in the matter of fact, shaped by neo-idealism, especially by its Diltheyan variety, makes one wonder whether this hybridization of ideas was truly unprecedented. Maybe it was the specific situation which made it appear in Poland earlier and in a more conspicious way than in other ethnic realms. The ten years that elapsed between 1960 and 1970 were a period when analyticists and phenomenologists (hermeneuticians) called for dialogue and demonstrated that the differences between them were merely an illusion. Those who were in favour of closer collaboration showed that the basic categories used by the thinkers of both schools, such as ordinary language, "naïve" consciousness and common sense, could be, for a variety of reasons, regarded as synonyms.[1] There is some evidence to suggest that this collaboration was possible, e.g., in the field of literary studies: investigations carried out by theoreticians who examined speech acts, especially those conducted by J. R. Searle, but also by R. Ohmann and S. E. Fish, have their roots in J. L. Austin's analytic philosophy of language, and they are unexpectedly close to the phenomenological approach.

As the above question lies outside the scope of our discussion of Ingarden's philosophy of literature, we cannot dwell on it here. It should, however, be dealt with in separate studies, which particularly ought to examine the general (ontological and epistemological) circumstances surrounding the dialogue between the two styles of thinking predominant in the twentieth century. As for Ingarden's position, it is enough to keep in mind the prospects it opens up. Our task for today is to show that it really does open them up. For this purpose, treatises which state the phenomenologist's approach to language and the role it plays in cognition are of vital importance.

Warsaw

NOTE

[1] On this synonymy, see for example: "Is not the ordinary language of daily life correlated with the ordinary phenomena of the life-world? Is it not true that the careful study of one must require the careful study of the other? Are not phenomenology (. . .) and linguistic analysis both approaching the same thing (concrete experience) from different angles?" J. Wild, "Is There a World of Ordinary Language?" in *Analytic Philosophy and Phenomenology*, ed. H. A. Durfee (The Hague: 1976), p. 192. See also: G. Küng, "Language Analysis and Phenomenological Analysis," in *Acten des XIV. Internationalen Kongresses für Philosophie*, Vol. 2 (Vienna: 1968); R. Schmitt, "Phenomenology and Analysis," *Philosophy and Phenomenological Research*, Vol. 23 (1962); and H. Spiegelberg, "A Phenomenology of Approval; On the Right to Say 'We,'" in *Doing Phenomenology* (The Hague: 1975).

GARY McNEELY

A HERMENEUTIC INQUIRY INTO HEIDEGGER'S GELASSENHEIT

Thought through Beckett's Waiting for Godot

It is essential to note at the outset that a condensed version of this paper was presented at the Second World Phenomenology Congress in Guadalajara, Mexico on September 12–18, 1995. This declaration is unsuspectingly critical. Although you, the reader of this text, will become familiar with the chiasmic method and the specific interests of this inquiry as you reflect on this excursion into Beckett's drama, I would have you consider that this meditation does not have a definitive point of closure. On the contrary, if this study has been successful as a chiasm, it will reveal that a continuous, yet evocative line of questioning extends from the performative dimensions of the conference through to the more conceptual interpretation imparted via a hermeneutic reading of this text. This extension is most acute when we note the curious affinity between a theatre performance and this conference. Delivering a conference paper is akin to performing on a stage, though for this conference our interests are in the main more philosophic than dramatic. Still, this similarity is instructive for it underscores the participation which is essential for both philosophic inquiry and dramatic theatre. The active involvement of participants should be stressed for, as you shall discover in Part I, it is this interplay which characterizes the essence of a chiasm. As you follow this text, then, I am hopeful that you will engage in the chiasm set up in the original meditation.

Before turning to examine the practice of a chiasmic method, it is important to highlight the two interrelated aims of this inquiry. Our general concern is to draw attention to the vital interplay between the "question of Being" and works of art.[1] As presented by Heidegger, and adapted by Merleau-Ponty and others, this illuminating exchange sets up a reflective space which helps clarify fundamental issues in existential phenomenology. Heidegger elaborates on this evocative possibility when he describes the phenomenal profusion of an artwork as one which deepens our perceptual gaze into "the constellation of truth after which we are *questioning*."[2] In the strange proclivity of its revealing, the work stirs our critical thinking by establishing itself as a site for and of ques-

A-T. Tymieniecka (ed.), Analecta Husserliana, Vol. LIII, 213–232.

tioning. In contrast to our current practice of categorizing an artwork either as a cultural artifact or as an example of creative genius, in this inquiry we are to consider that a work may help to recover and disclose dimensions of the world otherwise obscured, refused or excluded by the oblivion of "nihilism." By adopting this mode of "open thinking"[3] which meditates through specific works, we are freed to ponder the profuse dynamic of our present milieu, but in a way which also invigorates our interrogative practices.

Our complimentary and more specific concern takes up this interrogative dynamic by reflecting on selected scenes from Samuel Beckett's *Waiting for Godot*.[4] In this way, we may apprehend Heidegger's mode of thinking as Gelassenheit, so that we may attempt the task of re-thinking the "question of man's nature" in our modern epoch.[5] Our inquiry into "waiting" will make special reference to the drama's decisive opening and closing scenes, where we encounter Vladimir and Estragon's querying about what they are to do. The significant pauses which frame these moments proffer opportunities to reflect on the protagonists' waiting for Godot. The existential import of this "waiting" is especially poignant in the play's concluding segment where we encounter Vladimir and Estragon's curious "paralysis" as they fail to depart from the stage. By calling for our participation in this impasse, the drama helps the audience and us in this hermeneutic inquiry to sense more directly the tenuous yet original promise of "waiting." By attending to the subtle urgency of these pauses, we come to apprehend that our habituated modes of thinking are so constrained by a rational ordering that we are nearly obtuse to this prospect of "waiting (which) lets re-presenting entirely alone."[6] By marking this hiatus in "waiting," however, we are able to appreciate the open flexibility given by "stepping back"[7] from our domineering presence as *homo rationale* and the prevailing subject-object relation of our modern metaphysics.[8]

PART I: CHIASM AND THE INTERPLAY WITH DRAMA

Before turning to our exploration of *Waiting for Godot*, it is necessary to present a more detailed introduction to the practice of thinking through art as a chiasm. I have drawn on Merleau-Ponty's articulation of this phenomenon as he provides the richest and most promising articulation as a mode for philosophical reflection. As elaborated in *The Visible and the Invisible*, he distinguishes that at a primordial level "every relation with being" is a chiasm.[9] At its greatest extension, this myriad of relations is the "universal dimensionality which is Being."[10] This appeal to the multi-dimensionality of Being is subtle but crucial. It reveals that our own individuated presence is included in this nexus as we stand out in the "flesh of the world."[11]

These descriptions are insightful for they help distinguish the specific locus of our inquiry. Our concern is to draw witness to the intertwining relation which extends between the drama and our philosophic question, and to interrogate what the former may tell of the latter. The disclosive possibility of this relation becomes more evident when we consider that we are to "ground" our phenomenological descriptions and subsequent analyses as concretely as possible in the drama. Rather than deliberately superimposing an abstract interpretation as often happens in art or literary criticism, this method "tries to describe the way we use, think, question and *are* such notions and issues"[12] which pertain to Heidegger's discussion of Gelassenheit. More specifically, this approach

> aims at letting these subsequent thoughts remain nestled . . . in the work from which they have arisen. It seeks to found and ground its notions, no matter how abstract and theoretical, in the materiality, givenness, sensuousness, and contingent reality of the artwork.[13]

The upshot of this mode of address is that by remaining vigilant and self-reflective in "bracketing" the phenomena of "waiting" as proffered by the drama, we will be better prepared to engage in a critical analysis of this phenomenon despite the constraints imposed by the representationalism of our modern philosophic practice.

As we shall discover shortly in Part II, the central dynamic of this exchange is the manner of "reversibility"[14] which extends between the dramatic performance and our philosophic inquiry. To appreciate the play of this relation, consider the graphicness of each chiasm as preserved in the Greek letter "χ." To recognize its filamentous twist, we gather a sense of what Merleau-Ponty means by this "reversibility." This reciprocating moment entails that "every relation with being is *simultaneously* a taking and a being taken, the hold is held, it is *inscribed* and inscribed in the same being that it takes hold of."[15]

This reversibility carries phenomenological significance. Rather than working towards an active intellectual or rational understanding as a form of "possession," we are to participate in this exchange as a "dispossession,"[16] as we would then be "bracketing" the phenomena given to our perception. This "dispossession" is not incidental. For according to Merleau-Ponty, we are said to "have at a distance"[17] even as we touch in the midst of a chiasm. To appreciate this respect of "distance" as a hint of the dispossession safeguarded by a chiasm, consider the way we might describe the conduct of a conference. One of the most common pictures is to note that the audience is talked to by the reader, where the ultimate aim is for the audience to comprehend what the paper means by understanding its logic in its entirety. While not diametrically opposed to this description, a chiasm focuses on the complementarity rather than the polarity of this relation. It respects the differences between the *relata* while yet stressing that this relation is held together by the possibility of their touching. A chiasm is delimited by the space between the reader and its attentive audience, while yet allowing the audience not to aim at subsuming the text via an inclusive and possessive mode of analysis.

By working pliantly with the drama in a way which preserves this space in the midst of the touch, then, our aim is to open a reflexiveness which both deepens and broadens our critical perceptiveness without falling back and securing even more rigorously the subjectivity of the modern epoch. By engaging in this "distancing," we enter into a chiasm which is distinct from the dichotomous subject-object relation or the "me-other rivalry" which typifies our traditional metaphysics.[18] In this manner, our inquiry sets a crucial but difficult task; it helps *prepare* us to experience a non-representational, that is, non-metaphysical thinking which is outlined in Heidegger's view of Gelassenheit. If this engagement is to work, then, it is fundamental that

> a chiasmic approach recognizes the independence of another thought and leaves it within itself in its concealed contextuality and hidden depths. . . . A member of a vibrant chiasm increases its independence and grows in its holding to itself because its own latent particularities and directions can often only be brought to light by the way it bends and fits itself to an other.[19]

As we shall discover, then, it is this "distance" which keeps a chiasm illuminating. Our interrogation remains authentic insofar as we acknowledge and safeguard both sides of the chiasm. Its disclosiveness remains reciprocating, evolving and non-static where its multifarious relations continue to bend toward, touch and disentangle from one another. It is with regards to our participation in this responsiveness that we must not mistakenly anticipate that our chiasmic "exchange" is to resolve itself, as if we were working towards an encompassing criticism or a final synthesis. On the contrary, what keeps a chiasm provocative is the intensity of its crossing-over such that its filaments give way for and respond to the other's presence, but never close absolutely.

To facilitate this intertwining – and we must recognize that not every work of art accomplishes this task – it is crucial that we refrain from ordering the drama as ready-made for our interpretive (subjective) handling. Rather, to participate in the drama's "interminable reinterpretation,"[20] we are to let the drama be, to unfold of itself in an incisive manner as a substantive site for questioning. In this play of emergence and re-emergence, and I include Beckett's drama in such a listing, we may come to experience the work as a "project of an infinite *Logos*."[21]

It is pertinent to anticipate, then, that our encounter with *Waiting for Godot* does not arise at randomly chosen scenes. Though certainly not exhaustive of the drama, the sites included in this study have been picked as they declare themselves as localized, disclosive moments pertinent to our inquiry.[22] Each scene "reaches" across the divide between theatre stage and seats to beckon the audience towards a "remembrance" of this *Logos*. Our chiasmic involvement calls for returning constantly to these conspicuous moments, genuinely recognized as decisive by those familiar with the drama, to retrieve the silent dimensions of "waiting" otherwise precluded by the representational ordering of our world.

To enhance this task of letting the drama be, it is helpful to envisage that we find ourselves within the space of a theatre, rather than to focus on the drama as a text where its linguistic or conceptual details are given preference. Although this "imaginative exercise" may appear arbitrary, it is crucial. The multifarious dimensions of the performance become effectually marginalized if we do not open ourselves to the drama *as* drama. To experience this dynamic of the chiasm, then, we must enter into the intimate "divide" which both joins and separates the audience and the dramatic performance. Our participation in this resonating flux between the drama, the theatre stage and the audience is guided by the perceptual play of the "theatre" itself (as *theatron*)[23] as it situates a space for our participation. We enter into the drama *as* drama (as *dran*) by apprehending the authentic "doing" or "working" established by the performance as it brings its original "spectacle" into the light before the audience. We join in the "self-showing" of our *humanitas*[24] by reflecting on the phenomena presented in the work.

To close this reflection on chiasm I would reiterate that I have chosen to work hermeneutically with *Waiting for Godot*, for as a site of *technè* – to apply Heidegger's term – it is less hampered by our contemporary mode of philosophic practice which has become closely associated with a scientific inquiry. It is because our entanglement and complicity in the rationalism and technologism of our present world has become so immediate and thorough that we must attempt to prepare ourselves to step back from the prevailing pragmatics and logistics of our philosophic tradition. I would have you consider that Beckett's drama establishes such a decisive site; it persists in offering clarifying glimpses of our estranged place within the world, of our sedimented and habituated inter-relations. In consideration of its illuminating prospect, we may begin to foster a more self-critical and reflective posture towards our existential rapport with the question of Being, which is not subsumed by the calculative and systematic ordering of our epoch.

PART II: CHIASM AND WAITING

Having examined our chiasmic method, it is possible now to move to the substantive issue of the inquiry. To commence this study, it will be extremely helpful to introduce the excerpted passages from Heidegger's *Discourse on Thinking* which will serve as the basis of our examination. As in the case with Beckett's drama, I have chosen these segments as they present many of the crucial dimensions of Gelassenheit, as "waiting." Consider this passage:

Scientist: Toward the last you stated that the question concerning man's nature is not a question about man.
Teacher: I said only that the question concerning man's nature makes a consideration of whether this is the case unavoidable . . .
Scientist: To behold man's nature without looking at man!
Teacher: Why not? If thinking is what distinguishes man's nature, then surely *the essence of this nature, namely the nature of thinking, can be seen only by looking away from thinking.*
Scholar: But thinking, understood in the traditional way as re-presenting is a kind of willing . . .
Scientist: Then the statement that the nature of thinking is something other than thinking means that thinking is something other than willing.
Teacher: And that is why, *in answer to your question as to what I really wanted from our meditation on the nature of thinking, I replied: I want non-willing.*
Scientist: Meanwhile this formulation has proved ambiguous.
Scholar: Non-willing, for one thing, means a willing in such a way as to involve negation, be it even in the sense of a negation which is directed at willing

	and renounces it. Non-willing means, therefore: willingly to renounce it. *And the term non-willing means, further, what remains absolutely outside any kind of will . . .*
Teacher:	But perhaps we come nearer to it by a willing in the first sense of non-willing.
Scholar:	You see, then, the two senses of non-willing as standing in a definite relation to each other.
Teacher:	Not only do I see this relation, I confess that ever since I have tried to reflect on what moves our conversation, it has claimed my attention, if not challenged me.
Scientist:	Am I right . . . ? *You want a non-willing in the sense of a renouncing of willing, so that through this we may release, or at least prepare to release, ourselves to the sought-for essence of a thinking that is not a willing.*
Teacher:	You are not only right, by the gods! As I would say if they had not flown from us, you have uncovered something essential.

Further on, the dialogue continues:

Scientist:	With the best of will, *I cannot re-present to myself this nature of thinking.*
Teacher:	*Precisely because this will of yours and your mode of thinking as re-presenting prevent it.*
Scientist:	But then, what in the world am I to do?
Scholar:	I am asking myself that too.
Teacher:	We are to do nothing but wait.
Scholar:	That is poor consolation.
Teacher:	Poor or not. We should not await consolation – something we would still be doing if we became disconsolate.
Scientist:	Then what are we to wait for? And where are we to wait? I hardly know anymore who or where I am.
Teacher:	None of us knows that, as soon as we stop fooling ourselves.
Scholar:	And yet we still have our path?
Teacher:	*To be sure. But by forgetting it too quickly we give up thinking.*[25]

What it most curious about the awkward rapport of the conversation's participants is that it betrays our own discomfiture when we inquire into Heidegger's discourse. It is telling that our initial response is often to side with the scientist, when we listen to his opening reflection: "If thinking is what distinguishes man's nature, then surely the essence of this nature, namely the nature of thinking, can be seen only by looking away from thinking." We concede as he does that, "With the best of will, *I cannot re-present to myself this nature of thinking.*" It is not surprising that we are "disconsolate" when called upon to renounce our presence as the *ego cogito* or to abandon the pragmatic efficacy of our representational perspective. We hear that releasement calls for "renouncing" our mode of subjective willing. Still, how we are to broach this preparatory mode of thinking as a "non-willing," which is to serve as a "keeping awake for releasement" and yet which also keeps open our critical reflection to the "nature of man"? This move seems imminently paradoxical when our intuition seems ready to declare that this renunciation of willing is something heretical. How is this releasement credible when our metaphysical mode of analysis necessitates that we explain the world and by extension our essence in terms of the principle of reason?[26] While our apprehension of "waiting" is bound to be obscure and hesitant, it is essential that we examine these critical questions. As we shall encounter by thinking through the drama, we can open our perception to this "waiting" which takes issue with the orthodox "form of transcendental-horizontal re-presenting" that pervades our epoch.[27] To set this context of our examination, let us turn our attention to the drama's disclosive moments of the phenomena of "waiting." Our encounter begins curiously with the work's first lines of dialogue:

Estragon sitting on a low mound, is trying to take off his boot. He pulls as it with both hands, panting. He gives up exhausted, rests, tries again. As before.

Estragon: (*giving up again*) Nothing to be done.
Vladimir: (*advancing with short, stiff strides, legs wide apart*) I'm beginning to come round to that opinion. All my life I've tried to put it from me, saying, Vladimir, be reasonable, you haven't yet tried everything. And I resumed the struggle. (*He broods, musing on the struggle. Turning to Estragon.*) So there you are again.[28]

Uttered first by Estragon and subsequently reaffirmed twice by Vladimir, the conspicuous declaration that there is "nothing to be done" is set to pique our interest. It stirs our curiosity, for in the context of the theatre we are quick to deny its plausibility. Rather, we expect that something happens in the performance just as we anticipate our own participation in the contingency of the world. Still, it is interesting how this refrain draws our attention to the strategically placed yet curiously open-ended question which pervades the entire drama. We hear Vladimir ask, "What do we do now?," after Estragon battles with his boots, and after both argue about the Gospels and whether this spot is the exact one where they were to wait for Godot. But, we listen to his query as though it were the original question which prompts both to answer that there is "nothing to be done."

The incongruence and repetition of these two lines anchor our involvement in the drama and our own presence in the world. We confront our own expectancy as we reflect on the interrogative, "What do we do now?" The question frames exactly how we chart our day-to-day affairs. And yet, we also sense that it marks something which is terribly amiss. It seems totally incongruous to ask this question when nothing appears accomplished save that Estragon or Vladimir wait and will wait yet again for the next day, despite their games and playful musings. Still this paradox is central to the performance. Their fitful play defies the rational expectations we have of our own activities and decisions. As such, the drama sets up a moment for the audience to reflect upon the habituated modes of its various enterprises, and to consider what it is to accomplish, if anything at all. The drama lets the audience attend to the logic of its rationality, a logic which requires that we describe and explain our actions in this schematic form, otherwise our activities stand as incomprehensible.

This initial disclosure is strengthened when we encounter the scenes which are most often cited in commentaries about the drama. The first case develops where both protagonists grapple with the reasonableness of hanging themselves but hold off on this decisive action when they decide to "wait till [they] know exactly where [they] stand."[29] The second scene arises where Pozzo commands Lucky to "think," to which Lucky proceeds to make his thoughtful pronouncement at a whirlwind pace until Pozzo, Vladimir and Estragon clamour to remove Lucky's offending "thinking" cap.[30] The third case ensues in the second act where we find the drama's "abuse scene."[31] It develops when, on the lookout for Godot, Vladimir and Estragon, "turning simultaneously," interrupt each other to ask: "Do you – [see anything coming]?" Their courtesy to have the other "finish (their) phrase" boils over to anger, until finally Estragon mortally wounds Vladimir with the quintessential epithet, "Critic."

Although these cases are meant to be bizarre, each illuminates what is "most extraordinarily interesting"[32] about our world. Each lets us reflect on the calculative scope of our *humanitas* where we have a *modus operandi* for our actions. These scenes describe the expectancy of our critical-rationalism, whether or not we appreciate and evaluate the repercussions of our actions. Concerning the first case, for instance, Vladimir's query about what they are to do is given so genuinely and innocently that even Estragon's reply, "What about hanging ourselves?" does not appear out of place. The hyperbole of Estragon's original enthusiasm over its auto-erotic possibilities is ridiculous, for he might have replied otherwise, but Vladimir's prompting question, "What do we do now?," is one we all ask. Even Vladimir's subsequent "indecisiveness" is not unexpected for we would hope that they "know exactly where [they] stand." Still, their "reasonably" discharged decision to hang themselves does seem misplaced, for their "decision" *is* indicative exactly of what they do not know; they do not know whether one might die in the attempt to get an erection. The strangeness of this scene is not determined so much by comic play on the sexual innuendo of their choice, but by the fact that in the audience we still gauge their "decision" by a measure of what is "reasonable."

Likewise, in our third scene, we find ourselves confronted with the task of assessing whether their personal attack is a measure of our authenticity. We share in the comic play where Estragon and Vladimir face off at "high noon" and then "make up" after the final parting shot. This scene works, for we are sensitive to their lack of sincerity. As they "make up," we recognize immediately their quick return to the critical matter at hand – "what are they to do now?" This abrupt shift in direction shakes our perception; it leaves open the question of whether their exchange marks a moment of authenticity or whether it highlights the everydayness of our preoccupations.

Although decidedly comical, these cases are telling in the way they focus the audience's attention to appreciate that even the most extreme of situations can be reconstructed and evaluated from a logical perspective.[33] What is peculiar here is that in all three cases we look for the underlying logic despite the apparent inanity of the actions. The scenes reveal the extent to which we delimit our presence within the scope of a rational decision-making process, rather than centre on the nebulous prospect of what it is to "wait." In this manner, *Waiting for Godot* prompts the audience to experience how "global" our rational ordering of the world remains, and to consider that we establish our way in the world through our "subject-centred" "decisions," much as Estragon and Vladimir do through their various decisions.

The descriptive play of these scenes is incisive in the manner in which it shifts the audience's attention back to the original and seemingly innocuous query, What do we do now?, which sets the drama in motion, and yet by doing so it also draws our interest forward to the discomforting nuance of "waiting." The pathos of these moments emerges as we recognize that something substantive happens in the performance of Vladimir and Estragon grappling with the "impossibility" of their situation, though it appears empty as a sequence of inconsequential happenings.[34] In the second case, for instance, the scene carries forward from the excitement of Lucky's tirade to the enduring dilemma faced by Estragon and Vladimir. Having bid adieu to the departing Pozzo and Lucky, they console themselves with:

Long silence.

Vladimir: That passed the time.
Estragon: It would have passed in any case.
Vladimir: Yes, but not so rapidly.

Pause.

Estragon: What do we do now?
Vladimir: I don't know.
Estragon: Let's go.
Vladimir: We can't.
Estragon: Why not?
Vladimir: We're waiting for Godot.
Estragon: (*despairingly*) Ah!

Pause.[35]

And as well, in the closure to the "abuse scene" we find Vladimir and Estragon apologizing with ingratiating hyperbole only to return to the imminence of "waiting":

They embrace. They separate. Silence.

Vladimir: How time flies when one has fun! (*Silence.*)
Estragon: What do we do now?
Vladimir: While waiting.
Estragon: While waiting.

Silence.[36]

It is conspicuous that Estragon's reiteration of "while waiting" is held by a moment of silence, as this moment in itself draws the audience into the meditative space of "releasement." We enter into this "waiting" as we reflect on the openness of the preceding question. The constancy of this question points to the attendant repose of "waiting," given in the moment of silence. We are held with them in the constancy of their "waiting" when the promise of their subjectivity is left unresolved.

Waiting for Godot draws the audience into this comportment with "waiting" by sharpening the distinction between the hesitance of "waiting" and the "prototypical" subjective queries undertaken by

Vladimir and Estragon. This experience presents the audience with a glimpse of the fragility of its "subject-centred" modes of reasoning and social relations. By presenting the reasonableness of their endeavours in a questioning way, the drama clarifies how readily we conduct our affairs as the *animal rationale*, in contrast to the open suggestiveness of "waiting."

Our perception of this repose of "waiting" deepens as we become sensitive to the dramatics of the decisions and actions undertaken by Estragon and Vladimir. Our response to their smelling, eating, sexuality, and affections is to say that we are aware of their personal space, but we also admit that these actions garner them little in the way of a tangible presence. This reflective stance is especially urgent where in the audience we ensconce and safeguard ourselves in the everydayness of the world much as the protagonists shelter one another from the vagaries and uncertainties of the road while waiting for Godot. Although we are familiar with the drama's recurrent query, "What are we to do now?," it is poignant that we do not reflect on the contingent particularities of our circumstance and our relations with one another, as an issue of our authentic possibility.[37] Rather we rely simply on the privilege of our rationality and its practice of representing. Our tact is to retreat unquestioningly to the assurance that our *humanitas* is inscribed by an endless series of choices.

Still, as the drama proceeds, we begin to experience the unbidden gesture offered in the unexpected promise of Estragon's unresolved question, "What do we do now?" The pause which pervades his question helps us encounter the compelling yet difficult prospect of "waiting," for the pause itself resists being accounted for in the way that asks, "what does this 'waiting' represent?" By engaging in this retiring gesture of "waiting," we broaden our perceptive acuity in a manner that turns our thinking towards a modality not transfixed by the systematic modes of analysis circumscribed by our metaphysical tradition.

These reflections point to one of the urgent yet telling facets of Heidegger's *Discourse on Thinking*. Our propensity to regard our critical reflections solely as modes of representing is so ingrained that we find ourselves pondering the same perplexity referenced by the conversation's three characters. We want to know the why and wherefore of "waiting." To listen to the scientist's comments, this expectancy of representation overshadows the entire conversation:

Scientist: With the best of will, I cannot re-present to myself this nature of thinking.
Teacher: Precisely because this will of yours and your mode of thinking as re-presenting prevent it.
Scientist: But then, what in the world am I to do?
Scholar: I am asking myself that too.
Teacher: We are to do nothing but wait.[38]

The scientist's puzzlement over failing to represent this thinking highlights our uncertain perplexity when studying this project of "waiting." It is precisely this problem that we encounter when attending to the open prospect of waiting as presented in the drama. We seem impelled to ask, "What does this 'waiting' represent?," but as we have found, this question is left open. As noted in the drama, then, to listen to the teacher's remark, "We are to do nothing but wait," we are easily waylaid by the suggestion of doing nothing but "waiting." This reaction is especially urgent when it means that we are to leave off our usual practice of bringing an object before our subjective gaze. We are distressed by this "letting go." It seems completely counter-intuitive to countenance relinquishing the pragmatic proficiency of our representationalism when we are so habituated to this practice.

Although this "waiting" seems impossibly ambiguous, as we become familiar with its gesture of entreatment we do begin to broach the possibility of thinking non-representationally. To sense this reciprocating reflexivity given in "waiting," let us return to consider Estragon and Vladimir's "in-decision" at the close of the drama:

Estragon: Didi.
Vladimir: Yes.
Estragon: I can't go on like this.
Vladimir: That's what you think.
Estragon: If we parted? That might be better for us.
Vladimir: We'll hang ourselves to-morrow. (*Pause.*) Unless Godot comes.
Estragon: And if he comes?
Vladimir: We'll be saved. (*Vladimir takes off his hat (Lucky's), peers inside it, feels about inside it, shakes it, knocks on the crown, puts it on again.*)
Estragon: Well? Shall we go?
Vladimir: Pull on your trousers.
Estragon: What?
Vladimir: Pull on your trousers.
Estragon: You want me to pull off my trousers?
Vladimir: Pull ON your trousers.
Estragon: (*realizing his trousers are down*) True. (*He pulls up his trousers.*)

Vladimir: Well? Shall we go?
Estragon: Yes, let's go. (*They do not move.*)
CURTAIN[39]

This scene plays on the sharp repartee between Estragon's tragic comment, "I can't go on like this," and Vladimir's sardonic retort, "That's what you think." Having followed through the protagonists' tribulations, it is difficult not to be empathetic with Estragon as he whimpers, "I can't go on like this." We laugh and cringe at the cruel reality of his predicament, especially when just prior to this finale we hear that both ponder the option of hanging themselves. By listening to their exasperated plight, however, we experience the dimensions of our own finitude and captive presence on this earth. We recognize the bitter reality that we too must "wait" for the next day when our expectations and desires go unfulfilled. When our endeavours are obscured or deflected very little seems amiss if we scream: "I can't go on like this!" This exclamation sounds natural under the circumstances in the drama. But, it also prompts the audience to question whether it is incapacitated by the pause at the close of the drama. For to consider that we are constrained by this "waiting" appears to be excluded given that, as subjects free to set our own agenda, we anticipate being able to embrace the world as our object.

This expectancy is especially poignant when we consider that Vladimir's response, "That's what you think," attests to our accustomed mode of rationality. We understand implicitly his corrective qualification even as we would distance ourselves from the drama's discomforting disclosure of our presence in this modern world. When he points out the obvious, that Estragon is "wrong" to think that he cannot go on "like this," we are immediately sensitive to the "poor consolation" that surfaces in this final moment of "waiting." Vladimir's simple phrase and the pause as the curtain falls combine to open a gap for the audience to assume a reflective posture. This space lets the audience broach the query left unstated by Vladimir and Estragon: if it is not to be "like this," then how is it to be? What is at stake in the drama is precisely that we are loath to "wait." On the contrary, as in our subject-centred world, we strive to realize our aims, wishes and desires, rather than enter into the reposeful space of "waiting." From an ontological perspective, the urgent and imposing necessity of our subjectivity betrays our own impatience as our presence seems to be delimited solely in terms of our actions and decisions.

Still, by following the indefiniteness left hanging in the last lines before the final curtain, we do begin to trace out this prospect of "waiting." Listen again to the final lines:

Estragon: Well? Shall we go?
Vladimir: Yes, let's go. (*They do not move.*)

To appreciate the import of this final pause, consider the dramaturgic effect if the last lines are drawn out slowly with a dim to complete darkness, a close fade of a spotlight on the hanging moon, and then a prolonged return to a fully lit theatre. The characters are to remain stationary until the lights are dimmed and the curtain is closed. The significance of this delay lies precisely in the essence of "waiting." While we are offered a moment to take up the question, "Shall we go?," we are held by this "waiting" as we stand mediately with Estragon and Vladimir. This pause disrupts our rational expectancy and our reliance on the efficacy of our questioning. But by leaving the focal questions "What shall we do?" or "Shall we go?" unresolved, the drama turns the audience towards the repose of "waiting."

As we attend to these moments in the performance, the "waiting" sets aside our usual demand for principled action and concrete answers to our questions. It reveals that we may attend to our queries, but without closing them in a resolute answer. To put the point even more strongly, it is crucial for the dramatic play of *Waiting for Godot* that it does not resolve these questions. On the contrary, by holding these queries open, it offers the audience and us here the space to appreciate the "letting-be" of "waiting" for Godot. In "waiting," we give up the expectation of identifying Godot, though we remain attentive to Godot's presence.

This review of the drama clarifies an essential facet of Heidegger's exploration of Gelassenheit given in "waiting." The disconcertedness we witness in this passage from the "Conversation" points to the strain we face when preparing to move into this non-metaphysical mode of thinking. Notice again the uncertain tenor of the conversation:

Scientist: With the best of will, I cannot represent to myself the nature of this thinking.
Teacher: Precisely because this will of yours and your mode of thinking as re-presenting prevent it.
Scientist: But then, what in the world am I to do?
Scholar: I am asking myself that too.
Teacher: We are to do nothing but wait.
Scholar: That is poor consolation.

Teacher: Poor or not. We should not await consolation – something we would still be doing if we became disconsolate.
Scientist: Then what are we to wait for? And where are we to wait? I hardly know anymore who or where I am.
Teacher: None of us knows that, as soon as we stop fooling ourselves.
Scholar: And yet we still have our path?
Teacher: To be sure. But by forgetting it too quickly we give up thinking.[40]

We sense this strain in the sharpness of the teacher's comment which plays on the scientist's insecurities, "None of us knows what we are looking for, as soon as we stop fooling ourselves." We are immediately aware of the emotive tension which cuts through this moment in the conversation. This exchange is crucial on another level, however. When we listen to the scientist's query, "what in the world am I to do?," it is insightful to recognize how closely it parallels the drama's refrain, "What do we do now?" This similarity highlights just how obvious it is that we are poorly consoled by this "waiting." Both cases aptly reveal how quickly we appeal to this instrumental view of our "subjectness" in order to address our concerns and interests. We acknowledge this appeal by appreciating how readily we accord with the scientist's question. We are loath to renounce the orthodoxy of our "willing" in favour of the resolve of releasement.

But, what is also provocative is that both works touch upon and yet leave open the issue of representation. This "holding open" is the most difficult move in the "stepping back" which centres Heidegger's description of "waiting" as it is juxtaposed to our usual approach of securing and developing our potentiality.[41] On the contrary, Heidegger describes this "waiting" as an entreating or as a holding out towards, where our inquiries remain mutually with but not closed by the gesture of this offering. Likewise, in *Waiting for Godot*, this "waiting" is revealed in the performative hiatus left at the close of the drama. The crux of this moment is that nothing presents itself to fill the "waiting." That is, the dramatic event lacks its own re-presentation. We experience this "waiting" in its emptiness by attending to Godot's presence though we cannot place Godot within our representational field. In this case, "waiting" is experienced as a holding out towards an absence. As such, the "waiting" is not resolved by securing what is absent through the prowess of our subjectivity.

This "waiting in releasement" is prepared for by "stepping back" from the prevailing order of our subjectivity. In this renunciation which leaves off representing, we attend to the reserve of "waiting" which "releases itself into openness."[42] As in the case of the openness which pervades the drama's recurrent query, "what do we do now?," we encounter the openness of "waiting" as an "expanse of distance," where "in waiting we leave open what we are waiting for."[43] In this sense, we appreciate that "we do awaken releasement in ourselves . . . [It] is not effected, but let in."[44] This "renunciation" is not achieved through an act of volition. Rather, it emerges as we relax from our subjective tendency of grasping every opportunity as a possessive means of knowing or accomplishing something. In contrast to our prevailing mode of understanding where the "subject" brings its "object" into its field of vision, Heidegger describes this reflexivity as a gesture which "lets beings be." As a "letting-be," releasement harkens to a "higher form of acting [which] lies – [. . .] – beyond the distinction between activity and passivity."[45] This openness to a non-willing, non-subject-centred posture prevails to open the primordial space of our essence without reverting to and reinscribing the traditional mode of a "transcendental-horizonal re-presenting"[46] which relies on the questions that dominate our habituated perspective: What is a human being? and, What ought we to do? Though these theoretical and practical questions underscore our presence as the *ego cogito*, or the *animal rationale*,[47] we have experienced through the drama that we can open a new approach to these concerns. This "waiting" calls for a mutual interplay between interrogating and answering where thinking sojourns as a reciprocating involvement and attunement to one's surroundings.

CONCLUSION

We are to recall from the introduction that this inquiry is bound to be finite, though the chiasm itself may continue to serve as a site for future reflection. In this sense, I would be remiss by not acknowledging that these reflections set up *a* pathway but not necessarily *the* way to follow through Beckett's *Waiting for Godot*. I am not claiming to present a definitive hermeneutic "reading" of the drama. I am hopeful, thus, that different perceptions from other perspective "readers" will serve as an impetus to keep this chiasm evolving, and to spark us to continue to reflect upon philosophic issues by way of artworks. If this inquiry has been at least partially illuminating, I would hope that those who have followed this excursion will be prompted to turn to a performance of Beckett's drama, both to reflect on what transpires in the work as well as to renew this meditation on Heidegger's notion of "waiting." As alluded to earlier, the "working" site of this inquiry is not restricted to the theatre *per se*. It is preserved contiguously in the performative essence of the conference, and also continued by those who take up this essay as critical readers. But, in both cases, we are to recognize that it is the drama which serves as the impetus for this chiasmic exchange.

Atkinson College, York University

NOTES

[1] Martin Heidegger, "The Origin of the Work of Art" in *Poetry, Language, Thought*, trans. A. Hofstadter (New York: Harper & Row, 1971), p. 86. Also see Maurice Merleau-Ponty, "Eye and Mind" in *The Primacy of Perception*, trans. Carleton Dallery, ed. J. M. Edie (Evanston: Northwestern University Press, 1964), "Cézanne's Doubt" in *Sense and Non-Sense*, trans. Herbert L. Dreyfus and Patricia A. Dreyfus (Evanston: Northwestern University Press, 1964).

[2] Martin Heidegger, "The Question Concerning Technology" in *The Question Concerning Technology and Other Essays*, trans. W. Lovitt (New York: Harper & Row, 1977), p. 35; Martin Heidegger, "The Turning" in *The Question Concerning Technology and Other Essays*, pp. 41–43. For Heidegger's discussion on the essence of truth as *Aletheia*, see variously, *Being and Time*, trans. J. Macquarrie and E. Robinson (New York: Harper & Row, 1962), Section 44; "On the Essence of Truth" in *Basic Writings*, ed. David F. Krell, trans. J. Sallis (New York: Harper & Row, 1977); "The Origin of the Work of Art"; and "Aletheia (Heraclitus, Fragment B 16)" in *Early Greek Thinking*, trans. David F. Krell and F. Capuzzi (New York: Harper & Row, 1975).

[3] Samuel Mallin, *Here to Zero* (Toronto: S.L. Simpson, 1983), pp. 7, 23–24, 29.

[4] Samuel Beckett, *Waiting For Godot*, trans. from the French text by the author (New York: Grove Press, 1954).

[5] Martin Heidegger, *Discourse on Thinking*, trans. J. M. Anderson and E. H. Freund (New York: Harper & Row, 1966), p. 58.

[6] *Ibid.*, p. 68.

[7] Martin Heidegger, *Identity and Difference*, trans. Joan Stambaugh (New York: Harper & Row, 1969), pp. 49–52.

[8] See Heidegger's discourse on the possible "saving-power" which is harboured within our epoch. See his essays "The Question Concerning Technology" and "The Turning," as well as his discussion in "Nihilism as Determined by the History of Being" in *Nietzsche IV. Nihilism*, ed. David F. Krell, trans. Frank A. Capuzzi (New York: Harper & Row, 1982).

[9] Maurice Merleau-Ponty, *The Visible and the Invisible*, ed. C. Lefort and trans. A. Lingis (Evanston: Northwestern University Press, 1968), p. 266.
[10] *Ibid.*, p. 265.
[11] *Ibid.*, pp. 143–147, 248.
[12] Samuel Mallin, "The Line of Performance: Pina Bausch's Dance-Theatre" in *Harbour* (Vol. 1, No. 1, 1990), p. 29.
[13] Mallin, "The Line of Performance: Pina Bausch's Dance-Theatre," p. 31.
[14] This term is a key description of Merleau-Ponty's notion of chiasm. See *The Visible and the Invisible*, pp. 142, 263–266.
[15] *Ibid.*, p. 266.
[16] *Ibid.*, p. 266.
[17] Merleau-Ponty, "Eye and Mind," p. 166.
[18] Merleau-Ponty, *The Visible and the Invisible*, p. 215.
[19] Samuel Mallin, "Chiasm, Line and Art" in *Merleau-Ponty: Critical Essays*, ed. Henry Pietersma (University Press of America, 1989), p. 223.
[20] Merleau-Ponty, "Eye and Mind," p. 179.
[21] Merleau-Ponty, "Cézanne's Doubt," p. 19.
[22] Mallin, "Chiasm, Line, and Art," pp. 225, 233–235.
[23] On the etymology of "theatre" which links to the Greek *theoria*, as "spectacle" or "contemplate," see for example: Bruce Wilshire, *Role Playing and Identity: The Limits of Theatre as Metaphor* (Indianapolis: Indiana University Press, 1982), pp. 31–33; Martin Heidegger, *Parmenides*, trans. André Schuwer and Richard Rojcewicz (Bloomington: Indiana University Press, 1992), pp. 102–104; Hans-Georg Gadamer, *Truth and Method*, 2d ed., trans. Joel Weinsheimer and Donald G. Marshall (New York: Seabury, 1975), pp. 124–128.
[24] Heidegger, *Parmenides*, p. 103.
[25] Heidegger, *Discourse on Thinking*, pp. 58–62. (My emphasis in italics.)
[26] Martin Heidegger, *The Principle of Reason*, trans. Reginald Lilly (Bloomington: Indiana University Press, 1991), p. 120.
[27] Heidegger, *Discourse on Thinking*, pp. 63–64.
[28] Beckett, *Waiting for Godot*, p. 7.
[29] *Ibid.*, p. 12.
[30] *Ibid.*, pp. 27–30.
[31] *Ibid.*, pp. 48–49.
[32] *Ibid.*, p. 9.
[33] See Heidegger, *The Principle of Reason*, pp. 127, 129.
[34] Bert O. States, *Great Reckoning in Little Rooms: On the Phenomenology of Theatre* (Berkeley: University of California Press, 1985), p. 74. Also see, L. Kane, *The Language of Silence* (London & Toronto: Associated University Press, 1984).
[35] Beckett, *Waiting for Godot*, p. 31.
[36] *Ibid.*, p. 49.
[37] Martin Heidegger, "Letter on Humanism" in *Basic Writings*, trans. Frank A. Capuzzi, in collaboration with J. Glenn Grey and D.F. Krell (New York: Harper & Row, 1977), p. 196.
[38] Heidegger, *Discourse on Thinking*, p. 62.
[39] Beckett, *Waiting for Godot*, p. 60. See also p. 44.

40 Heidegger, *Discourse on Thinking*, p. 62.
41 *Ibid.*, p. 80.
42 *Ibid.*, p. 68.
43 *Ibid.*, p. 68.
44 *Ibid.*, p. 61.
45 *Ibid.*, p. 61.
46 *Ibid.*, pp. 63–64.
47 Heidegger, "Letter on Humanism," p. 196.

WOLFGANG WITTKOWSKI

CAN WOMEN REIGN?
Schiller's Maria Stuart *and the Nemesis*

Born in 1542, the Catholic heiress to the Scottish throne lost her father very early. She was brought up at the French court by her mother's relatives, the powerful Guises, was married to the heir to the throne of France when she was 16 and soon became Queen of France. Widowed two years later, she returned to Scotland at the age of 19 after her mother's death and "thoroughly enjoyed the business of ruling."[1] At 23, she married one of her English cousins; and when he made himself unpopular enough to be murdered, in 1567, she married his powerful murderer. She was defeated on the battlefield by the rebellious nobility led by her half-brother, was taken prisoner, escaped, was defeated again, sought protection in England with her cousin Elizabeth who, however, had her taken prisoner and tried in 1568, and nineteen years later, in 1587, decapitated. Elizabeth's reasons were that several coups had been attempted on Mary's behalf, and the threat her presence posed to the Queen's rule whose legitimacy was being denied by the Catholic world. In an act of cowardice, Elizabeth later accused her staff of being responsible for this monstrous transgression, the first in a series of regicides leading up to the beheading of Louis XVI in 1792.

Schiller, grappling with such burning political issues of his own time, granted "imagination a liberty over history" in his tragedy of 1800. His dramatic representation of Mary's last three days makes use of inventions he found in earlier Mary Stuart plays: a meeting of the two women (which never took place); presenting them as young, attractive and sexually very much alive, aged 25 and 30 years respectively, and both of them challenged to become competent rulers; and their relationships to such men as the fanatical renegade Mortimer, and Lord Leicester, Elizabeth's lover and formerly Mary's prospective husband. Both men are trying to save the Scot, although Elizabeth believes them to be her helpers in finishing off the rival queen.

According to current readings of Schiller's play, Mary, the feminine one of the two, is a loser and Elizabeth, the masculine one, a winner, supposedly illustrating the fact that under patriarchalism women cannot combine femininity with political leadership[2] or cannot be rulers at all.[3]

A-T. Tymieniecka (ed.), Analecta Husserliana, Vol. LIII, 233–248.

After all, isn't that how it is in real life? At least, Schiller must have thought so, and he cemented and perpetuated this sad situation with his play. Actually, the number of able women leaders of Elizabeth and Mary's as well as of our own time (and Schiller's!) should throw considerable doubt upon this view of reality and Schiller's thesis respectively – if it was his thesis at all. And that, indeed, is the question I want to raise; or more precisely this: does Schiller really present Mary, Elizabeth, or both, as being unfit to reign?

Whichever conclusion they support, however, most current interpretations proceed to diminish Mary's qualities as a ruler and a woman while uplifting those of Elizabeth and excusing her obvious ethical shortcomings. They advise us to drop the traditional solemn admiration for Mary's last transformation, be it into a beautiful or a sublime soul. Either way, Mary merely escapes the political realities she has proven unable to manage,[4] but in so doing is glorified for her Christian self-sacrifice; while Elizabeth, though often morally repulsive, carries out her royal duties ably and successfully – the only way to improve world conditions.[5] Most critics allow for her accomplishments as recorded by history. That is, so we learn, why Schiller places the victorious queen at the end of the play and not as its title figure. For the same reason, some add, Mary's defender old Talbot, and his position of political ethics and justice are presented as old-fashioned, antiquated, the implication being, of course, that politics, by its very nature, has nothing to do with morality.

Perhaps this is the wisdom of shrewd Germanists who claim that they, at long last, have come of age in matters of politics. It is hardly the way most ordinary people feel. And we know well that the 18th century in particular applied strict standards of morality to everything and not least of all politics. Of course, nobody was believed capable of being perfect. If the evil individual exhibits some better traits, they are poisoned by a main vice, – in Elizabeth's case, her hypocrisy.[6] Conversely, Mary's weaknesses are ennobled by her outstanding honesty. Leicester's egotism leads him to misuse and betray both queens. Mortimer's sensuality, which goes to the extreme of violence, overshadows even his generous self-sacrifice. While Mary's generosity and genuine sympathy with others stand out, these traits are utterly foreign to Elizabeth and Leicester. Finally, Mary is the only one besides Talbot who insists on justice and repudiates the use of evil means.

This may be easier for the convicted defendant facing execution, but it does not change the fact that one true virtue attracts or joins with others:

honesty, generous care for others, justice. Current Schiller scholarship seems to overlook the fact that fulfilling this triad of the Golden Rule is exactly and exclusively what could improve world conditions and preserve peace – at least according to the *Natur- und Völckerrecht* of Christian Thomasius (1705). Schiller learned about the so-called "Father of the German Enlightenment" (May 29, 1799 to Goethe) when he was studying material for *Maria Stuart* (1800) and just in time for *Zum Antritt des neuen Jahrhunderts*, when (as this poem deplores) France and England were wrestling for world domination, trampling on peace, freedom and justice everywhere. His *Maria Stuart* was to tackle the hottest political questions around 1800.

In Schiller's works, the force organizing tragic action – the *nemesis* – plays a decisive part. As in contemporary thinking cause and effect are firmly linked, this antique goddess of justice and revenge presents the dire consequences of major transgressions of the universal moral order without sparing the innocent. She sends a strong warning signal to the audience and also to the figures during the action of the play and she does so early on, before striking, thus offering the protagonists an opportunity to escape harm by stopping, turning around and seeking a new and better beginning.

Herder had shared his ideas concerning *Nemesis: Ein lehrendes Sinnbild* (1786) with Schiller the year following its publication (August 8, 1787 to Körner), and the playwright (himself author of two voluminous historical narratives) took to heart Herder's claim at the end of his learned study: "If mankind is to learn from history, teacher-scholars should dedicate the latter to nobody but *nemesis and fate*." From this time on (i.e. in his *History of the Thirty-Years War* 1790/92, his *Wallenstein* 1799 and all dramas to follow), Schiller made powerful use of the *nemesis* as the correcting force of action, even though he did not share Herder's personal belief in it as a real historical agent.

We learn that prior to the start of the action both queens have provoked the *nemesis* plentifully. Mary, for her part, has agreed to the killing of her misbehaved husband and has frivolously exalted his murderer by marrying him. Elizabeth, on the other hand, has taken her royal sister prisoner. In the play, we are repeatedly reminded of this gross violation of international law and human rights – the burning political problem around 1800 as Napoleon was forcing his new order upon Europe. Elizabeth's act evoked a series of unwelcome reactions: the hostility of the Catholic countries, France and Spain, the queen's excommunication, assassination attempts, all of this exacerbated by harsh British reprisals against Catholic citizens.

This is my first point of thorough disagreement with current readings which acclaim Elizabeth as a shrewd, pragmatic and successful politician. Moreover, the events of the play as well as those directly preceding it – the law made singularly on account of Mary, the farce of her trial, the hiring of the secret murderer, and her eventual judicial murder – only serve to top off a policy marred by fear and shortsightedness, law breaking, lies and a total lack of imagination concerning new and better ways of proceeding. Whether historically correct or not, Schiller's play presents the British policy towards Mary Stuart as a thorough disaster. On the other hand, it might be said, Elizabeth became Mary's *nemesis*. But the British queen's feeling is just the opposite when she complains,

> She is the fury of my life! Chained
> To me by fate as my tormenting demon
> [. . . a hellish viper . . .] Mary Stuart is the name
> Of every calamity which knocks me down!
> (IV, 10; 3230ff)

Clearly, Elizabeth perceives Mary as "almost a superhuman figure."[7] Yet she never comes close to realizing that she may be speaking of an instrument and signal of *nemesis*,[8] nor does she feel forewarned to consider a new course of action. There is an even more sinister side to this insensitivity, and this is my second major point of disagreement with current *Maria Stuart* scholarship: Elizabeth, who follows one crime with another even worse one, in the process destroys her lawful, constitutional government, the basis of her popularity, transforming it step by step into the tyranny for which she really longs and which will be complete by the end of the tragedy.[9] "In her end is her beginning"; the freedom she desires – total unaccountability – will turn out to be her own and her nation's punishment. Mary will make this known to the world at her own end, thereby realizing the motto – it was hers in history – in her own way.

For Mary, on the other hand, does expect her participation in her husband's murder to be avenged by the *nemesis* (I, 4). But it is different in the end, when her fight for Elizabeth's pardon is lost and she, who has fought so tenaciously for her crown and her own life, accepts her unjust execution as an atonement, granted by God's grace, for her crime of blood for which she has done penance long ago (V, 7; 3571). Having heeded her years of captivity as a warning to repent, to try a new beginning, she has, in Schiller's eyes[10] and those of his audience, effectively atoned for her crime and now confronts us practically free of guilt and almost faultless. No *nemesis* pursues her any more; her autonomous acceptance of death endows her with moral and tragic dignity. Trying for another new beginning, Mary in Act I had, through Mortimer's mediation, accepted Leicester's offer to free her; he was to use his influence with the queen because "Elizabeth's free will alone" could set her free (I, 6; 665). Mary holds on to this belief, rejects Elizabeth's device of murder and requests a meeting with her.

Her request is supported by two of the queen's three advisors. Leicester, however, does not agree, as does Talbot, neither out of humanity, lawfulness, political far-sightedness, nor in order to win Elizabeth over to such motives. Rather, he plays the *nemesis* of his "virgin queen" who is on the brink of betraying her not so secret lover and replacing him with the young successor to the French throne. Both of them know that meeting Mary will oblige Elizabeth to pardon her. But Leicester also knows that Elizabeth expects, albeit erroneously, that Mortimer will solve the problem by killing the prisoner in secret. Playing on the queen's vanity, cruelty and sly secretiveness as well as her reveling in the anticipation of seeing the royal petitioner and pretender defeated, humiliated, broken, Leicester easily persuades her to agree to the meeting.

This twofold foul play, however, backfires against both the doer and the instigator. Both women are trying to impress Leicester; Elizabeth allows herself to be carried away by her arrogance and jealousy, calling Mary a whore only to provoke being called a bastard, as Mary closes in triumph: "If justice ruled in Britain, you would be laying at my feet. For it is I who am your King" (III, 4; 2450ff). Thus Elizabeth is defeated by her own means, and we understand that no other than the *nemesis* is at work here.

Of course Mary must pay for her victory with her life when the accomplices of her would-be savior, Mortimer, provide Elizabeth with a legal excuse by their attempt on her life. As in history, Elizabeth admits her guilt by scrambling to shun all responsibility for signing the execution order. Therefore – worse than in history! – she feels well served by Burleigh who admonishes her to heed "the people's voice, the voice of God Himself" demanding Mary's head (IV, 8; 3068). And great is her apprehension at the news that Talbot, who has already saved her life this day, moreover has now dispersed the crowd by speaking reason. For it illustrates his claim that she does indeed have the freedom to do what is right and reasonable; that she may refrain, at least for the moment, from giving in to her irritation and anxiety. Yet the essence of his words does not get through to her at all. Nor did it to the Germanists (my third major point of disagreement), even though Talbot's great political speech deals with a topic not entirely irrelevant to German Studies – the alternative between constitutional government and dictatorship – and despite the fact that the ending of the tragedy proves Talbot's prediction exactly right. He also identifies Mary metaphorically with the *nemesis*. Excerpts of what he says in IV, 9; 3114ff:

> Now, the living Mary makes you tremble. [. . .]
> However, the beheaded Mary
> Will make you tremble even more.
> Resurrected from her grave,
> As a goddess of disunion, as a demon of revenge,
> She will walk about your country
> And turn your people's hearts away from you.
> Now the British fear and hate her.
> They will avenge her when she's dead,
> Granddaughter only of their kings, victim of hate
> And jealousy. Soon you will recognize the change, [. . .]
> A transformed England and a transformed nation.
> For with you is no longer the grandeur of justice
> Which won you every heart. Fear,
> The horrible accomplice of dictatorship,
> Preceding you, will clear each road you travel.
> You will have done the ultimate extreme.
> Whose head is secure, when this one fell!

Talbot will add in the last scene:

> I am too old, my hand too straight and stern
> To seal the newest deeds of yours. [. . .]
> I have not saved your better part. Live
> And rule with happiness! Your adversary is now dead.
> From now on, you have nothing to respect nor fear. (4023–31)

Tyranny has been achieved – the freedom Elizabeth has yearned for: to rid herself at least of the monarch's shameful, slavish service to the people whom she so utterly despises (IV, 10; 3190).[11] The obvious parallel to this is her resentment of women's alleged slavery to men in marriage (II, 2; 1155–1184).[12] What she wants, in both cases, is to become openly unaccountable and uncommitted to any other human individual or shared value system.

As clear as this is, another question is no less urgent: Why is a policy which ineluctably led to such an end being appreciated by critics, appreciated as proof of Elizabeth's political competence? Which inevitably leads to another question: what else could Elizabeth have done? Besides the many steps she never should have taken, there is one thing she positively should have done. At their meeting, the Queen of Scots reminds her:

> O, had you listened to me when I sought so urgently your eye!
> Then never things had come to this!
> Had you declared me your successor,
> As you do owe it to the law,
> My gratefulness and love had won you
> In me a faithful friend and relative for ever. (III, 4; 2366ff)

Nor did the historical Mary ask for more. Her son did become Elizabeth's successor after all and accomplished that which had been Mary's dearest wish: "To unify two nations, noble, free and cheerful,/Under the shadow of the olive tree" (I, 7; 830f). Instead of naming Mary her heir, and in pointed contrast to uniting the two neighbor nations, Elizabeth pompously announces the half-hearted and ill-fated effort to connect "the crowns of France and Britain" (II, 2; 1223f). France is Britain's enemy and supports attempts to overthrow Elizabeth. The latter despises marriage. More to the point, her would-be husband assures her that he will respect England's religion and hold Mass in private. As reigning queen, Mary had made exactly the same promises and not only kept them but even earned papal censure for ruling in favor of Scottish Calvinism rather than the Catholic minority.

That is the reason, unrecognized throughout and my fourth major point of disagreement, why Schiller introduces Elizabeth's betrothal to the French crown prince in Act II. Historically it was negotiated much earlier and marked a low point in Protestant-Catholic tensions – much lower in fact than shortly before Mary's death. Nothing then could be further from the truth than Burleigh's highly emotional prospect of another violent conversion of faith in England under Mary's rule (IV, 3156ff). It is by far the most impressive speech of this advocate of *raison d'état* and power politics. In light of Talbot's great speech and of the betrothal in Act II, however, it should be read as a blinded man's honest trick to pressure Elizabeth into his wrong line of reasoning. The historical validity of this argument is not represented in the drama, but neither are Mary's notorious tolerance and mildness towards Calvinists set in contrast to Britain's oppression of Catholics. The author of the *History of the Thirty-Years-War* was too aware and weary of the tremendous devastation caused by the bloody clashes of religious denominations.

What is important here is that the Catholic Mary had a strong following in England. Eventually, an alliance between the two cousin or "sister" queens might have been hailed by the entire nation (see Mortimer I, 6; 556ff). To arrive at that, not even extraordinary political far-sightedness would have been needed, but certainly a strong dose of self-confidence and trusting generosity on Elizabeth's part, which is lacking here, as in history. "Do to others what you wish others should do to you," Thomasius requests with Jesus (Luke VII). Not at all! Instead of this (and far worse!), Elizabeth violates the maxim of the *Justum* again and again: "Don't do unto others what you don't want others to do unto you."

Of course – and this constitutes my fifth major point of disagreement with current scholarship – Elizabeth could and should have learned better when she was young, far from the throne and being kept prisoner herself. To express this, Talbot states with courtly euphemism,

> the gracious father of this country
> Raised you by misery to your primal duty. [. . .]
> From early on and undistracted by worldly vanity and noise,
> Your mind was trained to collect itself, to turn inside
> Its thoughts and value the true treasures of this life.
> (II, 3; 1382ff)

On the other hand, the French court had never allowed the spoiled teenage princess and queen Mary a chance for serious reflection and – as Talbot puts it – had indeed "blinded her by the vices' splendor" (1393). Here, critics tend to sympathize with Elizabeth who, they claim, was forced to resign herself early on and to deny her femininity by contrast to her more fortunate rival.[13] However, Schiller[14] and the historians see it the other way around. So, too, does Talbot, speaking of "poor Mary." Elizabeth herself validates his arguments officially by using them partially to her own advantage when she hypocritically proposes to step down (IV, 9; 3139ff, esp. 3155ff). Of course, those arguments are for her an excellent preparation to reign. All the more striking are the results: England's ruler is incompetent, vicious, violent, and a resentful hypocrite who claims to cherish her virginity above all while entertaining an intimate affair with Leicester quite openly (II, 2), and while offering herself in secret (and only in pretense) to Mortimer in an attempt to use him for Mary's secret murder (II, 4).

Mary, on the other hand, having grown up under adverse circumstances, has learned her lessons through her ordeals in Scotland and in English captivity, has become a woman who fascinates by her femininity and honesty and who leaves not the slightest doubt as to her competence as a ruler (my sixth major point of disagreement). Despite having had no practice for years, she aptly demolishes Elizabeth's claims (III, 4) and Burleigh's untruthful official arguments, exposing their unlawful character (I, 7) just as she actually did with her accusers during her historical trial. At the same time, she calls for faithfulness to the constitution and separation of powers in the vein of benevolent despotism:[15] for political structures which Elizabeth's government cynically ignores, turning England into a virtual tyranny (Schiller's contemporary, Kant, was highly critical of the fake republic, England.) Clearly, as a religiously tolerant queen who refuses to equal Elizabeth's criminal means and bigotry, Mary embodies the ideal of a ruler as established by

> The founders of this country. Endowing women
> With the authority to reign as well, they indicated
> That this nation's sovereigns should excel not in severity,

but in gentleness and generosity (II, 3; 42ff): virtues held high during the Enlightenment. The text invites us to take Talbot's and the founding fathers' opinions for Schiller's. It is my seventh point of disagreement: Nowhere does the text share the belief of modern critics that in the framework of patriarchalism no woman can reconcile a political with a sexual role; in fact, the contrary is true.[16] Conflicts are, of course, a part of life. But Schiller's Elizabeth fails as a queen not because she is *a woman* but because she is *this woman*; while Mary Stuart presents herself as the perfect queen, indeed, *because* she is *a woman*, namely *this woman*.

In Schiller's *Don Carlos*, somebody says: "What this office [of a reigning authority] requires is nothing more than a human being."[17] And Mary is exactly that, a human being, i.e. a woman, with strengths and weaknesses – just as men (and especially most of Schiller's men of power) have them in abundance. A weakness shared by both queens is that they entrust themselves to Leicester and try to impress him. But there is a difference: while Mary is deeply ashamed of her mistaken choice, Elizabeth calls (in vain) for the elusive accomplice even at the end. Mary's advisor could be none other than Talbot – "the chorus" (Körner to Schiller 7.9.1800) who distances himself from Elizabeth, the tyrant.

For a long time, Mary's final moments, as presented in the drama, were admired as a model of moral sublimity or beauty. During the last decades, however, doubts about the credibility of her pious transformation have emerged.[18] By dealing with them, we will indeed arrive at a stunning revelation about Schiller's representation of his historical heroine – my last and strongest point of disagreement with *Maria Stuart* scholarship to date.

On her last walk, Mary enjoys the splendor of royal pomp once again. She consciously displays the joyous composure of a royal heroine and martyr. Her obvious theatricality reflects two traits attributed to the historical Queen of Scots which are dear to Schiller's anthropology: 1. her rise to the highest human position – handling her situation like an artificial one and commanding her posture at her will (*On the Sublime*); 2. her fierce determination to demonstrate on and to the "theater of the world" (Fraser) what kind of ruler Elizabeth is.[19]

It is unthinkable that Elizabeth's ending would be mourned by her servants in the way that Mary's is who consoles them all and bids each of them good-bye individually. Afterwards, having confessed in private to a Catholic priest, she turns away from her lamenting subjects, away from the crowd which thirstily awaits the bloody spectacle, and up towards heaven: in Schiller's terms, a "beautiful" soul.

However, as has rightly been pointed out, in the process she has to reject repeatedly the earthly realm. These are "sublime" efforts which alternate with that beautiful harmony she attains with herself again and again. Far from contradicting one another, these two lines of emotions together are indispensable and highly indicative of Mary's process of self-perfection.[20] Firstly, hers is no small achievement; she must make powerful efforts to overcome her lively earthiness, her love of life and her anxiety of death. Secondly, since she indeed possesses the necessary moral strength to do so, she does not need to struggle convulsively, or even visibly while reaching a new level, as Elizabeth must at the end. In this way, graceful beauty and sublime dignity alternate and form a union which elevates Mary to a perfection – beyond gender differences –[21] which Schiller found in representations of ancient gods and goddesses. Might she be on the verge of being transformed into the representation of such a goddess?

Fittingly, that would have to be the goddess of justice and revenge, *nemesis*. In all his dramas from *Wallenstein* on, Schiller endows the instrument, representative and human incarnation of the Nemesis with more or less superhuman traits which he indicates, however, so tactfully that this artful touch has escaped most readers.[22]

During the course of the play we have become used to the aspect of the divine being present on earth – at first from a fanatically distorted point of view such as Mortimer's but then, at Mary's end, in accordance with her own "Catholic belief in the immanence of the divine."[23] There have also been colloquial allusions to ancient goddesses. Mary is spitefully called Helena (a half-goddess), the goddess of misfortune, Ate. Mortimer idolizes and indeed confuses her with the heavenly Mary as he admires especially her rightful violent outrage over Elizabeth (III, 6): "I have thrust the knife into my adversary's heart, at least enjoying the moment of revenge and triumph!" (III, 5; 2458ff). In I, 6; 588, 605, Mary speaks unequivocally of justice in demanding revenge for her treatment in England. Elizabeth feels (and Talbot says) that she may appear as a "goddess of revenge."

But how in the world (and in heaven) might such a role be reconciled with Mary's elevation to saint and martyr? Poorly enough, several critics have found, pointing to the striking contrast between her devout confession on the one hand and the stern, resolute final position she takes towards Elizabeth and Leicester on the other. It is indeed to these passages that we must turn. And a look at the sources reveals that forgiveness and revenge were indeed inseparably united in the thinking of Mary the Catholic – and that Schiller used them approvingly.

In order to receive God's forgiveness and communion, Mary herself must forgive. Thus, for the sake of her personal reconciliation with God and herself, she repents and apologizes for failing to forgive her adversary out of hate and vengefulness (V, 7; 3676ff). Later she confirms it publicly: she apologizes to the queen, her "sister," for her toughness and even forgives her "with all my heart," – her own death (V, 8; 3782ff). What a resounding reminder on and to the "theater of the world" that *sub specie aeternitatis* it is Elizabeth who needs forgiveness for what she has done to her sister queen!

She continues: "May God save the Queen and grant her a happy tenure as regent!" The real Mary even expressed the hope that Elizabeth might eventually serve God at some point in the future! Such an obligatory wish can obviously be uttered without Mary (or anybody else) worrying that God might stop the historical process, i.e. the law of causal connection of all things, which He Himself has established: "World History is Judgement of the World," this famous Schiller line in the spirit of the *nemesis* implies not so much the reward for good deeds as the unfailing retribution for evil ones.

And Mary, to her most deeply felt satisfaction, uses her chance to become instrumental in promoting this process upon "the theater of the world." Her words, separated from her subjectivity and in full harmony with God's objective intentions, are indeed designed to heighten the world's repugnance against Elizabeth to the utmost. In sharp contrast to the conventional scholarly wisdom that Mary escapes from political reality, these words show her as intentionally constituting a highly effective *politicum*.

Mary reserves her last words, however, for Leicester. To be sure, in her confession she did not repent of her scorn for his treason. Rather – with a stinging double edge – she blamed herself for her own sinful (and foolish) falling in love with that undignified man who has so "unfaithfully left and betrayed" her (V, 7; 3684ff). Still, shaking off this love was her hardest struggle. Really? There is more ahead, as she is reminded by her confessor:

> You have to undergo another hefty fight.
> Do you feel you have sufficient strength to
> Defeat your hate or bitterness?
> MARIA: I am afraid of no relapse. I
> Have sacrificed my hate and love to God. (V, 7; 3758ff)

Has she? In this instance, "sacrificed" very probably carries the additional meaning of "surrendered, transferred." In pointing to what appears to be a relapse on her part, some critics seriously underestimate the formidable calibre of the historical Mary as well as the thorough sympathy with which Schiller represents her. Exactly as it was in history, she asks Jesus to receive her with open arms as she takes her final leave. But then she meets Leicester's eye and trembling, her knees giving way, she sinks into his arms, putting him in the Christ-like position she has just invoked. It is an almost comical paradox, which nevertheless leads to the climax of the work. Clearly, Mary is dealing with Christ and the Christian God in an unexpected, unconventional way. I am venturing this reading: She will be embraced by Christ only after this man has been taken care of once and for all and by her!

The emotional shake-up takes its physical toll on Mary, but apparently without effect on her spirit which, she claims (and Leicester concedes wholeheartedly afterwards), is in the process of exchanging the earthly realm for heavenly bliss. Above un-Christian subjective disgust, she verbalizes the indignation which is objectively felt by humanity and Christ, God and the universe.

With a calm and composed voice, but at the same time with biting irony, she announces the final judgement against this villain. The most egotistical, ungenerous man present, he has ruthlessly put Mary, Elizabeth and Mortimer at their utmost risk and now falls under Christ's verdict: you shall be measured by the measure you have used yourself, i.e. you will be treated the way you have treated others. This is not only most authentically Christian, but is also in full accord with the Old Testament, with the Golden Rule of Thomasius, and especially with the way the *nemesis* proceeds: it is by treason that the traitor is felled. And this is exactly what Mary, her instrument, is doing.

Leicester is present at the execution because he is suspected to have tried to lead Mary out of prison; now Mary confirms before witnesses that he promised to do so. Devastating as this is, he has deserved no better, and by not letting him get away with the fruits of his villainy, the *nemesis* reaffirms our trust in the moral government of the universe – an indispensable element of any tragedy in the traditional sense.[24] "(*He stands devastated. She continues with a mellow voice*)." Again, she does not accuse him directly, but rather her own foolishness for having expected freedom and genuine love from such a man. As far as he and his queen are concerned, she warns: "Live happily ever after, if you can!" "Kneel at the feet of your Elizabeth! May your reward not be your retribution!" By implication she humbly leaves, surrenders all this to God and the *nemesis* – and surely she can do so with the fullest trust and confidence in eternal justice.

With "retribution" as her last word before taking leave for good, Mary has functioned for the *nemesis*, the goddess of order and just revenge who warns, punishes, and in so doing warns characters as well as readers and spectators. This is demonstrated and confirmed by the rest of the tragedy. All Talbot's prophecies are proven right: that tyranny was the form of freedom Elizabeth desired; she brought it about by taking action against Mary. And she herself causes the companions of tyrants, loneliness and isolation, to catch up to her. Now that she has thrown out her servants or jailed them in the Tower, she calls for none other than Leicester – and is informed that "The Lord asks to be excused. He went per ship to France," i.e. to Britain's enemy now more than ever!

In testimony to Schiller's unfailing sense of history, his *Maria Stuart* may leave us with the retrospective view of the two rival queens as presented by a present-day woman historian, Antonia Fraser (553–555). Although Elizabeth cruelly denied her cousin "the French burial she craved," James I of the United Kingdom honored his mother "by the last of the royal monuments [in Westminster Abbey] and the most imposing of them all." Elizabeth is resting in her tomb, raised earlier by James, alongside her elder sister, the "Bloody" Mary Tudor. "Barren in life, the two of them had been left to lie alone in death. Mary, however, lies [in her tomb] amid her Stuart posterity." From her, "who never reigned in England [. . .] and who died at the orders of an English queen, [. . .] every sovereign of Britain since her death has been directly descended, down to the present queen," Elizabeth II. As Mary Queen of Scots exclaimed: "In my end is my beginning."

Schiller could not know all that. But he congenially honored her motto in this sense: "My way of dying will be my triumph in history." Or again: "World History is the Judgment of the World."

University at Albany
State University of New York

NOTES

[1] Antonia Fraser, *Mary, Queen of Scots* (New York, 1968), p. 255.

[2] Gert Sautermeister, "*Maria Stuart*," in *Schillers Dramen: Neue Interpretationen*, ed. Walter Hinderer (Stuttgart, 1979), pp. 175–216, esp. 181–188.

[3] Julie D. Prandi, *Spirited Women Heroes: Major Female Characters in the Dramas of Goethe, Schiller and Kleist* (Frankfurt/M etc., 1983), pp. 29–114.
[4] Terence J. Reed, *Schiller* (Oxford/New York, 1991), p. 91.
[5] Todd Kontje, "Staging the Sublime: Schiller's *Maria Stuart* as Ironic Tragedy," *Modern Language Studies* **22** (1992), pp. 88–101. Lesley Sharpe, *Friedrich Schiller* (Cambridge etc., 1991), p. 273. According to the "feminist analysis" by Karl Lokke ("Schiller's *Maria Stuart*: The Historical Sublime and the Aesthetics of Gender," *Monatshefte* **82** (1990), pp. 123–141, esp. 138, 143), in Mary, Schiller's concept of the sublime reveals "the fanaticism of complete identification with the father and destructiveness."
[6] In his letter to Goethe (July 30, 1799), Schiller characterizes her as the hypocrite queen. For further clarification cf. G.A. Wells, "Villainy and Guilt in Schiller's *Wallenstein* and *Maria Stuart*," in *Deutung und Bedeutung*, ed. Brigitte Schludermann (The Hague, 1973) pp. 100–118.
[7] Sharpe (n. 5), p. 269f.
[8] Therefore, I disagree with the translation of *Furie*, "fury" by "nemesis" (although I applaud the insight that the author means "nemesis") by Sophie Wilkins, *Schiller: Mary Stuart* (Varron's Educational Series, 1959), p. 130.
[9] Kontje (n. 5, pp. 88, 98) defends Schiller against Paul de Man's "perverse" allegation that Schiller is siding with tyranny. But incredibly he himself claims that Schiller shifts the blame for Elizabeth's dictatorship onto Mary (thus honoring today's fashion of exchanging the roles of victim and victimizer).
[10] *Erläuterungen und Dokumente: Friedrich Schiller, Maria Stuart*, Christian Grawe, ed. (Stuttgart: Reclam, 1978), p. 66. Similarly p. 89: William Robertson, *Geschichte von Schottland* [. . .] 1762, vol. 2, pp. 184–186).
[11] Kontje (n. 5, p. 97) maintains that Mary's execution was motivated by Elizabeth's "desire to continue her reign as an otherwise just and even beloved ruler." In fact, she embraces the execution as her liberation from such hated necessity.
[12] Hildburg Herbst in her highly recommendable ethical interpretation "Herz, Seele und Gewissen: Aktualisierung eines historischen Stoffes in *Maria Stuart*," in: Richard Fisher, ed., *Ethik und Ästhetik: Worte und Werte [. . .], Festschrift Wolfgang Wittkowski* (Frankfurt, 1995), pp. 235–246, esp. 243, nevertheless ponders the "global proportion" of Elizabeth's timely association of marriage with subjugation, sacrifice of liberty. However, instead of pioneering the women's cause, Elizabeth hurts it severely since, although she seems to speak for it, she does so motivated by an egotistically narrowed concept of freedom, totally analogous to her anarchic and despotic concept of political freedom. Schiller felt encouraged to picture Elizabeth in this way; by Friedrich von Gentz, *Maria: Königin von Schottland* (Braunschweig, 1799), whose view is essentially corroborated by Fraser (n. 1). Cf. Ferdinand van Ingen, *"Maria Stuart: Friedrich von Gentz und Friedrich von Schiller,"* in: Fisher, loc. cit., pp. 247–257.
[13] Karl S. Guthke, *Schillers Dramen: Idealismus und Skepsis* (Tübingen, 1994), p. 214f.
[14] Schiller, *Über die notwendigen Grenzen beim Gebrauch schöner Formen* (1795), last pages.
[15] As Mary illustrates, benevolent despotism may provide a better government than one limited by parliament, as was the case in England as well as in Scotland at that time, the latter being torn by rivalries between members of the aristocracy, including Mary's relatives. For favorable comparisons of Mary's "absolutism" with Elizabeth's "par-

liamentarianism" cf. Ferdinand van Ingen, "Macht und Gewissen: Schillers *Maria Stuart*," in: W. Wittkowski, Ed., *Utopie und Verantwortung* (Tübingen, 1988), pp. 283–309; F. J. Lamport, "Krise und Legitimitätsanspruch," *Zeitschrift für deutsche Philologie* **109** (1990), *Sonderheft*, pp. 134–146.

[16] Sautermeister (n. 2), pp. 185, 187.

[17] Cf. Stephanie Kufner, "Utopie und Verantwortung in Schillers *Don Carlos*," in: Wittkowski, *Verantwortung* (n. 16), pp. 238–255.

[18] Harry Loewen, "The End as the Beginning: The Nature of Maria Stuart's Transformation," *Seminar* **15** (1975), pp. 165–180. Although Loewen has read Fraser's biography with its convincing description of how Mary combined religiosity with nemesis-like behavior, he sticks to the dilemma of Schiller scholarship, chosing this as a way out: he shifts Mary's famous step from this world to the other ("In [my] end is my beginning") from before the beginning of Act V to her very last line or even to after her death.

[19] See Fraser (n. 1), pp. 506–555.

[20] David Pugh, "Schiller as Platonist," *Colloquia Germanica* **24** (1991), pp. 273–295, esp. 285f.

[21] Helmut Fuhrmann, "Revision des Parisurteils: 'Bild' und 'Gestalt' der Frau im Werk Friedrich Schillers," *Jahrbuch der deutschen Schillergesellschaft* **25** (1981), pp. 316–366, esp. 362–364, although Mary has been excluded from all references!

[22] That is obvious with the heroine in *Die Jungfrau von Orleans*. For *Wallenstein* and *Wilhelm Tell*, cf. W. Wittkowski, "Theodizee oder Nemesis-Tragödie? Schillers *Wallenstein* zwischen Hegel und politischer Ethik," *Jahrbuch des Freien Deutschen Hochstifts* (1980), pp. 177–237, esp. 223f; W. Wittkowski, "Weimarer Klassik," in: Ehrhard Bahr, ed., *Geschichte der deutschen Literatur* (Tübingen 1988), v. 2, p. 219.

[23] David Pugh, "Enlightenment and Absolutism in Schiller's Classical Drama," *Weimar Classicism: A Reassessment*, Gerhart Hoffmeister, ed. (Bristol, 1995), pp. xx–xx, esp. 14. The point has been made by others: Dieter Borchmeyer, *Die Weimarer Klassik*, Vol. 2 (Kromberg, 1980), p. 294 reads it as aesthetic publicity; Ilse Graham, the first one, in *Schiller's Drama: Talent and Integrity*, London, 1974, pp. 166–170, reads it as an expression of Mary Stuart's spirit reconciled with her sensuality and transfigured into the wholesomeness of perfect humanity and transcendence.

[24] Lessing, *Hamburgische Dramaturgie*, 79, Hauptstück.

PART FOUR

FINE ARTS AND THE CONVEYING OF THE AESTHETIC EXPERIENCE

MIHAI PASTRAGUS

PHENOMENOLOGICAL AESTHETICS AND THE CONTEMPORARY ARTS

1. THEORETICAL PREMISES

Throughout the centuries the different aesthetics that have followed one another have conditioned artistic production, stressing either their object or the creative subject. Aesthetics of the first type have followed the principles of ontology, and those from the other type cast their lot under the sign of gnoseology and psychology. Where the nature of the object and aesthetics, but also the creation process, are concerned, things started to change with the reflections of Kant. In his "analysis of Beauty" he considered the object of art to be found in the play of the spiritual faculties, because according to the presence or the absence of that "play," the object will appear beautiful or ugly. This opinion, which is more spontaneous than elaborated, quickly found itself a place in aesthetic and artistic conceptions that appeared after Kant, simply because the object of knowledge, namely things without distinction, is conditioned by the active interference of the object. Kant did not make any distinction between empirically real things and beautiful things. This opinion will later be found in Croce, Bergson, Lipps, Husserl and Geiger. Hartmann was of the opinion that the most important path that aesthetics follows in our century is that which proceeds from phenomenology "because nothing can help us here other than a tendency to come as close as possible to the phenomena themselves, to observe them more precisely than we have done up to now and to learn to see them in all their variety" (Hartmann, pp. 34–35). Phenomenological aesthetics was much influenced in its formation by psychology, and that is why sometimes confusion occurs between the empathic tendency and the phenomenological one. Moritz Geiger is the one who was credited with separating them. Concerning the relation between the object and the act, phenomenology lays stress on the latter, but not to exclude the former. Neglecting the object does not mean ignoring its objectivity. The two poles of phenomenological analysis, noema and noesis, virtually guarantee that analytical properties of phenomenological aesthetics incline toward an intuitionist view in which the concept of intentionality is the key

A-T. Tymieniecka (ed.), Analecta Husserliana, Vol. LIII, 251–269.

not only for the analytical method, but also for the entire aesthetic system.

The idea that the beautiful is part of our existence is accepted and sustained as an axiom. Existence is considered to be a state which is attributed to Being, its first general way of appearing. But the beautiful is a determined state of existence and it involves the fact that Being exists in a certain way for the phenomenological consciousness. Therefore, since it is a way of existing for Being, the beautiful is a phenomenological state. The beautiful is a state, essentially a state of harmony, but it can be also an object and even a feeling as long as it conjugates with essence. But the beautiful cannot be perceived through an act of living. The beautiful – in any hypostasis – appears in the sphere of consciousness, but this is possible only so long as consciousness gets some hyletic data about the qualitative states of the objects. An analysis of the qualities which offers hyletic information to consciousness was made in detail by Harold Osborne, who is cited in the work of a Yugoslav aestheticist, Michael H. Mitias: "Osborne starts from a distinction between three types of aesthetic qualities (descriptive, evocative and emotional or expressive) having a special ontological status – they are sort of dispositional emotions. . . . When Osborne talks about aesthetic qualities, he does not mean emotional qualities, but feelings and moods. . . . Expressiveness does not denote a relation or a certain aspect different from aesthetic qualities. Expressive objects or works of art are the meaning and contents of expressiveness. Works of art are expressive thanks to their aesthetic qualities; they express the interior of their creator's nature, meaning that we cannot judge them according to the artist's qualities" (Mitias, p. 38). Therefore, the existence of the really beautiful, conceived in a phenomenological way, which is the basis of what we can perceive at present, means there is a positive possibility of seeing the appearance of subjective phenomena in the field of existence. Nature, and so the world of the beautiful, comes to us in a series of appearances or subjective phenomena, as possibilities that one consciousness has to mould the physically given into perceptible images. The sum of these operative possibilities of consciousness forms the object of aesthetic and artistic experience. The meaning of forming and transforming acts appears at the level of aesthetic experience. The aesthetic value of a work of art which presents itself to our perception is only an ideal, that experience tends to accomplish if it presents some interest. But we always perceive the artistic or aesthetic object in a continuous chain of perceptions, each of them being an experience. The naturalistic aesthetic which advanced the "mimesis" theory of creation considered the beautiful to be a gift or a passive process in which the subject just "registers" outside developments of things, the appearances of outside conditions, or, in the best case, the copying of a given "model." In contrast, phenomenological aesthetics reserves the term *art* for hypostases of cognitive and emotional activity, for consciousness always comes close to something from the outside. In these states or hypostases of cogitational activity our ego defines itself and becomes more and more complete as it totalizes the many results of intended acts in the field of artistic experience.

If we abandoned ourselves to applying objectivistic and especially objectivist clichés in which the subject is only a simple data gatherer or a simple passive registrar of a positivistic type, then phenomenological aesthetics would be just the expression – at an artistic level – of a certain solipsism in which a subjective universe, filled with empirical data, would be enough for itself, as in a monadic universe. However, the act of artistic setting-up has its objectivity, though, of course, in a way other than the realist-naturalist way. The sources of their kind of objectivity could be at least three: a) the validity of the values that were created by the arts; b) the transcendental unity of their apperception among different subjects; c) the emotional apriorism which shows the immanence of the artistic and aesthetic value in the human disposition (intentionality). Having such a type of objective existence, the beautiful can lay the foundations of a science. In Husserl's opinion, it is known that there are only two types of science (the sciences of facts and the sciences of essences). Aesthetics according to its object, roles and specificity, is considered to belong to the sciences of essences, for at least two reasons: it permanently watches and constantly views, using specific means, essence, that is, a type of essence, and further, its expressiveness belongs to its essence. Therefore, we can easily say that phenomenological aesthetics is neither empirical nor simply deductive, but eidetic, cognitive-affective. As observed in our introduction concerning what is specific to phenomenological aesthetics, this important current of our times refers to the understanding of essence. This is a quite difficult process. It can be reduced to three cardinal moments:

a) the suspension of judgements by using the operation called epoche, that is, the elimination of those elements which belong to the physical given, which do not concern this method;

b) the eidetic reduction, that has as its purpose the elimination of the individual existence of the analysed object, now subordinated to attention or to the creation process, by eliminating from the field of attention all the experiences, hypotheses and theories which are specific to an object;
c) the transcendental reduction, which means eliminating everything that does not match the schemes and principles of our consciousness concerning the respective given.

After this last stage, the object arrives at its pure identity or "eidos" (the ideal image). But is not this eidos exactly the artistic substance we are looking for with so much effort? At least the arts of our century are moving toward this conclusion. The vital condition of the phenomenological view of the beautiful consists of harmonizing object being and subject being (noema and noesis). From the phenomenological point of view, the beautiful expresses itself in three hypostases: as an existent, as an essence and as a phenomenon. As an existent, the beautiful consists of the structural harmony of the essence; it is a motion, a way to an end which can be found in a structure. This process takes place at the conscious level, having the form of intentionality when followed.

As a relation of consciousness to the object which exists in the form of a model that has been projected by a subject, intentionality, exists therefore in all forms of consciousness. As for essence, phenomenological aesthetics emphasizes the harmonious structure of consciousness and its object, namely, with what is called the essence of intentionality. As a phenomenon, the beautiful also includes in its structure its ground, namely the object itself as a sum of the elements which form the noema and give the hyletic data, but also the sum of the predicates of the object. All these form the noematic nucleus, the stuff with which noesis will operate at the subjective pole. But when the sum of the predicates is called on to express the intentional content, namely, the essence of the object which the consciousness tends toward, then the form loses its material characteristics and becomes the meaning of the noema. This is what M. Petrisor, one of the Romanian specialists in phenomenological aesthetics, said: ". . . what gives a meaning to the phenomenology of structure is intention, the meaning which becomes an object for intuition" (Petrisor, p. 38).

The relation between the noemic and noetic, between the objective and subjective, between real and unreal, is explicated in a stronger way in the analysis that Mikel Dufrenne has made of Wagner's "Tristan and

Isolde." This story has intrinsic virtues which make it easy to analyse from the artistic point of view; that legendary halo, the Homeric simplicity of the story, the violent purity of the passions, all these give it a dramatic and poetic characteristic which emphasizes the genius of the composer. But these aesthetic virtues are only contingencies, possible states which can be found in the subject; they are in the positive core, the hyletic data which form the meaning of the work. They can become present only at the level of the accomplished work. In all artistic creation, we, creators or receivers, accept a series of conventions, and as a matter of fact, this is exactly what we shall do with "Tristan and Isolde." Sometimes we accept even the unbelievable. For example, although Tristan is dying, he speaks in a loud voice, the shepherd is a very good musician, the actors mime conventional movements in a way we can hardly find in reality. For the sake of the work's truth, we accept a lot of conventions. "If Isolde," said Dufrenne, "is not real, she will not be unreal in her relation with history, but with the truth that the work has a mission to reveal and to settle, a truth that it cannot reveal but through music. . . . This unreal is also neutralized, which means that we do not really take it as being unreal" (Dufrenne, pp. 50–57).

Therefore, everything is as if during the play, the real and the unreal are balancing and neutralizing each other, as though the eidetic neutralization did not come from me as a subject, but from the objects themselves; what happens on the stage invite me to neutralize what is going on in the concert hall and vice versa. On the other hand, the facts that are taking place on the stage invite me to neutralize the actors and so on. As a matter fact, all I perceive after these reductions is neither the singers, nor the heroes; they are who is singing, but it is the song and not the voices, music which the orchestra has not created, that I perceive as this sum of words, music, stage, movements, and this is what I come to see and hear, this is real for the subject. A good show, whatever it may be, is an aesthetic event with an internal and necessary meaning, because what can be felt and what gives this meaning offers me the real aesthetic object. From this reasoning concerning the method of eidetic reduction, but also perception, we must deduce that the viewed element does not consist of the given that can be felt at the surface, but the internal meaning which comes from the movement of a noematic nucleus within the noetic space of intentional consciousness.

Moritz Geiger and Roman Ingarden, the most well-known aestheticists of the phenomenological movement in the first stage of its turning to art, were aware of the fact that introducing the phenomenological method into aesthetics means not only trying to find a solution for the crisis that aesthetic judgement had to pass through at the beginning of our century, but especially bringing art and aesthetics to the level of a cultivated and mature outlook, something which came more and more into view among the public of this twentieth century.

This method is given different accents by every aestheticist or artist who uses it, but the results have always been the same and they have come immediately. At the First World Congress of Phenomenology, convened in 1988 in Santiago de Compostela, there was analysis of the inheritance of Husserl and also the creations after Husserl, and it was proven that the phenomenological method is being applied more and more practically to all forms of artistic creation, from all the fields of philosophy, to literature, plastic arts, music, architecture, choreography and the theatre. In a famous dialogue between the Norwegian poet Emil Boyson and the phenomenological philosopher Roman Ingarden, on the occasion of the latter's visit in Oslo in 1966, Boyson said that "artists, especially painters and poets have, since Descartes, the habit of being ruled by the idea of nothingness. But they have also learned to blame the way in which they were treated as mediocre people, shadows of nothingness. Feelings, colours, sounds, the stuff on which they work, do not belong to the world, they have different roots in existence in the feeling of the poet. The poet (the artist) is a person who works with and inside himself, with his own sensations and ideas, which appear inside himself. By tasting the dignity of aesthetics, he is proud of his own independence, calling himself a king, or exactly like Baudelaire who considered imagination to be the queen of human faculties, but a King or a Queen without a country . . ." (Äsbjorn, p. 169). This opinion expounded by Ingarden and Boyson, seems to become more and more important in the world of twentieth-century art, an opinion that has as a basis the transcendental "ego" of the artist. And since we must be sincere to ourselves, this belief, this situation is not a deviation from the progress of knowledge and artistic creation, on the contrary, it symbolizes the fact of the artist's becoming mature, acquiring the power of tackling reality not – as he had done up to now – in an objectively physical manner, but in an objective-noetic way, insisting upon the creative ability of imagination, in correlation with the paradigmatic axis that has been offered by eidetic axiology. The new method offered by phenomenology, known to this century and adopted, even assimilated by the arts, cannot be made responsible for the playful artistic experiments in some artistic fields, experiments which in a few cases still provoke doubts concerning their affiliation to art. But, apart from these failures, the vital renewal of the arts was a beneficial one for the arts, creators and the public, because these arts, these poetic mutations have created the premises of a new civilization, a "whole" or integralist one. Using the principle of verifying in the spirit of scientific requirements we shall try to confront philosophical and aesthetic assertions with the cliches on art's evolution during this century, especially concerning the methodological mechanism of creation and the reception of the work of art.

2. MUSIC – THE MOST PHENOMENOLOGICAL ART

The comprehension of contemporary music, as well as of classical music, is not entirely possible except by a comparison with an aesthetic vision, even a summary one. That is why we can say that aesthetics is placed before a double object. The eidetic method must naturally enter into a relationship with one or another of two realities, the work and the spirit which creates the work. It has to probe the relation between these and not consider the work of art as an isolated thing. Aesthetics likewise has to explain two functioning realities, each in relationship with the other. If it is concerned with the work, then it will have to consider the spirit which creates it; if it looks first to the spirit which creates, then it will also have to take into consideration the created object. This relation has an aesthetic nature, and any other relations which sustain the work, such as other spiritual realities, are considered non-aesthetic. In music, though, aesthetics is the meeting place for all the values of spirit which contribute to the act of creation. The two realities are exactly like two vast fields, with many ramifications and extreme complexities. Speaking about the creative spirit in music as a noetic act; we may say that the entire human spirit goes in different directions, being applied to fluid and different gifts, but that it especially becomes real when it comes into relationship with a musical work that seizes it. When the spirit is animated and penetrated by the intention of the creation of an existent sound, then it starts to move with all its intentional complexity and mobilizes its entire spiritual energy. Any intention which is not connected with an artistic purpose remains non-aesthetic, but as soon as it refers to the creation of a work it becomes an aesthetic intention. But an intention cannot be conceived from the object it refers to. However, it is that object's necessary complement. This complex relation will be better understood as soon as the second term of the aesthetic comparison is analysed. It must be also remembered that all of the spiritual potentialities of the creator – composer or singer – can be found in the artistic intention. Therefore, the creating intention is an act of will which refers only to the work, an act sustained by conceptual considerations, emotional considerations, but also conditioned by the very exigencies of the spirit. The creating intention is determined by thinking and sensation, by will and affect, connecting to the essence of the spirit in all its dimension.

The second term of the artistic creative act, in aesthetic relationship with the first one, is the musical object, a concrete given (noema), a basic existent on which a quantity of possible determinations can be applied. It can enter an infinity of logical systems, but in each of them it loses something of the richness of its determination. In other words, we can say that the musical object is polyvalent, so it can be suspected of emphasizing an infinity of values. In the same way, a series of qualifying determinations can be applied to a symphony or a concert. Because the object in sound is a concrete one, all qualifications are possible, but it is not easy to see which of these are essential to a particular musical object under discussion. So, the problem is to find, among these values of the object, those which give it unicity and essentiality at the same time.

Saying, for example, that music can give joy or pain means, as a matter of fact, denying the essence of the musical object, because other artistic methods can easily express these emotional states. This general manner of making defining statements is contestable, because it does not hit on what is essential to the musical object; joyful or painful expression is neither specific, nor exclusively musical. The desired particular value, essential to the musical object alone, can be found in the constitutions of music itself. For in fact it is made of sounds and successive intervals. This is the musical essential. In other words, every sound and every interval contains its own aesthetic value and depends especially on the way it accomplishes the artistic intention. This manner, which is connected with the cultural level of the artist and with his technique of controlling sounds, but also with his inspiration, calls on the whole spirit to give the composition or the interpreting their inherent aesthetic content.

The well-known Swiss aestheticist, Jean-Claude Piguet, rightly considered that in music, the aesthetic relation establishes connections between two worlds, each of which has its specificity and richness and lives in its own life, and he went on to explain that neglecting the material essence of the work of art would mean accepting some abstract generalities and we would then risk falling into vagueness. The aesthetic method is placed between the two positions: it considers the material aspect as well as the spiritual aspect of music; it also gives importance to the spirit, but only as a creator of a well-defined work. It devotes itself to a double complementary object: to a sound form and to an inseparable creative intention. Therefore the aesthetic relation which links organically and specifically the world of spirit and the world of sounds, which unites the intervals and the sounds with their creative intentions and prevails over the various influences which would polarise the act of creation, rules the whole musical work. The fine arts and literature have, most of the time, a clear and direct meaning; they reveal natural elements, human actions, etc. Their meaning comes from the comparison between the art and the nature which they imitate in a certain way, but it is impossible to give a natural meaning to, for example, a fugue by Bach or to the Fifth Symphony of Beethoven. The musician works with a material entirely different from that of other artists. Here, the specific must be looked for in the order which he tries to establish in this vast field of sounds which does not exist for the simple human being or, if it exists, does not say much. Order and combination, in fact, are what give a meaning to the work. This notion of order, which supposes arranging, combining, selections, valuing, measuring, emphasizing elements, also supposes the presence of an ordering consciousness which concentrates and reaches the notion of technique understood as a system of abilities, methods and principles out of which the creator gives life and meaning to a form, to a sonorous existent. In the arts, but especially in music, we discover this double significance and belonging of the meaning to both the material order (the noemic) and also the spiritual order. The beautiful belongs to the technique only in a certain way; technique is just one of the conditions – a necessary but not sufficient condition. If it belonged only to technique or to it in a decisive way, then art would identify itself with science. But art means so much more: it is science with beauty. An artistic work which is not beautiful does not exist.

The beautiful, as an essential product of art, does not proceed entirely from the act of knowledge; knowledge is just a condition for obtaining beauty. The beautiful is the truth of art, it can be found in the pleasure that this kind of knowledge gives. Knowledge and pleasure in music, but in the other arts too, are suited to a sort of mutual support; they constitute two different aspects of one and the same spiritual activity, and do not exclude, but support each other. But knowledge in music always depends on valuation in which spirit must make the effort to choose some sounds, some scales, some techniques, rules, principles etc. which can form and transmit a meaningful message. It could be said that knowledge is the basis of a valuation, even a hypothetical one, of which beauty is the measurement, the more or less total engagement of the creator. In a way, the joy of beauty has its beginning in knowledge of musical existence, and that is why we consider that aesthetic feeling is a constitutive element of beauty. It is, in other words, an "Einfüllung," a gift of the ego to other people, an almost religious fusion with the sonorous existence we shape. But for reaching the aesthetic judgement of beauty we cannot be satisfied only with the supersession of the ego by others.

The creative ego uses, in an unconditional way, the rule and the order of sound in the name of the envisaged intentionality. The beautiful starts to exist only as I accept the rules and the emotional value of sounds deliberately. A difference can be operated between two categories of beauty: definite beauty and hypothetical beauty (the latter appears as only a possibility of the beautiful based on the values of sound). Between the two forms of beauty in music there can exist an infinity of shades which are psychologically marked by emotional degrees, in comparison with each other, feelings which change from the simple attraction of the gift, to the supersession of the creator. But beauty in music is always based not only on knowledge of musical value, but also on the condition of understanding among people (emotional contamination). If in order to judge the beauty of a work I make efforts to know these values, the feeling is based on an objective given, one knowable, on which aesthetic judgement is also formed.

But in order to be valuable, aesthetic judgment must accomplish the condition of the universality of aesthetic judgement, which can be obtained only in two ways. There is the naïve realism which existed till phenomenology appeared and which pretended that beauty was in the work and you only had to find it. On the other hand, in a Kantian sense, beauty does not belong to the object, but to a subjective harmony felt by the subject but lived by all subjects who feel and live. In keeping with this latter variant, which comes from the paradox of objective subjectivity, recognizing beauty can be done by searching for the subject in the object, where it remains pure as long as it gives aesthetic meanings to the object. The universality of aesthetic taste does not belong to feeling, but especially to the connection between the subject and a value or the transfer of values into subjects. Definite beauty is not really universal, because it belongs to my personal adhesion to a value, just as long as I recognize the value in a work confirmed by a burst of the feeling. Only then does the hypothetical beauty become a universal one. Therefore, the beautiful in music can be found in the relationship between the work and the human being, especially its creator (composer or interpreter) and then the public. So, this is in a few words, a demonstrative way from which we learn that beauty in music belongs definitely to the play of the subjective faculties of the human being, to the complex relationship between noetic content and the noematic nucleus.

Hence the idea that value is an "objective and valid dimension of the noetic structure which escapes the destruction in the real existential field challenging time and the perspective of an everlasting condition, which can be compared to that of the universality" (Boboc, p. 166).

3. FROM THE ASCERTAINED VISIBLE TO THE PROFOUND INVISIBLE

The visual forms of art, especially painting, sculpture and architecture, have a great importance in the development and the understanding of aesthetic types and the artistic styles. As a matter of fact, the entire artistic movement is a continual process of creating new forms and styles and, at the same time, an effort to pass – in specific ways – from the ascertained, seeming visible to the profound invisible, to the inferred possible. The ascertained visible that is found in macroconfigurations and spatiotemporal microbuildings, as well as in the possible worlds in their pure forms conceived from the rational point of view, constitutes the three fundamental movements in the evolution of art. We shall try to tackle the problem of art not by taking into consideration its evolution, but through understanding art's object, applying the phenomenological method specifically to the fine arts.

As a premise, we consider every work to have a matter – the sensible proper or the object. But this sensible must be separated from the material which produces or sustains it. The matter of music, as we have seen, is the sound, not the instrument which produces it, and the matter of poetry is the word, not the voice which pronounced it. A material instrument is a sort of matter. Therefore, the material and the sensible are, from the phenomenological point of view, two different things. But there are some arts in which the sensible cannot be separated from the material which produces it, e.g., architecture and sculpture. Stone, marble or bronze are definitely meant for this purpose, for their sensible qualities, for their strictness and elegance. From the aesthetic point of view, the material does not appear as material only, but also as a support of the sensible. It is used to form an object which is meant for contemplation and not use. The artist does not confront the material so that it may disappear in his eyes as a material and yield space to the manifestation of the sensible. The material is what is first submitted to phenomenological reduction. The material does not aestheticize itself but reveals its entire sensible richness. It denies itself through its qualities. Its virtues become aesthetic only when, through reduction (creation), the images of the sensible start to appear as forms of a possible inferred within the ideal. The beauty of a mass will become the beauty of a bust, the beauty of a line will be that of a profile, the granulation of the stone will symbolize the brilliance and the softness of the skin. That is how we arrive at judging the beauty of a work through the beauty of the object which is represented by the ideal and referred to intentionality.

But this mechanism can also hide some faults. The self's accepting the illusion of the represented object can decrease the importance of the artistic effort and implicitly, the acceptance of some works just because the represented object becomes so great from the subject's point of view. The conquest of the sensible can be made more systematically in painting and in music than in other arts, this for at least two reasons. The represented object seems to solicit the entire creative being, and further, the material is so much present that it seems to represent the object more than its visual or sound qualities. The revealing of the sensible is made by removing material till the material (object) which is willingly represented, is obtained. This is, surely, a complex and difficult enough operation, which is connected not only with effort, but especially with reason, inspiration, the faculties of the a priori emotional. Simple cutting, cutting out or geometrisation is not enough for making the aesthetic object intentional. A curve or a column, if they are made with artistic sense, can be displayed to eyes in such a way that they will become as sensitive as before the most elegant combination of colours. Concrete architecture has prepared us to enjoy the beauty of artistic mass, especially if it is associated with some older or newer mythological connotations. Geometry is not enough to bring out the aesthetic. This is proved by the fact that a series of pieces meant for use, made by technique, in spite of their engineered structure, are not artistic works. They may be influenced by art, but that is all. "A cathedral is beautiful," Dufrenne thought, "because every perspective breaks and overturns symmetries and regularities, because the light, which is filtered and spread through the stained-glass windows, makes the stone sing, because the geometry is just a continuous bass singer amid the orchestra. In sculpture, geometry is also at least broken by the artistic imagination or by the exigencies of the model" (Dufrenne, p. 405).

Therefore, we understand once more that the sensible distinguishes itself by rejecting legal rationalism but also through having an abundance of spontaneity and the unforeseeable, sometimes even strangeness, in the construction of form. In all the arts the sensible must be composed as to be perceived without ambiguity and difficulty, by using different schemes of harmony and rhythm. Through these schemes the artist elaborates the specific language of his work, in connection with the desired meaning. The schemes can have a double function: they can define and classify the elements of the aesthetic language either of a tradition, of a school or of the current to which they belong or which they initiate. The schemes give essence and wholeness to the work. The tonality of sounds, the scale of colours, the scale of choreographic movements, the scale of words in a literary work, the scale of lines and surfaces in architecture or sculpture and even the scale of characters in theatre are the schemes which give to each work in this enumeration a meaning and a significance, a certificate of belonging to the artistic genre.

But eidetic reduction does not stop here in remaking the material, it proceeds to the structure of the object because, as a matter of fact, this is the purpose of every work. Being capable of genesis, a work can be structurally analysed by taking into consideration the movements and the elements which form it from the aesthetic point of view. These may be harmony, colour, rhythm and even melody. People often speak, not without reason, of musical painting, as well as of pictorial music. The matter of every art, in order to acquire aesthetic valence, must be connected to a structure with a meaning. Speaking of painting, there is a dispute over the relative importance of the two parts of its plastic structure, drawing and colour. During the long discussion, which continues even nowadays between the supporters of colour and those of drawing, the latter group has an advantage: the rational way of thinking, sustained by the authority of the great rationalists and personalities of the Enlightenment. But the supporters of drawing also have against them the accusation – serious enough – that drawing is guilty of imposing imitation, and that this "mimesis" stops imagination. This is a limitation. But drawing also has the merits of suggesting form, style, of making representation more accurate and of directing perception straight to the object. It is also true that if we want to choose a more sensible aspect of the object, than we shall naturally choose the outlines, the drawing, the part that is easier to understand. It is not only the sign which is easiest to read, but also the sign that can be considered to represent the essence of the object, the essence in connection with which the colour is just an accident, a minor quality. So it is obvious that as long as the pictorial object is confounded or identified with the represented object, as long as painting maintains the practice of giving significance to a rational truth, drawing will continuously hold the first place and it will tend to outshine colour.

If, as we have seen, drawing is a prisoner of form, then colour, according to the present level of understanding of the hierarchy of values, determines the contents and refers to the essence, in the highest degree. We can say that contemporary painting is a gift of colour. To acquire this status, it had to await a revolution for the opportunity to dispossess the drawn form, to create an object that asks to be perceived and not interpreted, an object before which any commentary becomes superfluous. But this is a new conception of art and a new way of perceiving art, a new conception of art because it must come to be through the will of creation, not through imitation; concerning perception, because it means that the new fine arts must admit that appearances can be hypostases of another truth no less important than rational truth. Perception has made an important contribution to the rehabilitation of colour in aesthetics and that is why it itself needed to be rehabilitated, at least as an element which is capable of revealing a certain truth. The phenomenologist Pierre Trotignon will consider that "perception is a manifestation of the phenomenon as an objectivity that has been entirely reduced to objective perception, for a subjectivity that is entirely invested in its role as a place of manifestation" (Trotignon, p. 52). As a matter of fact, the perceived is united with its context, having within it an element which prevents all spatio-temporal and numerical ensembles from articulating themselves in terms that are easy to identify. Present sensation and images or past experiences never appear except on a horizon of meanings. Merleau-Ponty, the most authoritative explorer of phenomenological perception, considered that the real problem of memory is found in perception, because perceiving does not mean judging a multitude of impressions which bring memories with them, a judgement capable of completing them; rather, it means seeing an immanent meaning without which recalling memories is not possible. Remembering does not mean bringing back before your consciousness a picture of the past surviving by itself; it means falling into past times and revealing step by step the intervening perspectives that it contains. "Perceiving does not mean remembering" (Merleau-Ponty, p. xv). Therefore, perception is what can help us represent the intentional object, it is what identifies the aesthetic in the work of art, immanent significations that rationalism considered very little. Since colour, which is the essence of painting, is not an object to be represented by thought, the purpose of colour is that of its being perceived, of impressing our eyes and of giving a basis to the spirit provoking in it certain states of dreaming, certain inner movements. But as it appears to daily perception, colour is not perceived because of itself, it's just as independent quality. However, the theory of colours will be what systematizes the experience of colour, especially as the painter must not dedicate himself only to the description of shades of colour. It is absolutely necessary in its subjective relationship with the viewer, for colour to become objective through an idea. In contemporary painting, it has become a truth that the functions of drawing have now been taken up by colour, the latter becoming the most important element of painting. Every work contains a key, a code, a secret plan. Every authentic painting is built on a certain harmony of colours, which sometimes passes from domination to obsession. Red for Rembrandt, green and subtle grey for Velázquez, the harmony between yellow and blue for Vermeer; the harmony among violet, green and orange for Pierre Bonnard, and so on. The harmony of colours needs a stronger shade which gives particularity in association with a prevailing colour that determines the entire scale of colours. Choosing his scale, each painter decides on the importance of colours and what they symbolize, on the style, but also on the destiny of the works. But the movement in the aesthetic object, which can be only a spiritual movement, oriented towards a special object, cannot be accomplished if we do not take into consideration the rhythm also. The logical rhythm which is at work in the way a work is composed does not exhaust the content of this notion. In the fine arts and in architecture, the rhythmic schemes which compose a work can also be named melodic. The "melody" of a fine arts object, but especially of an architectural object, consists of the rhythm that rules their composition. The rhythm can be recognized in the compositional scheme of the form in which the clear and the obscure, the system of contrasts, the arrangement of the plants, the use of perspective foreshortening are distributed. We can say that rhythm is the way through which a work which has become total reveals its semantic significations. Through rhythm we are led to the privileged lines, to the paradigmatic axis of the construction, to the secret of the chromatic field. The work accomplished through these identified elements has the ability to awaken the receiving subjects' participation in empathically decoding these meanings in order for the sensible to reveal itself with its true meanings.

The artistic movements of our times, especially those of the last decades, will confirm the tendency to essentialize artistic expression in the many methods and approaches to which artists appeal (see Brâncusi, Moore, Dali, Miró and so on). In discussing painting we shall briefly treat abstract expressionism, which is the approach most connected to the phenomenological method. One of the critics of the trend said that: ". . . the affiliation with political revolution has also demanded the permanent abandonment of the old philosophy and the adoption of a new one, as a basis for interpretation of the new art's existence" (Smith, p. 25).

The artistic motif of this movement, supported by a series of great personalities after the First World War (Ernst, Matta, Dali, Masson and so on) has been that of dream and reality, so that, following the focus on anxiety in modern art, it might extend art's interest into the inward universe, into the deposits of experience which dream reveals. The cognitive efforts of impressionism, futurism and cubism which look at the reality of the object, were then completed by the surrealism that adds, from a perspective which is close to symbolism, a vision of subjective inferiority, one separated from external objects. Imagination is cultivated and explored in vast dimensions, as the most productive function of the human spirit. In art works, strange, unusual, unreal elements, risk is cultivated. Risk becomes the first principle of composition, asked to lead the artist to the revealing of the truth. Some techniques, as collage, fragments of images, numbers, figures face to face, frontage, the evasion of surfaces, haze, x-ray (the hazard of photo), decals assemblages of ready-made objects in irrational compositions, all these become the new forms in which expressionism projected the subjective universe of the ego's objectivity. In its repertory we recognize the recurrence of fantastic motifs from the gallery of oriental exotic mythologies or from the Europe of the Middle Ages, similitude with the world of the mystical nightmares of Hieronymus Bosch, Pieter Brueghel, Grünewald, Goya and so on. In the final paragraph of his article: "The Secrets of the Magic Surrealistic Art," André Breton describes: "These summer roses are blue, and the forest is made of glass," and his metaphors were constructed far from the huge territory of this reality.

4. TOWARDS A LYRICAL FUNCTIONALISM

The philosophy of Husserl, with his inspired method, followed by an axiology based on emotional apriorism, could not fail to reach the architects, whose important efforts follow the line of contemporary civilisation. During the twentieth century they have managed, under the banners of the new philosophical conceptions, to synthesize in a surprising form both Cartesian rationalism in its formalist variant, connected with mathematics, and geometry (which had been declining in influence) with emotional apriorism. The way was thus found to a new constructive lyricism. This lyricism does not come from any superfluous formality in the scheme put out front, but from the functionality of the new architectural forms. The aesthetic conception of the most modern advanced architectural current of our century, led by Le Corbusier and Frank Lloyd Wright, proceeded from the idea that something is beautiful if it corresponds best to the human needs, and that is why five architectural principles referred to the functions of the building, the front, the roof, the interior, functionality. For these architects and their disciples, architecture is the complete art, which reaches the Platonic magnificence of mathematical order provoking spiritual speculation, of the perception of harmony wherein rough emotional connections are placed within the limits of a phenomenology of emotional connections. Hence the idea that, in architecture, "the artistic (aesthetic) constructions are placed within the limits of a phenomenology of the emotional life" (Pastragus, p. 147).

The taking of emotionalism into consideration will not only give a lyrical tint to the new architecture, but will humanize it. The modular structure of Corbusier is the emblematic symbol that warrants the evolution towards this humanism. "In the new architecture, the beautiful can be obtained only by rationally satisfying the spirit, through strict connections; through inventive genius – plastic and geometric – it can make order and unity easy to measure, organizing them according to clear rules satisfying visual sense, producing different evocative sensations, producing the architectural emotion, using implacable means" (Melicsow, p. 216).

In contemporary architecture there is a new plasticity, a synthesis of all arts, the total art, as those of the Bauhaus School said, or in other words, here is the place where all the arts meet, starting with music and with theatre. The architect, by uniting the inside with outside, noema with noesis, creates a new perspective for living space, develops the taste of civilization, with space, time, movement, thought and feeling receiving absolutely new dimensions easily comparable to those of surrealistic visions. Modern architecture confirms the tendencies of this complex art revolution changing conceptions about space, culture, home, traffic, work, health and values. Among the countless examples of architectural creation we should mention the work of Kiyonori, Kikutake, "Ocean City" (of 1962), a surrealistic creation with strong lyrical accents. This creation imposes itself through two elements of clear originality: the attempt to use the ocean by building homes on floating platforms and some original solutions for the construction of these platforms. In all these expressions, we can draw the conclusion that contemporary architecture, in its most original tendencies, has managed to convert the geometric line, strict and tough, into lyrical fibres able to produce and sustain the emotions of a new poetics, one that feeds on both reason and feelings. To this great work of civilization and functional culture, phenomenological aesthetics had made a very important contribution, one born out of the genius of Husserl, Scheler, Ingarden, Geiger. . .

Translated by Raluca Vârgolici & Gheorghe Stan.

Iasi
Romania

REFERENCES

Asbjorn, Aarnes, "A Poet's Life and Work in the Perspective of Phenomenology," *Analecta Husserliana* **XXXVII** (Dordrecht: Kluwer, 1991).

Boboc, Alexandru, *Etică şi axiologie in opera lui Max Scheler* (Bucharest: E.S., 1971).

Boyd, Robin, *New Directions in Japanese Architecture* (New York: Braziller, 1968).

Dufrenne, Mikel, *Fenomenologia experientei estetice*, Vol. I (Bucharest: Ed. Meridiane, 1976).

Hartmann, Nicolai, *Estetica* (Bucharest: Ed. Univers, 1974).

Melicsow, Marcel, *Architecture Modernă* (Bucharest: E.S.E., 1975).

Merleau-Ponty, Maurice, *Phénoménologie de la perception* (Paris: Gallimard, 1970).

Mitias, Michael, *The Ontological Locus of Aesthetic Qualities*. The Eidos of Art, vol. II (Crakow, 1985).

Pastragus, Mihai, "Aesthetic Concepts of a Phenomenological Origin," *Analecta Husserliana* **XXXVII** (Dordrecht: Kluwer, 1993).

Petrisor, Marcel, *Curente estetice contemporane* (Bucharest: Ed. Univers, 1972).

Piguet, Jean-Claude, *Découvertes de la Musique* (Neuchâtel: Edition de la Baconière, 1948).

Smith, Edward-Lucie, *Movements in Art since 1945* (London: Thames & Hudson, 1969).

Trotignon, Pierre, *Les Philosophes français d'aujourdhui* (Paris: P.U.F., 1970).

YNHUI PARK

THE ARTISTIC, THE AESTHETIC AND THE FUNCTION OF ART

What Is an Artwork Supposed to Be Appreciated For?

Every artwork is simply an object like other objects, but even if it were physically and perceptually indistinguishable from other objects, it would be an object of a peculiar kind. While the value of all other objects is intelligible only instrumentally and thus extrinsically in terms of their specific functions, the significance of an artwork can be explained in terms of its predisposition to be intrinsically enjoyable. Whatever else an artwork may be thought to be, it is an object primarily for appreciation. Paintings, sculptures, songs, plays and poems exist necessarily as objects to be appreciated. As Wollheim points out,[1] part of the meaning of an artwork is it's being "an object for appreciation." We visit museums, go to theaters, and read literature, it is assumed, primarily and ultimately in order to appreciate these works.

It is not clear, however, specifically what we are supposed to appreciate in all artworks, and only in them. This question assumes that all works of art share a common and unique property, or properties, to be appreciated because they are not found in other kinds of objects. This claim would be quickly rejected on the grounds that different people appreciate the same artwork objects differently. Thus, most recently philosophers like Eaton and Davies take as self-evident that what we value and appreciate in an artwork is neither universal nor fixed, but relative and changeable because it is culturally, historically and traditionally determined.[2]

However, if that were true, then the same word "artwork" must in reality mean quite different things to different cultures, in which case even though we might be saying "artwork," we would be in reality talking about quite different things in different cases, and different values according to the different culture, historical period and context. This implies that if we want to discuss "artwork" in a classificatory sense meaningfully, we must find common criteria for the expression, i.e., "artwork" *as* artwork.

What, then, *should* we look for when appreciating an artwork? What particular value do we expect from it? A satisfactory philosophical theory

A-T. Tymieniecka (ed.), Analecta Husserliana, Vol. LIII, 271–286.

of art must answer these questions. In this paper I want (I) to argue that all the theories of art now available to us fail to answer these questions, (II) to show that the distinction between the concept of what I call "the artistic" and the concept of what I characterize as "the aesthetic" is essential if a better answer is to be found, and finally (III) to propose my own answer.

I. A CRITIQUE OF CLASSICAL THEORIES OF ART

In his recent book, Davies distinguishes between the functionalist and the proceduralist definitions of artwork. The functionalist defines an object or an event as artwork in light of the specific role it plays, whereas the proceduralist defines a thing or an event as artwork it if has undergone a certain procedure deemed necessary to obtain the status of artwork within a particular tradition. According to Davies, there is a sort of dialectical relationship between the two processes.[3]

As we all are aware of, one of the most important philosophical debates which has taken place in recent decades, more specifically within the tradition of analytic philosophy, concerns the definition of artwork. Our interest here, however, lies not so much in the definition of artwork itself, as it does in the nature of the values assigned to that class of thing already defined *as* artwork. In order to pursue our interest here, we must determine the proper function of an artwork, because the value of a given artifact can be made intelligible only to the degree to which its function is fulfilled.

It might be suggested that *the* search for this function is misguided because an artwork has many functions and different artwork has different functions. Of course, there is some truth in this observation. Nevertheless, it does not address the actual intent of our question. We are not denying that an artwork, just as any artifact, could and in fact does, function in different ways for different purposes. However, to recognize any artifact, say, a bicycle, is to have perceived the unique function particular to that class of object which distinguishes itself from other classes of objects. The reason why an object is called a "book" must be related to its purported proper function, so it is only in light of this function that the value of an artifact within a particular class of objects can be assessed. Since our problem is to know for what we are supposed to appreciate an artwork and thus assess its value, we must know what the function of artwork should be.

Every theory of art purports to define "artwork." And all, with perhaps the exception of "institutional theory," which illustrates the proceduralist definition, are designed to provide a functionalist definition. For the functionalist it is only in light of its specific function that an artwork can be distinguished. For the convenience of our reflections, I want to divide different theories of the function of art into categories of either *cognitive* or *performative* theory. The theories of art I would call "classical" theories such as imitation theory, expression theory, and formalist theory all fall in the cognitive category, while the theory I would characterize as "phenomenological" – as exemplified in Bachelard's "phenomenology of poetics"[4] belongs to the performative category.

a. *Cognitive Theory*

For some time, it has been customary to distinguish between cognitive and non-cognitive theories of art. Imitation theory is said to be cognitive because this theory assumes that the function of art is primarily descriptive, while expression theory and formalism are non-cognitive because the former takes the function of an artwork to be a subjective expression of the artist's emotion, while the latter finds the essence of an artwork in those formal properties which are capable of arousing a certain emotion in the viewer.

A little reflection shows, however, that both the expressionist and the formalist theories are in reality two different kinds of cognitive theory in disguise. On the one hand, taken literally, the expressionist theory is absurd because the expression of an emotion can properly be attributable only to human beings. Therefore, if an artwork is said to be expressive, it must be on the grounds that somehow it conveys some kind of human emotion, so what is meant by "expressive" means in reality "descriptive" or "representational," hence cognitive.

On the other hand, in a formalist account, an artwork has no cognitive function because what we appreciate in an artwork is such formal and material properties as shapes, lines, colors, and texture of the artwork insofar as these formal qualities generate "aesthetic emotion." The function of an artwork would therefore consist in creating formal properties which are capable of causally provoking such an emotion. However, to explain why and how they create "aesthetic emotion" in the viewer, Bell, the foremost critic, must appeal to what he refers to as a "metaphysical hypothesis," wherein our aesthetic emotion derives from realizing that we are in touch with the eternal and ultimate nature of reality through the formal properties of an artwork. In his view, "Whatever the world of aesthetic contemplation may be, it is not the world of human business and passion, in the chatter and tumult of material existence it is unheard, or heard only as the echo of some more ultimate harmony. . . ."[5] Art that transports us from the world admits that aesthetic emotion is derived from our cognition of the ultimate nature of reality. This view in turn implies that it is not the aesthetic emotion itself but cognition of the ultimate metaphysical reality that counts in artwork.

Mistaken as it is, the cognitive theory of art is more deeply rooted and widely accepted than it appears to be. For most of us non-experts, it is hard to explain how modern artwork in general, and abstract paintings in particular could show truth, if, indeed, they do. Yet, modern artists, critics and historians of modern art, as well as philosophers, and laymen like us, with no theory at all, almost unanimously believe the cognitive function to be the primary function of artwork.

For Rilke, one of the greatest poets of the early twentieth century, "Poetry, (and) creative literature, is nothing but the elementary emergence into words, the becoming uncovered, of existence as being-in-the-world. For others who before it were blind, the world first becomes visible by what is thus spoken."[6] According to Kandinsky, a revolutionary Modernist painter, "the abstract painting leaves the 'skin of the nature' but not her laws, cosmic laws."[7] In explaining Miró's style in painting, Muller, an art critic, writes that "it (the style) allows a man to be aware of something grotesque within himself."[8] As philosopher of art, Heidegger writes about Van Gogh's *A Pair of Shoes*: "It is only in the picture that we notice all this (i.e., the mutual belonging of world and earth at the core of the tool) about the shoes."[9] In his account of the increasing tendency toward abstraction in modern painting, Lyotard, a postmodernist-philosopher, writes: "I shall call 'modern' the art which devotes its little technical expertise to present the fact that the unpresentable exists. To make visible that there is something which can be conceived and which can neither be seen nor made visible: that is what is at stake in modern painting. . . . This abstraction itself is like a presentation of the infinite, its negative presentation."[10]

All of the above quotations either assume or state that the fundamental function of all artwork is cognitive, namely, descriptive, representational, and is thus informative, concerning the world, society, and metaphysical reality, presumably in a manner not attainable or describable by any other means than artwork. No one would question that through contemplating and studying artworks, often we *seem* to get information about the world and ourselves, and seem to gain insight into metaphysical reality. Nevertheless, the cognitive theory of the function of artwork must face the difficulty of accounting for at least three crucial and obvious facts about artwork.

First, artwork is appreciated only insofar as it is looked at, contemplated, listened to and read attentively, constantly, repeatedly, even after we have already understood well what it is intended to mean, show, and reveal and thus inform. We read and read *Hamlet*, and go to a theater to see it performed even if we have memorized all the lines by heart. We hang Picasso's *Guernica* or Jasper Johns' *Flag* on the wall of our living room and display Duchamp's *Fountain* or one of Pollock's works in the corner of our study even though we all have understood, and agreed on what all these artworks try to mean, show and reveal. These facts indicate that the ultimate destiny of all artwork is not to be merely understood but to be appreciated. If the point of artwork is merely cognition, it is impossible to make sense of artwork's destiny. If the function of artwork is truth, all that we need in dealing with an artwork would be to understand it. If the significance of an artwork is knowledge, it is difficult to explain why we are interested in hanging paintings, in displaying sculptures, in rereading a poem, and in watching the same play many times, even when we know all of these works by heart. This demonstrates that even though all artworks have cognitive import in some loose sense, their function is not like that of newspapers or academic books and that, even if their ontological status must be taken as a kind of statement, their statement is not like any other statement in newspapers and scientific books. This means that if an artwork has any cognitive function this is so only in a special and very broad sense of the word "cognition."

Second, the cognitive theory of art implies that the reason for which a particular artwork is more highly valued than others must be its greater cognitive import. This view in turn implies that we must have already known the truth or truths of the artwork we appreciate, admire and evaluate. An undeniable fact about most art lovers, however, is that most generally they neither know nor intend to know the putative truth of their artwork. And yet they admire, evaluate, enjoy and thus appreciate them in museums, theaters, concert halls, and libraries.

Third and finally, even among the most committed cognitivists there is no clear and definite agreement on the particular nature and kind of truth a particular artwork is assumed to reveal. And most people have never understood, cannot comprehend, and will never see with clarity what the putative truths of *Guernica*, *Hamlet*, *Flag*, *The Waste Land*, and *Fountain* are supposed to reveal and mean, even when provided with endless explanation by art experts. Furthermore, even among the most convinced cognitive theorists of art, what truth a particular artwork is supposed to reveal is a matter of infinite debate, if indeed these artworks have some sort of truth and also if indeed all artists intend primarily to show a certain kind of truth in their products. If that is the case, then the value of an artwork can neither be assessed nor appreciated according to its indeterminable putative truth. Still, artwork was, is, and will always be appreciated, assessed, admired and criticized by many people in different societies. And modern artwork and any artwork cannot be thought of as being appreciated simply for its cognitive contribution alone. The untenability of the cognitive theory of artwork's function brings us to an alternate theory which I have characterized as "performative."

b. *Performative Theory*

By "performative theory" of the function of an artwork, I mean a theory that holds that the proper function of an artwork is not so much to reveal truth as to give rise to a special emotion in those who contemplate it as artwork. Bachelard's phenomenology of poetic experience best illustrates what is referred to as "performative theory."

From this perspective, the point of an artwork is not to inform, but to make us happy. The aim of a poet is to create certain kinds of verbal images which would enchant the reader. Successful poetic images attract us, enchant us, and render us pleased. Poetry would consist in creating verbal images that would provoke an enchanting, pleasing experience in the reader. Because of these effects, this experience is rightly called "felicitous." Such experience is generated by poetic images because they are associated with real and concrete conditions under which most people, if not all human beings, would be happy, or with things or situations to which all human beings aspire universally. Images of the house, hut, shell, or nest, for instance, are poetic because they are subconsciously associated with our mother's womb where we were most protected and satisfied. Images of the cellar, wardrobe, or drawer are also poetic because they are intuitively associated with the place where some of the happiest parts of our childhood were spent. Images of the infinite cosmic space are poetic because they symbolize the freedom we all aspire to enjoy.[11] The same kind of explanation would apply *mutatis mutandis* to all kinds of literary artwork, and to all forms of non-verbal artwork as well.

But, however attractive this account of poetry and by extension artwork in general may be, it has a major obstacle to overcome. Bachelard's theory fails to distinguish between real things and their images: between a real nest and a verbal image of nest, between a real sunset and a painted sunset, to be more precise, between the experience derived from real things and the experience derived from their images in artwork. If the charm and felicity of poetic experience are explainable in terms of the psychological association of certain perceptual objects with certain past positive experiences, then the real sunset, the real nest and the real shell would provoke stronger poetic, i.e., pleasurable experiences than their images as created in artworks, for the real things are closer to the real conditions or situations where we were most happy – for example, when we were in the wombs of our mothers. And, if Bachelard is right, it is hard to explain the universally shared human need to create artwork and to enjoy it since the satisfaction we expect to get from artworks could easily be obtained through real things, natural or artificial.

Indeed, Bachelard's theory overlooks the linguistic nature of artwork. Images of the nests, shells or sunset in an artwork, and artwork itself, are linguistic in character, while a real nest, shell, or sunset are not. What makes the visual experience of images in an artwork "poetic" is not simply their psychological associations with the conditions under which we have had the most happy experiences in the past. In a fundamental way it has something to do with linguistic features of the images. An artwork is necessarily a kind of statement or discourse in the broadest sense possible, and is thus "ideal" in character (even though it is always physically embodied), while mere objects and mere images, natural or artificial, are not.

As a physical object, an artwork must be causally connected with us, but as language, it cannot exist without meaning something. To know what an artwork is, is not simply to be stimulated by it, but primarily to understand, in some sense, its meaning. This implies that the function of an artwork has, to some extent, a cognitive dimension.

Therefore it seems as though the value of an artwork and the special satisfaction we get from it are essentially cognitive. In other words, to borrow an expression of Goodman's, an artwork provides us with "enlightenment."

Here we appear to be contradicting ourselves because we have already rejected any cognitive theory of art. For, if an artwork is appreciated as language, why is it that languages are not simply equally appreciated? If an artwork is a sort of cognitive statement or discourse, how can it be differentiated from other such statements as are typically found in, say, a scientific context?

There are two ways in which an artwork as language can be distinguished from languages in other discourses. As I have tried to argue elsewhere,[12] the difference between them lies in the difference in their distinctive modalities. While the modality of a statement or discourse in a non-artistic context is *assertive*, their modality in an artistic context is *hypothetical*. When in a scientific or perceptual context a sentence like "The snow is white" is uttered, at least for the speaker, it is assumed that the statement is factual and true. But when the same sentence is written in a poem or a fiction, the statement is only a proposal to be imagined or conceived and thus at best probable or possible. The logical implication is that with art the question of truth in a proper sense of the term cannot arise. If to appreciate an artwork as artwork is tantamount to appreciating it as language, we are not appreciating its cognitive significance. If all artwork has some sort of cognitive dimension, this is so only in a very figurative sense. Contrary to Goodman's view,[13] the delight we get from an artwork which we usually call "aesthetic" cannot be cognitive in a strict sense. Art is not obligated to show either what there really is or how things actually are. The language of art is concerned with showing what kinds of things are possible and how they could be conceivable.

On the other hand, the language of art is different from the language of all other discourses other than art insofar as the former tends to be concrete, i.e., figurative while the latter tends to be abstract, i.e., general. Each artwork as a linguistic discourse is necessarily inclined to be concrete because it intends to discover the possibility of representing, symbolizing or expressing certain unique features of an object, event or experience. But the very notion of a concrete language which represents, symbolizes, or expresses a unique feature of an object, event or experience is contradictory, because language is possible only when it generalizes diverse features of an object, event or experience. Insofar as something is called "language," it does not merely exist, but it means something. Insofar as language means, what it means is necessarily universal, abstract, hence "ideal" instead of being particular, concrete, hence sensory. The meaning of a language requires sacrificing the concrete and particular aspect of what is meant. The language of scientific discourse tends to be increasingly abstract and aims to show the universal aspect of the particular thing it represents. Science fulfills our need to manipulate the natural environment. However, scientific success in this regard is possible only at the expense of sacrificing the world and things in it as we actually and concretely experience them. Art was born at the moment when humanity became conscious of this aspect of the scientific representation of reality and, more generally, of cognition. Cognition means representation, representation means conceptualization, conceptualization means abstraction, and abstraction means distortion, insofar as it is different from concrete reality. The goal of cognition is to represent things as they are concretely. But this goal turns out to be contradictory, because distortion is a necessary condition of cognition. Cognition implies a necessary gap between the knower and the known. Art was invented as a way of resolving this contradiction, and thus overcoming its attendant frustration. Therefore, as Bachelard puts it, "the poet does not describe, he exalts things."[14]

But it is only in the light of the proper distinction between "the artistic" and "the aesthetic" and of the clarification of their relationships that the goal of art and its value can truly be understood.

II. THE ARTISTIC AND THE AESTHETIC

In the context of discourse on art, the concepts of "the artistic" and "the aesthetic" are very often used interchangeably as if they were identical. Moreover, "philosophy of art" is often understood as a synonym of "aesthetics." The evaluation of an artwork is often understood as an aesthetic evaluation of that work. More generally speaking, even when they are not taken synonymously, the meaning of each of the two concepts has never been clearly defined.

No doubt, both the notion of the artistic and that of the aesthetic can be predicted in an artwork. An artwork, however, is "artistic" only insofar as it belongs to a certain class of objects, and "aesthetic" merely insofar as it is predisposed to be appreciated not merely through the intellect but also necessarily and primarily by means of sensory perception. The concept of "the artistic" is thus classificatory and ontological, while the concept of "the aesthetic" is valuation and psychological. "The artistic" refers to how a piece of artwork is intended to function, whereas "the aesthetic" refers to the way in which the intended function should be fulfilled.

As has already been said, artwork belongs to the class of language in the broadest sense of the word, and exists semantically to the extent that not only verbal artwork such as poems and novels but also nonverbal artwork such as paintings or sculpture or music or ballet, including the most abstract kind, are intended to mean or signify something. Therefore each artwork is something to be interpreted and thus understood. Needless to say, not all language falls into the class of artwork.

All language is used for different purposes, either descriptive, expressive or performative purposes. Yet, contrary to some prevailing views, and despite contrary appearances, an artwork as language seems to have cognitive function because it is descriptively used either to represent or express or exemplify something, all in Goodman's sense of cognition.[15] But, just as not all language is used descriptively, not all descriptive language is necessarily artwork. Only a special kind of descriptive use of language is qualified to be under the category of artwork. It is at this juncture that what can be referred to as "the artistic" in an artwork reveals itself as necessarily having something to do with what can be referred to as "the aesthetic" even though the former does not mean the same as the latter.

If the artwork as language is necessarily relevant to the aesthetic, then the reason is that it is presented as an object at least partially to be grasped by sensory perception. All artwork desires to be perceived, and for this, it appeals not to the intellect but to the senses. The most abstract painting, verbal artwork and even "conceptual artwork" is no exception. Although all artwork insofar as it is open to sensory perception is "aesthetic," its does not follow that all objects open to sensory perception, i.e., aesthetic experience, are artworks; for all natural and cultural objects, which are not artwork, are also open to sensory perception. Therefore, if all artworks are aesthetic objects, they are a very special kind of aesthetic objects.

Although "the artistic" relegates an artwork to a class, and "the aesthetic" refers to the way an artwork is perceptually accessible to the viewer, they are inextricably intertwined. Insofar as an artwork is a kind of language, its meaning can only be understood through the intellect. However, the meaning of an artwork as language is designed to be grasped and understood, but by means of the senses, namely, aesthetically. Accordingly, as Eagleton states, "(a) poem is in this sense a perfected form of *sensate discourse*."[16] If the concept of the artistic includes the concept of the aesthetic, the converse is not true. There is a necessary relationship between the two. But their relationship is asymmetrical.

It is only at the end of our reflections on the relationship between the artistic and the aesthetic that we are able to understand better the failings of the classical theories of art we discussed above. And it is in light of this distinction between the artistic and the aesthetic, and their relationship, that we realize that the theory of art referred to as "cognitive" placed too much emphasis on the side of the artistic and ignored the crucial importance of the aesthetic, while the theories of art referred to as the "performative" committed just the opposite error. From Plato to Goodman, the prevailing theory of art, particularly concerning a theory of modern art, has been "cognitivist," and the fundamental problem of this theory of art is that it at least marginalizes, if not totally omits "the aesthetic" dimension of artwork. It is only when the aesthetic recovers its rightful importance that a better theory of art can be constructed.

III. VALUE OF AN ARTWORK AS ARTWORK

How can the distinction between the artistic and the aesthetic and their relationship just described help us better understand the function of artwork and, as a result, the nature of our appreciation, and hence the value of an artwork?

From the perspective of "the artistic," an artwork certainly gives fresh insight as to how to think and look at the world, objects, and human life. But the cognitive insights it intends to provide are unique and new, so its putative truth claims have to be described as being so unique, so "sensory" (i.e., aesthetic), that the language or the claim's meaning can hardly be clear or commonly accessible, or "true or false." Since a decision about the validity of any truth claim logically presupposes, first, the understanding of the meaning of the claim, and second, that the criteria of a particular truth must already be publicly available, the implied claim of an artwork can never be judged to be either true or false and remains cognitively dubious, perhaps even meaningless.

Granted, the intent of every artwork is to represent a concrete reality or experience irreducible to any concept, and to try to describe it in an adequate language. Since such a language is not available, each artwork must invent a new language. For, as Blocker has convincingly argued,[17] the fundamental assumption of the traditional cognitive/performative debate is misguided. What is meant and represented in an artwork is inseparable from the language in which it is meant and represented. It is for this reason that the language of art demands to be judged "aesthetic," i.e., concrete and sensory, and in the process, ultimately unique and singular.

But the very notion of a language as singular, unique, hence non-conceptual, is contradictory. A language has necessarily to be universal, repeatable and public. Thus the intent of an artwork as a discourse defeats its own realization. It is for this reason that inevitably, in all artwork there is a tension between the artistic and the aesthetic. And, it is in light of this tension that the value of an artwork as artwork and its attraction can be understood. What then does the value of an artwork as artwork consist of? For exactly what reason or reasons are we moved by a piece of art when we contemplate it as artwork?

It is generally assumed that an artwork's value is primarily "aesthetic" and that the aesthetic value is *intrinsic* in the sense that the value does not refer to any desire or need. Certain colors, shapes and other sensory features of an object are said to be aesthetic because mere contemplation of their intrinsic qualities delights us, thus, by identifying an artwork's value with its aesthetic value. Eaton writes that "aesthetic value can be explained in terms of properties of things valued in and of themselves."[18] Following this line of thinking Davies says that "The primary function of art is to provide enjoyment."[19] But without reference to any human concern, the notion of intrinsic value and its intrinsic enjoyment are unintelligible. The problem is therefore to determine what specific need an artwork fulfills.

Yet, despite our implicit assumption, and Goodman's claim,[20] as was noted earlier, the ultimate function of an artwork cannot be thought of as satisfying our cognitive interest; for originality in content, form, and subject matter, and technical excellence are repeatedly pointed to as what constitutes the proper value of an artwork, hence as the cause or reason of the power an artwork exerts on its viewer. For example, Muller, a French art critic, cites "the firmness of the design of van Gogh's paintings, the originality of their harmony, the correctness of their composition" which he refers to as "something artistic,"[21] while Golding, an art historian, cites technical innovation as that which constitutes the greatness of Picasso. According to him, "[Picasso] establishes a free, variable system of perspective."[22] Herries, a philosopher, assumes that the significance and the value of an artwork derive from the fact that artworks at any given time reflect the ideological conditions of a given period of history, when he states: "Modern art has its origin in the disintegration of the traditional order of values which once assigned to man his proper place. Knowing this place, man knew what to do. With the death of God, this order lost both founder and foundation."[23] Levi-Strauss, an anthropologist, is more speculative about the reasons why an artwork moves us. He states that an artwork does what it does because "[it] arrests in time the disintegration of the contingent in favor of the pretext and incorporates it into the work."[24]

However attractive these diverse explanations of why an artwork moves us may appear at first glance, they all have some problems.

In the first place, linguistic creativity and technical originality can also be found and valued in discourse other than art discourse. If the metaphorical or figurative use of language is characteristic of literature, such features of language can also be found as frequently in newspapers or scientific writing as in certain poems. Yet this language outside the context of art is not artwork; only poems and other media of art are appreciated as artwork. In the second place and more importantly, if the reasons for which the linguistic innovation or the technical originality of artwork is valued and thus appreciated are ultimately cognitive, as we have pointed out earlier, it is impossible to understand why we enjoy reading the same poem repeatedly, or to explain why we hang paintings on the walls of our living rooms, while we are not interested in repeatedly reading a paper on Einstein's relativity theory or hanging a picture of the North Pole that is immensely informative. Third, if we value artworks only insofar as they are cognitively significant, only insofar as we gain new insight about metaphysical eternity, the world, people and events through an artwork, and consequently are delighted by this insight, how then can we account for the fact that we delight in reading the same poem repeatedly, or in contemplating the same painting we have hung on the wall or in listening to a piece of music over and over again? For it would be absurd to imagine we are repeating the insight. No theory of art can be satisfactory unless it provides satisfactory answers to all these questions, and all the theories we have considered have failed to answer them. We must therefore look for an alternative theoretical solution.

A possible answer to the question can be found once again through considering the relationship between the artistic and the aesthetic. If both the aesthetic, i.e., the sensory, and the artistic, i.e., the conceptual properties are essential components of an artwork, then what we appreciate an artwork for, cannot be reduced to either of them.

If an artwork is a form of discourse, then it is something to be intellectually understood, and if it is also to exist in a unique, singular, concrete and, finally, sensory fashion, then an artwork is something to be aesthetically experienced. Thus, on the one hand, the fact that an artwork is an object to be understood explains why people talk about the meaning of an artwork, and on the other hand, the fact that a painting will be hung on the wall for long periods of time and that people will read a novel whose story they know by heart implies that ontologically an artwork is a language whose meaning must be found and interpreted, and is at the same time psychological, it being a sensory object whose qualities are experienced and appreciated. That such features as color, shape, and form in a painting, or such properties as metaphor, freshness of image, and style in a piece of literature are the concrete grounds we use for evaluating an artwork should suggest that they are different devices for rendering the language of a given artwork as aesthetic as possible.

An artwork can neither describe, represent, nor express in any strict sense any actual things or facts either objective or subjective. However, as we have already seen, the ultimate and proper function of an artwork is to propose possible or felicitous facts or worlds. Although these facts or worlds cannot be *known*, they are disposed to being *understood*. Due to the uniqueness of these possible facts and worlds, they have to be proposed in sensory (i.e., aesthetic) terms. And, since an artwork is a set of sensory terms (i.e., aesthetic language), its meaning can be understood only, as it were, sensorily (i.e., experientially). Finally, since an artwork can be only comprehended aesthetically (i.e., experientially), it is necessarily an object provoking emotional experience, often rightly referred to as "aesthetic" effect.

This aesthetic delight can be accounted for in terms of the excitement stemming from discovering a new and unique language, which in turn opens up possible facts and possible worlds differing from and transcending established ways of seeing, and thinking about, the actual world, which is, on the one hand, cognitively distorted as a necessary result of its linguistic, hence, conceptual representation, and, on the other hand, emotionally oppressive because of this linguistic distortion. The universally shared human impulse to create artworks is a manifestation of our desire and our attempt to break the established, that is to say, distorted and oppressive mold, and thus to transcend the mundane. In order to be faithful to concrete reality, and thus authentic to itself, art has to break the established mold of expressing our thought and representing the world we know. In order to achieve this goal, art must be as sensory, as concrete, in other words, as "aesthetic" as possible. And it is only through art that the gap between language and its representation, between the conceptual and the perceptual, between the abstract and the concrete, between the objective and the subjective, between meaning and reality, and ultimately between man and the world can be closed. Thus, in art, we wonder, as Bachelard puts it poetically: "when a dreamer speaks, who speaks, he or the world."[25] Thus, art is also an expression of our existential attempt to find metaphysical or religious happiness.

This means that the function of an artwork is at once to liberate us from the constraint of the present habit as well as past conditions of life and to find an ideal world where we can be cognitively more truthful, where we can be ontologically more in harmony with the world as it is. It is through art that we strive to realize these metaphysical ideals, and these ideals, embodied in an artwork, are that which excites us, and what we appreciate the artwork for.

Simmons College, Boston – Emeritus

Pohang University of Science and Technology
Korea

NOTES

[1] Richard Wollheim, *Art and its Object*, 2nd ed. (Cambridge: Cambridge University Press, 1980), p. 160.
[2] See Marcia M. Eaton, *Basic Issues in Aesthetics* (Belmont: Wadsworth, 1988), p. 143 and Stephen Davies, *Definitions of Art* (Ithaca: Cornell University Press, 1991), p. 221.
[3] Davies, *ibid.*
[4] See Gaston Bachelard, *The Poetics of Space*, trans. Maria Jolas (Boston: Beacon Press, 1964) and his many other works on this subject matter.
[5] Clive Bell, *Art* (London: Chatto and Windus, 1914), p. 26.

[6] Rainer Maria Rilke, quotes from, "The Notebooks of Malte Laurids Bridge," in Martin Heidegger, *The Basic Problems of Phenomenology*, trans. Albert Hofstadter (Bloomington: Indiana University Press, 1988), pp. 171–172.
[7] Vasily Kandinsky, quoted in Joseph-Emile Muller, *L'Art moderne* (Paris: Librairie générale française, 1963), p. 101. (My translation.)
[8] *Ibid.*, p. 96.
[9] Martin Heidegger, "The Origin of the Work of Art," in *Poetry, Language, Thought*, trans. Albert Hofstadter (N.Y.: Harper & Row, 1971), p. 34.
[10] Jean-François Lyotard, *The Postmodern*, trans. Brian Massumi (Minneapolis: University of Minnesota Press, 1989), p. 78.
[11] See Bachelard, *The Poetics of Space*, *op. cit.*
[12] See Ynhui Park, "The Function of Fiction" in *Phenomenology and Philosophical Research*, March 1982; also Ynhụi Park, "The Ontological Modality of Artwork" in *Contemporary Philosophy*, October 1986.
[13] See Nelson Goodman, *The Ways of Worldmaking* (Indianapolis: Hackett, 1978).
[14] Gaston Bachelard, *The Poetics of Reverie*, trans. Daniel Russel (Boston: Beacon, 1969), p. 163.
[15] See Nelson Goodman, *Languages of Art* (Indianapolis: Hackett, 1976).
[16] Terry Eagleton, *The Ideology of Aesthetics* (Oxford: Basil Blackwell, 1990), p. 15. (My emphasis.)
[17] See H. Gene Blocker, *Philosophy of Art* (New York: Scribners, 1979).
[18] Eaton, *op. cit.*, p. 145.
[19] Davies, *op. cit.*, p. 221.
[20] See Goodman, *The Ways of Worldmaking*, *op. cit.*, pp. 139–140.
[21] Muller, *op. cit.*, p. 175. (My translation.)
[22] John Golding, *Cubism* (Cambridge: Harvard University Press, 1988), p. 11.
[23] Karsten Harries, *The Meaning of Modern Art* (Evanston: Northwestern University Press, 1968), p. 53.
[24] Claude Levi-Strauss, *The Savage Mind* (London: Weidenfeld & Nicolson, 1967), p. 29.
[25] Bachelard, *The Poetics of Reverie*, *op. cit.*, p. 161.

ANDREW JAY SVEDLOW

REVERIES ON AESTHETICS

> Beauty is the clearest manifestation of perfection and the best evidence of its possibility. If perfection is, as it should be, the ultimate justification of being, we may understand the ground of the moral dignity of beauty.
>
> George Santayana

This essay puts forward questions concerning the critical and poetic encounter with art. Facets of this intent are expressed most commonly by questions such as: What is art? Why do we make art? How is art an indicator of historical and sociological trends? What is thus about the aesthetic experience?

Santayana assessed the encounter with art as a manifestation of a sense of coming to terms with beauty. This understanding of art and beauty has been attacked, ridiculed and relegated to the pleasantries of a gentler philosophy. Beauty, as such, has seemed to have lost relevancy in a century dominated by lost generations and annihilistic tendencies. In his time, however, Santayana brought aesthetic education to a broad American audience interested, as we are today, in coming to an understanding about the valuation of art and the objects and ideas we encounter on a daily basis.

This assessment of quality and criteria by which to make such conclusions about art and objects has been paramount in the realm of aesthetics. It is all too often that there exists little if any public forum by which aesthetic questions can be shared, except, of course, in professional philosophical journals and conferences. But the general dissemination of aesthetic concerns has been relegated to the contemporary critical analysis of Siskel and Eberts.

Students, consumers and reformers all too often sidestep issues of quality and value for non-questioning conventional standards of political and educational rhetoric. The significance of aesthetics as a form of study is in its power to question basic tenets of accepted lore and fact. The goal, perhaps, of bringing forth philosophical questions on quality, value, beauty and importance is to integrate such queries into

A-T. Tymieniecka (ed.), Analecta Husserliana, Vol. LIII, 287–291.

the activities of everyday life. At its best, aesthetics in the educational setting can be a stepping-off point to a greater understanding of shared and personal values. In practice, it appears that aesthetics takes into consideration the processes and products of our viewing, valuing and judging of objects and ideas by which we confront and are confronted. Dominant concerns of aesthetics are attempts to define art, the language by which we share our valuations and the phenomenon of the experience with art, objects and ideas. These activities make up a world of aesthetics.

The wonderful and mysterious thing about aesthetics, and its educational significance, is that it begins and ends with the formulation of insight-gaining questions. Every day we encounter and are bombarded with mass amounts of visual imagery, information, and the products of a post-industrial consumer economy. This transaction carries with it a continual need to sift, choose, and act upon choices of quality and value. The study and discussion of the process of valuation and prioritizing is one that aesthetics deals with on both an intimate and a universal level.

At times, this process of inquiry into the formation of a sense of value is one of non-rational and tacit behavior. At other times it is a purely intellectual delineation of choices. No matter to what degree of inquiry, the pursuit of understanding the process of valuing and sorting through the material and intellectual creations of our world is one of many shifting ideas and queries.

There are numerous philosophers, critics and aestheticians who have grappled with this problem in the past and who continue to attempt to deepen our understanding of the encounter wit art, objects and ideas. It is the mission of art educators and educators in general to continue the dialogue for young people and adult audiences.

This dialogue might follow the pattern of beginning with the question, "What is art?" The very coming to terms with the word "art" and its referent is of concern to any aesthetic discussion. It is particularly interesting to discern the impact of art on individuals and art's role within a historical and contemporary context. An attempt to extend personal views of art beyond ethnocentric blinders is an enlightening consequence of exploring the definition of art.

As this question of definition gets reduced, educators can begin to ferret out philosophical viewpoints on what art should be and social scientific views on the psychological, sociological and anthropological aspects of art. "Why is art?" as expressed in social science inquiry is a natural follow-up to the "What is art?" question.

These two concerns can then be followed by "What has been and is thus about art?" Of course, a good deal of aesthetic inquiry takes an historical tack by looking to art as a documentation that carries inherent explanations of its own origins. The student of aesthetics travels, therefore, from "What is art?" to "Why do we make art?" to "What is the origin and historical significance of art?" Added to this sequence is an all important notion that the inquiry into art also holds the possibility for poetic invention to take place. This process joins the inquiring mind with the art and elicits co-creations that interpret essential qualities of the encounter with the specific work of art.

From questions of definition, validity and historical importance, the student of aesthetics can leap into the phenomenon of living out the art through the making of art and artistic statements. Aesthetics, in this case, has the potential to be truly transforming.

Gaston Bachelard, the French philosopher, wrote that

> We must recognize the fact that new experience says "no" to old experience, otherwise we are quite evidently not up against a new experience.

If the aesthetic experience is to be, through inquiry, a transforming one, it must say "No" to the old experience with objects and art encountered in the world. This "No" saying it not so much to deny the pre-existence of worthy experiences or to prophesize a material culture anarchy, but to co-create, within an enlivened world, the transaction of cherishing, caressing and beholding the newly seen objects themselves. The invention of such a world view is the creation of a phenomenological intuition, a world different from positivist analysis and different from existential philosophy, because this is not a terminal utopia.

To say "No" is to take action in a continually changing world; the observer participant is forced to keep on transforming the vision of objects which themselves are changing. This continual questioning, inquiry and evolution gives birth to new, fresh experiences. To take each encounter as a selective transaction, wherein an object, such as a piece of pottery, is examined and dreamed about wholly anew, makes such meaningful experiences possible for creating a progression into the future of a totally new and expanded horizon.

This philosophy does not stop to validate the existence of pottery, or to explain its origins, its creator's intentions, or its contextual significance. Nor does it stop at concretizing the meaning of its Being. Rather, it expands into a giant spiral that grows as you follow the pottery downward into its ever changing nature. This is a philosophy of Becoming.

Time, the quality essential to most transactions with objects, is replaced with a time continuum which has no beginning and no end. In John Dewey's view of the aesthetic experience, an encounter with an object has realizable value in one's reflection upon the experience with it. Thus upon reflection, the individual brackets the experience with a beginning and an end and within that temporal experience recognizes the event as that which transformed the individual's understanding of the world of that object. In other words, the temporal encounter with an object, upon reflection, can produce a transformation in the subject. But if that temporal quality is expanded, and the pragmatic aspect of the physical world is suspended from belief, then the spiral of non-graspable time surrounds you and draws you ever more intimately towards the sacrifice of the stasis of the object.

Therein the projected evolution of the encounter with an object is one of perennial, cyclical possibilities, not one of linear progression towards a transcendental ideal. As we progress through and with the spiral, we have the possibility of enrichment. The farther the spiral moves, expands from the realism of mechanics, the closer it flows to a creative questioning of essential qualities.

The pottery is no longer merely a plastic container for temporal practical activities. It is boundless in its possibilities. It is vessel, women, Gaia, earth mama archetype, it is still waters, it is death trapping us under its silence, it is the great darkness before the light, it is "No," a slow, intimate query into reality. Instead of a pure element, reducible to its observable presence, the pottery is born anew on each visionary leap into the imaginal world of the subject, whose caressing hand acts as catalyst to infinite realms of creation.

This form of behavior, or dreaming with objects, is beyond the normal analysis of science and psychology. There is no indication of positive identification and classification in this process. No snapshot of reality is revealed as Truth in this process. The past is denied, only to be reified in an unfolding enlightenment.

This philosophy is a celebration of expressing "No" to formulas, standardizations and conventions. This reading of objects allows us to reveal and unravel ourselves as the object. Instead of reading into the object or taking out from immature indications of reality, the action of "No" allows us to break the immediacy of the object as ordinary. We revel in the ordinary through careful selection of inquiry. If "No" is the counter to conventionalized, plastic consumption, then the possibility of true personal quality can take place.

Why is it that we purchase cheap, shoddy mass-produced objects when with a resounding "No" we can explore the hand-made, the sensual, the unique, the ecologically and ergonometrically appropriate? The answer is in the educational process of aesthetics.

A society that possesses the courage to face an unknown future will support a free art, knowing full well that often the return for such support will be questions and provocation, or even nonsense.

Karsten Harries

New Hampshire Institute of Art

REFERENCES

Bachelard, Gaston, "A Philosophy of No," *ETC* **XXVIII**: 2, 1956, p. 244.

Harries, Karsten, "The Ethical Significance of Modern Art," *Design for Arts in Education*, July/August, 1988, p. 11.

Santayana, George, *The Science of Beauty* (New York: Dover Publications, 1896) (facsimile).

ROBYN GANGI

MUSICAL AND VISUAL ENCOUNTERS

An Investigation of the Aesthetic Experience

This paper focuses on a study conducted by Mihaly Csikszentmihalyi and Rick E. Robinson which is described in *The Art of Seeing: An Interpretation of the Aesthetic Encounter*. Csikszentmihalyi and Robinson interviewed and surveyed curators of art museums – people whose schooling, training, and professional experience dealt with the visual arts – throughout the U.S. in an attempt to understand the nature of the aesthetic encounter. Through these interviews and surveys they formulated a model of the aesthetic encounter in respect to the visual experience.

My thought after reading the study was to see what would happen if I held the model up to my own work with my choral students: What would cancel out? What would apply to both? Would there be some qualities unique to each area, the visual and the musical, that might alter the model? How so or to what extent? Is not an aesthetic experience an aesthetic experience no matter what the object is, or is its nature different according to the object being engaged by the perceiver?

The model proposed by Csikszentmihalyi and Robinson was based on the feedback of experienced and highly trained "see-ers" who were employed by some of this country's finest museums. Their responses were based on their encounters with tangible objects – objects of the plastic arts. But of the temporal arts – dance, music theatre – which only exist for the time they are being created, what would be the feedback on encounters with these intangible objects from a very different group of artists: my high school students? Could there be anything in their experiences that might find a parallel within the same model constructed by Csikszentmihalyi and Robinson? How wide an applicability does the model have?

My spring concert took place in late May with all of the choral groups (a 160 voice concert choir, a 40 select-voice chamber choir, and a 16 select-voice camerata) performing in one evening. The program consisted of some of the finest works in the choral literature: one of the *Vespers* by Sergei Rachmaninoff; the *Ave Verum Corpus* of William Byrd; *The Dark-eyed Sailor*, a folk song arrangement by Ralph Vaughan Williams; *My Sweetheart's Like Venus* by Gustav Holst; John Dowland's

A-T. Tymieniecka (ed.), Analecta Husserliana, Vol. LIII, 293–321.

Can She Excuse My Wrongs; the round *Old Joe Has Gone Fishing* by Benjamin Britten from the opera *Peter Grimes*; *Tota Pulchra Es* by Maurice Durufle; from Leonard Bernstein's musical *Candide*, *Make Our Garden Grow*; and, for double chorus, Franz Biebl's setting of *Ave Maria* – these are some of the highlights of the two hour concert. It was truly a great evening of music and musical performance. After giving us a standing ovation, parents, and students and administrators openly shared how deeply moving this concert was. For days later the feedback was equally moving. The performers, my students, were also deeply affected by what occurred. While the evening was still fresh somewhere within them, I thought I would have them get some of their thoughts down on paper. Could they put into words the transcendence of the previous night's performance? Would their responses somehow correlate to those of the museum curators? The initial questionnaire to my students, who answered using pseudonyms, was as follows:

1. Can you put into words your thoughts/experiences from the spring concert?
2. Did you come away with something more than or changed from when you went into the evening?
3. Were there certain activities or occurrences in rehearsals that prepared you for this experience, i.e. would you have had this experience without the rehearsal procedures you have been through? What were those activities and how did they enable you to have these experiences?
4. An aesthetic experience can possibly be defined as an intense involvement of attention in response to a visual or aural stimulus for no other reason than to sustain the interaction. The experiential consequences of such a deep and autotelic involvement are an intense enjoyment characterized by feelings of personal wholeness, a sense of discovery, and a sense of human connectedness. But to make it possible there must be a challenge contained in the object and the viewer or listener must have certain skills. Choosing one or more specific examples, what was the challenge in the music and what skills did you possess so that your interaction with the work was heightened?

After receiving their answers to this first round of questions, a week later I asked them to write again, still maintaining their anonymity:

1. Did you ever sense some sort of connection with those of the time frame of the songs you were singing or as someone of that time and/or place? That is to say, "Yes, I know that (feeling, thought, or desire), too."

2. Did you ever feel you were where the songs took place? Did the song or the singing of it transport you somewhere? This may not have been a specific place or time. But was there something unreal or surreal experienced during the singing?
3. Did the music ever make you ask questions? For example: What was it like there? What was it like then? Why does this text or music cause a reaction to me?
4. Do the thoughts enter your mind while you are singing or after? But even though the words weren't right there, did that "orb" of thought affect your singing? How has your encounter with the music changed over the several times you have sung it? Choose one or two pieces and be specific.
5. Did you ever wonder if you knew more or less about a song before learning it that this would alter how you eventually experienced it? In other words, would you have sung the song with more ____ or less ____ had you known when the song was written or for whom it was written, or under what conditions, either political or religious or geographical, the composer wrote it?

Sifting through the responses of 160 students, I found that there were indeed many places within their responses that were parallel to those of the curators' studied in *The Art of Seeing*. There are some places in the model where their answers do not fit simply because two very different art forms are involved. Csikszentmihalyi and Robinson focus on the visual arts – static, fixed in time and space, always there to be revisited. The performing arts, however, are dynamic, ever moving in time and space; they must always be re-created in order to be re-heard. Re-creation is susceptible to the circumstances of the event: the performers may be tired or emotionally keyed up; the performance may be the last concert on a tour; the first piece on the program may have a different energy from successive pieces; concert dresses or costumes may fit too snugly; the varied skills, the training, and the experience of the performer(s) profoundly affect how the music will be transported from the two dimensional page to aural and time-bound space.

Going through the responses, I began to jot down my own questions or thoughts about the work I was getting into. For example:

1. Training a musician may be akin to training a painter to be a painter of paintings: to translate or transport and transmit the symbols set forth by someone else, i.e.:

[A] Painter paints ≫ work is completed ≫ perceiver (viewer) encounters the work (does not need to be a trained painter in order to look at it *which is what is meant to be done with the work*) ≫ this engagement creates a new work unique to the perceiver.

[B] Composer composes ≫ work is completed ≫ perceiver (musician) encounters the work (*does* need to be trained enough to be able to translate symbols into sound in order to listen to it *which is what is meant to be done with the work*) ≫ this engagement creates a new work unique to the perceiver (musician) ≫ perceiver (listener) encounters the work (does not need to be a trained musician) ≫ this engagement creates a new work unique to the perceiver *but which is dependent on the perception of the performer*. For a fair example, listen to the two recordings, separated by twenty-some odd years, of Glenn Gould playing the Goldberg Variations of J. S. Bach. Same music, same performer – two distinct renderings of the work. Which one *is* the work written by Bach? Can the performer be unthreaded from the final work enough for us to know what the composer meant? In doing so, though, has the work been rendered impotent and dismantled enough so there is nothing left to perceive? Can the performer get totally out of the way of herself in order for the creator's intentions to come through? Or is this an impossibility? Even if the painting was meant to hang in the closet for none to see except the artist, it is still there to commune with someone. The music has not fulfilled its life if it has not been brought off the page and given life, if for no one else than the performer. The creation is meant to be. To commune and breathe with something beyond itself.

2. Might it be that teaching is the art of making education an aesthetic experience for the student, no matter what the subject matter?

3. Does the seven period day, with the repetitive, rhythmic ringing of bells (an enlarged effect of TV with all of its darting images), disrupt the focus of the students, reminding them of their time-full status (as opposed to the timeless one experienced in an aesthetic encounter), hemming in their learning experience – educational strokes moving in contrary motions to those of the aesthetic experience?

4. Why do we want to experience something that removes us from our present state? "I really pity people who are not artists most of all, for they have no refuge from the world," Vanessa Bell (Virginia Woolf's sister), wrote in 1939, ". . . I often wonder how life would be tolerable if one could not get detached from it, as even artists without much

talent can, as long as they are sincere" (*The New Yorker*, June 5, 1995). C. S. Lewis sees this yearning for something beyond ourselves as a desire for our original and intended state: Our home is of a timeless realm and not governed by the physical bonds we live in daily. We are literally fish out of water:

"How time flies!" as though the universal form of our experience were again and again a novelty. It is as strange as if a fish were repeatedly surprised at the wetness of water. And that would be strange indeed; unless of course the fish were destined to become, one day, a land animal. (138)

The above are some of the questions I have.

The following is an outline of Csikszentmihalyi and Robinson's study. From the responses they received from the museum curators they were able to surmise that the aesthetic encounter could be described as having four dimensions that are distinct yet inseparable from each other: the perceptual, the emotional, the intellectual, and the communicative. They also concluded that common to these four dimensions of the aesthetic experience there appeared to be other qualities that were also related to Csikszentmihalyi's description of optimal experience or the flow model: a focus of attention, a limitation of the stimulus field, a loss of ego or transcendence, and that one's skills are adequate to overcome the challenges found in the work. Adding to these they notate the process of discovery experienced by the perceiver and the inherent human quality and temporal element of all art. Within the outline of the study I am placing pertinent excerpts of the students' responses where parallels exist between the visual experiences of the curators and the musical performance experiences of my students. Because of the different nature of the experiences resulting from interaction between the object and the perceiver inherent to the two arts dealt with here, there are areas where the students' responses fall outside of the model. I have created an extension of the model offered by Csikszentmihalyi and Robinson proposing that, indeed, there is a difference between the aesthetic experience of the visual and music-performing arts. Those responses of the students taken from the first set of questions are notated (1-page number) and those responses of the second set of questions are notated (2-page number). Page numbers listed as (33) refer solely to the text *The Art of Seeing: An Interpretation of the Aesthetic Encounter*.

Outline: *The Art of Seeing Juxtaposed with Students' Responses*

I. Perceptual dimension (29):
  A. What was the physical nature – what is seen?
    1. Considerations:
      a. real v. reproduction;
      b. size or scale;
      c. form, line, color, surface.
    2. Curators' responses reflect traditional principles of order, harmony, balance.
    3. Sometimes their perceptions extend beyond the visual to other senses.
    4. Other physical features in the work provide access to the creative process by which the work came into being.

II. Emotional dimension (33):
  A. What was the emotional nature – upon the seeing . . . ?
    1. Considerations:
      a. The work creates emotions;

> I felt as though I was no longer performing, but that everything was happening inside of me. It was an experience that has never, and may never again be equalled. It was amazing to me that within only half an hour, I could feel so many emotions. I felt as though I was expressing every feeling I've ever experienced, just laying it all out in my own little world. (1–67)

> I learned new emotions; and that's the greatest power ever to be learned. (1–87)

> I was amazed that I was able to feel such passion for a work of art. (1–94)

      b. There is a connection to personal feelings, past associations;

> It flowed out of me, representing my emotions from the past: the hurt, the pain, the love, the hopes and dreams. . . . It was special to me because my grandfather had just died and I felt like I was singing it for him. (1–133)

      c. There is a physical or visceral reaction ("it grabbed me");

> I was shaking after the end of [the concert]. (1–73)

> We were all working so hard I felt the *electricity* going through our music. (1–175)

2. There can be a development of reaction over a period of time:
   a. some said an initial reaction must be positive before the work is given any serious thought;
   b. others said there need not be an initial response of any kind before engaging with a work;
   c. an initial dislike need not deter one from the work.

> In order to sing *Slava v vish ni Bogu* [one of the *Vespers* by Sergei Rachmaninoff] I needed to feel the music. At first I was not into this one at all but I then began to listen to the sopranos and altos and I could not believe everything that was going [on]. Every time we sang that piece I was mesmerized, I was put into a trance. (1–94)

B. Emotional appprehension (40):
1. It is a salient feature of the aesthetic experience.
2. It is not homogenous.
3. The quality of the emotional dimension is dependent on the time spent with the work.

III. Intellectual dimension (41) – engendering thoughts:
A. Considerations:
1. Characterized by a need to sleuth out . . .
   a. the work's inner or hidden message, or . . .

> I ask many questions when we sing. Like why does music make us feel so bold yet frightened and exposed and joyous? And how is it that just one song can uncover unfelt feelings, things you never would've imagined thinking or things that you never felt were possible to feel? It makes you wonder what were these composers thinking of when they wrote this and what kind of people they were? Did they possibly understand the powers that they possessed by bringing this music to life? (2–2)

   b. the work's history or nature, or . . .
   c. an attempt to label, to place in an historical setting.
2. It impels one to search out new ways of seeing a work, to discover another dimension to it.
3. Its historical quality evokes an atmosphere of its era to the person in the present.

B. For some this came into play after the perceptual, emotional, or visceral:

> I think going into the music blindly is better because you need to just learn it first – but when you know it, when you start or you are at the point where

> you can become it, then if you learn the background and/or the time frame – you know the little details. Yes, I would sing the song with *more* intensity if you can relate it to something, if you have knowledge of where it came from. The more you know about something I think the more personable you can make it, to yourself, to your audience, and to you and each other. (2–3)

IV. Communications dimension (62) – discourse ensues:

A. Considerations.

1. The work gives you something in return; you engage with it:

a. The emphasis could be on differences between past and present ideas, era to era, culture to culture.

> . . . Why have we become what we have? We have taken our musical virtues and talents and replaced them with crashing vocals and screaming windpipes. These songs make me ask, "Why was I born here, and not back then?" But then I realize that I appreciate it much more now because I realize the difference between real music and imitation. (2–66)

b. The emphasis could be on the continuities found between ideas past and present or between ideas all found in the present – boundaries of space and time dissolve, allowing an interpersonal exchange to occur: "With the help of information, imagination, and empathy, the viewer can in fact share the dreams, the emotions, and the ideas that artists of different times and places have encoded in their work." (71)

> (Here is an example of communication with a work of the past set forth in a present medium. A kind of aural pentimento.)
>
> In *Make Our Garden Grow* [by Leonard Bernstein, from his musical *Candide* based on Voltaire's work] my personal performance – as I think [were] many others' – was greatly affected by a story Mr. Gangi had told us one day in chorus. I was then able to relate the words of the song to *my* experiences and regrets allowing the song to become more personalized. The little anecdotes we get during chorus are very important, because they stick with us and influence us more than you could imagine. (1–97)

2. The work can be a vehicle for stimulating fantasy and imagination.

> In the past four years I've been a monk, walking and praying; an angel; a young lover rolling in the grass; an arrogant young beau, and the hand

of the elegant young belle that slapped him! All through the music and your focusing lens. (1–112)

3. This can lead to reflecting on one's present state.

I realize more than often how hectic the world is, after being so devoted to one particular activity. It is sometimes difficult to get back "in the swing of things" in the whole complicated process of life, after being so involved and dedicated to singing; it feels almost as if my head were in the clouds. (1–85)

4. The work can stimulate associations (it looked like . . . ; it sounded like angels . . .).

I could picture great battleships on an ocean – just sailing off into the sunset. Corny isn't it, but it was like the setting sun. Angels awakened from heaven and sent their blessings, sang down their warmest regards, . . . holy but ancient as if part of a ritual or religious ceremony. (2–1)

5. It may have a transcendent quality – it removes one from the present and transports them to a timeless realm.

I don't know if there are words to describe the way I felt the night of the concert. When I left the concert that night my parents were asking me why I was so quiet but I couldn't describe to them the way I felt. On the ride home I just sat and stared out the window not paying any attention to anything but my thoughts. I remember looking up and realizing I was home. It seemed like I was in the car for only five minutes. Basically the feeling I got from being in the concert was like being in a dream. (1–83)

Other Formative Qualities of the Aesthetic Experience

V. Attentional element – the focusing of attention (119):

A. Object directness – "I was so wrapped up in it a bomb could have gone off."

I find myself feeling like I am no longer where I actually am. There seems to be no one and nothing else in the world except for me, Mr. Gangi, and the music. It is a feeling that is hard to describe, but it is one that I experience nowhere else but while I am singing. I seem to put every bit of my energy and concentration into the singing; it is like being "taken over" by the music. (1–85)

B. Limitation of stimulus field.

1. The specific circumstances wherein a work is engaged enhancing the interchange between the perceiver and the

object thereby altering the perceiver's orientation of time and space – the lighting and room size where the paintings were displayed or seeing *Guernica* in Spain with Franco's guards outside the museum v. in New York at MOMA.

> I remember one practice, a class day when it was pouring outside and you could hear the pitter-pattering of the raindrops. We sang *Slava v vish ni Bogu*. It was as if the angels were crying for us. All I could picture were the gods in heaven rejoicing and praising us for the miracle we had given them – the lights were low, everyone was still and quiet. It was moving. I don't think moments like that are made for words. (2–2)

C. Loss of ego or transcendence – losing the most human of qualities: self consciousness (122).

> [Mr. Gangi] made me see that music wasn't black and white things on paper written by a person who died 200 years ago. Music is inside you, it's a living, breathing thing that just takes you over. That was my challenge, to let the music take me over, and when it happened, I was changed forever. (1–88)

VI. Skills and challenges:

A. Skills.

1. Each of the dimensions (perceptual, emotional, communicative, and intellectual) are found to be stronger in some people than others.
2. These continue to grow over time.
3. They also grow by interaction with other people of similar interests and by continually exposing onself to more works as well as returning to already seen works.

B. Challenges (127).

1. "Cracking the code" requires knowledge in other areas.
2. The challenge is to apply this information to what is seen.
3. There is the challenge of rediscovering the same work or reusing the information at one's disposal, having a tenacious curiosity.

> I think the challenge for me was to look inside myself and say, "What am I doing here? Am I good enough to be here? Do I have what it takes?" The challenge for me was to answer those questions. (1–88)

VII. Discovery:

A. Usually there is the discovery of one's potential or unrealized skills.

1. "The content of the various discoveries closely parallels the four major foci of the aesthetic encounter: perceptual, emotional, intellectual, and communicative." (128)

> Music is so often the language that my thoughts flow in, I believe. Hence, I often find words hard to come by. Leaving this setting makes me frightened to think that no one's going to be able to interpret my thoughts – the thoughts that can only be expressed through music, and not words. (1–47)

B. Discovery of the work can lead to discovery of one's self and/or one's relationship to the world around them, either specifically or generally.

> I remember the end of the song when I looked at everyone. They were like shaking or . . . quivering, my eyesight couldn't adjust. It was like their souls were being released or let go like they were fighting to be released, struggling to break free from the chains that had kept them intact for so long. They had found somewhere better and they didn't want to return and in effect part of us remained behind, but the place isn't forgotten, it's just unrecognizable. A fight between heaven and earth. (2–2)

C. The process of discovery contributes to a significant encounter with the work.

> Near the end of our last Chamber Choir rehearsal before the concert, just as we were about to sing *The Dark-eyed Sailor* [arranged by Ralph Vaughan Williams], you said something that truly made me think. You said we should "keep a little bit of what you love in everything you do" (or something to that effect). That made me realize how wonderful music could be if you truly love it. I think that this enhanced my experience of the concert, and has greatly affected the way I will live the rest of my life. (1–53)

VIII. Human quality:

A. It connects us with others.

> A connection . . . like a complete understanding of what those people went through. Yes . . . time frame – I can say I did with *Slava v vish ni Bogu* – I guess it was like I belonged in that era, as if some part of me reawakened when we sang that song. Like a past life or something. I could imagine it just so vividly: a life, a death, a sweet offering from heaven. (2–1)

> *Slava v vish ni Bogu* often made me ask myself what it would be like to sing that song when it was originally written – not for a high school chorus concert in 1995, but for the purpose for which it was originally written. (2–14)

B. Its imprint of humankind is upon it.

1. "I am communicating with someone."
   a. "The aesthetic encounter inevitably involves some realization that humanity is communicating with humanity." (132)

> The music never made me ask questions, but it *did* make me realize just how little human nature changes over time; for a teenager in 1995 to be able to relate so much to something written 400 years earlier is pretty amazing. (2–12)

IX. Temporal element:

1. The time in history when the work comes into existence.

> . . . certain songs that we sang would remind me of, or make me think about what went on in the time when a certain piece was written, who sang it? How did it sound when they sang it? And why was [it] sung in the first place? And furthermore, . . . (2–18)

2. The time in history when the viewer lives.

> . . . if they heard it now, how would they say we were doing it? (2–18)

3. The time over which the viewer sees the work (how long do they look at it, how many times do they see it?).

> That one time when we tied our shoes over a five minute period of time. I could've sat there and done that for hours. Just to sit there and experience time. It's such a real and human experience. Certainly nothing you would do on a normal day to day basis. (1–110)

> Although we have been singing *Ave Maria* [by Franz Biebl] since the beginning of the year, every time I sing it, I sing it for a different reason, but each time I do sing it I feel more whole. (1–182)

> When singing *Slava v vish ni Bogu*, we transformed from Masuk High Concert Choir to a group of people singing praises to God in a distant land. It was an incredible experience. This cannot happen unless you are comfortable enough with the music that you can enjoy it. I began feeling this once I was comfortable with the music. The most intense feeling of this was Thursday night at the concert. The song wasn't just notes anymore, it was music. (1–92)

Here I depart from Csikszentmihalyi and Robinson. I am developing qualities in this next section based on the students' responses and my experience which are unique to the choral aesthetic experience.

Areas Contained in the Choral Music Performing Model Not Found in the Visual Art Model

X. The one work as experienced by many people simultaneously, the group experience.

A. The performer's experience via a four-fold relationship:

1. The performer's relationship to the group with whom they are performing.

I have never been in a chorus like the one at Masuk [High School] – it amazes me. How we can all be in our separate conversations with completely different things on our minds at one minute, and at the next, it seems we are one voice, creating a beautiful musical masterpiece; our differences seem to momentarily fade into nothing. It is a feeling like no other that I have ever experienced. (1–86)

150 some odd people stretched across the stage and into the pit, and for the first time I understood fully "a chain is as strong as its weakest link," because there were no weak links. Without any physical connection we connected. It was not me alone crescendoing, . . . or flowing . . . it was the energy between us all. That soprano or tenor ten feet away moved my sound as I in turn did for someone else. How could a group of people this size, come together for an hour of every intense emotion existing, every day, 180 days a year, sharing and trusting – without bonding into one soul or energy or life or something? I have shared more of my inmost feelings with this group than anyone else in my life. (1–111)

2. The performer's relationship to the music they are creating and perceiving.

The *Ave Maria* was probably the most beautiful piece of music of any kind that I have ever heard. Somehow, I didn't realize that I hadn't given my all for the rest of the concert until we got to that piece. From the first note of that piece I guess I could say I was captured by the feeling. And when I sang I gave it everything I had. I almost shut my eyes when I was singing. I felt like I was a part of the music. When I hit all those incredibly high notes I felt like I was soaring. And at the very end when it was all over I collapsed on my friends. I just couldn't stand anymore. I was exhausted and it hit me the minute the concert was over. During the concert, if I was tired I couldn't tell. Though the whole concert was the cause of my exhaustion, it was mostly the *Ave Maria*. That piece has such power and beauty. (1–201)

3. The performer's relationship to the conductor from whom they are receiving direction as to how to go about performing the music.

> We had so many meaningful rehearsals, it would take a book to hold them all, but chorus isn't a class, it is an experience. It had made me realize what an art music truly is, and Mr. Gangi has instilled in me a desire to experience life, instead of just going through the motions. (1–57)

> It wasn't until the other night that I realized something. One thing that I should have known earlier. During *Ave Maria* everything was just like normal rehearsal, when I took a quick moment to look at who was standing in front of me. Naturally, I was in the front row, so *you* were the only one in front of me. The thing I realize was how grateful I should be to have such an opportunity to spend a year, and hopefully three more, working with such a talented man. Every time I try to describe you, I can only think of two words: brilliant genius. This one man in front of me took an enormous group of kids, some of them who had never sung a note in their lives, and turned them into such an incredible group of performers. So for another brief moment I just looked and could see how much you appreciated what we (the choir and yourself) had accomplished. I see you as a role model . . . I'd base my life on you and what you've taught me. (1–89)

4. The performer's relationship to the audience for whom they are performing.

> I feel that I did come away with something more that evening – I always do – even sometimes after class . . . perhaps that's when we are truly ourselves. There's no one to impress or put on a show for. We are in touch with *us*. And maybe we can't understand it because we're always trying to cover it up – to protect it, so no one will hurt us. We're afraid to be hurt. When we sing it's as if we're naked, free of everything we're normally afraid to show. I think maybe we are free and naked when we sing. Totally pure and naked, and innocent like a newborn child. In touch with ourselves and you and each other. We communicate. (1–109)

XI. The performer as both perceiver and creator (is not the viewer of the painting also creating the work?).

A. The activity of singing as a creating process is due equal attention to that of the perceiving process.

1. The activity of singing as a creative process is transformative. Singing involves the preoccupation, the attention and the awareness of the work along with the intense physical and imaginative engagement with it as it is being created.

a. This results in an enlarging of the self.

> I think musicians feel so much more than regular people. They see things that no one else can see. (1–65)

Many people will never know what it's like to feel things so deep inside themselves. I think others don't even know it's there. For me to be able to find out so much about myself, I think I am blessed. I left knowing that if I never do something like this again, for four years of my life I was a real musician because I gave what was in me. (1–77)

b. This is due to the natural prerequisite for this activity which is for the individual, either consciously or unconsciously, to release the self: allowed to roam freely amongst this complex array of sensations, the self is able to expand.

In a sense I lost a part of my soul to that concert, but I also gained a new perspective on who I am, . . . I experienced a wholeness that I never felt before. (1–79)

c. In giving the self over to the music the performer becomes the vessel through which the work finds its venue, its fleshing out, its skin.

During *Ave Maria*; my God, . . . I wasn't me anymore. It was like I *was* music and realized that I wanted, *needed* to feel this exhilaration every day of my life. (1–87)

The music was a part of my whole being and generated sort of a light or glow around me. (1–61)

d. The creating and the perceiving are one activity: one perceives as one creates and one creates (puts "it" together, understands it, arranges the symbols into an individually coherent message) as one perceives.

You guys are singing *Ave Maria* right now and it still makes me want to cry. How can *we* make such beauty? (1–19)

e. Sometimes in the act of singing a performer puts on the work as if it was clothing and fills it with his/her self.

You help me to come to reality sometimes, . . . and enjoy the music *we* are making. When I sing, I sing with my heart and soul. I put me in it. I let *me* shine through my voice. (1–59)

2. The activity of singing is more than an activity, a mere doing. It can go beyond this and become a process of experiencing or reliving.

Also, a big challenge for me was to worry less about having each note right, and instead be able to make each song my own personal experi-

> ence. Thursday was one of the few times I was able to stop saying, "Why the hell are we singing about a garden?" and see the real meaning of what I was singing. You helped me to do this when you'd give examples, sometimes from your own life, of what the song meant. Then, instead of *doing* it, I was *experiencing* it. (1–68)

XII. Preparation for the occurrence when the work can be perceived.
   A. Performance skills are developed.
      1. This facilitates a person's ability to perform the work in such a way as to illuminate it for those listening and, also, increases that person's ability to engage with the work.

> The exercise that helped the most was lying on the floor and letting the music draw you into a world you never even dream of, carrying you off and becoming one with the entity that captured the soul. (1–79)

   B. Since the performer is both the perceiver and the creator, the rehearsal period may be seen as the time for culling through the work's nature. As the sculptor moves the clay in the hand, warming it, shaping it, letting the Muse find its way through the fingers and into the work, a singer sifts through the sounds in the body and on the tip of the tongue, lets the Muse enter the breath and seek out its shape in the aural space surrounding the singer or instrumentalist: "True singing is a different breath, about nothing. A gust inside the god. A wind" (Rilke, 231). But the axis of Csikszentmihalyi and Robinson's study is the visual aesthetic encounter of the perceiver in relation to the work seen as a separate entity, complete in itself. They do not explore the visual aesthetic encounter in its fullest sense: the triune relationship of the painter/creator to the painting/work to the perceiver/*creator*. The idea that the perceiver is also an integral part of the work in its entity due to the fact that the perceiver is creating – or re-creating – the work as they see it is touched upon in one passage where it is discussed how beneficial it is to come to a work with some knowledge about it:

   Viewing art is not only a receptive process, he [the curator] exclaims, but a creative one as well:

> and the one way you become creative as a viewer is to keep going to exhibits, going to the galleries, going to the shows, read the art journals and magazines, prepare yourself as much as possible, and you'll be amazed at how much more you'll enjoy looking at art if you have that kind of background. (119) (153)

In music, as in the other temporal arts, the aesthetic encounter involves both the perceiving and the creating experience since the work is always subject to the perceiver's creative process before it can be perceived at all as it was meant to be – music to be heard or dance and theatre to be both seen and heard. Beethoven's sonatas were not meant to remain as ink markings on paper. They were meant to be translated into sound so as to be heard. Enter the pianist – the re-creator. The preparation the re-creator goes through before presenting the work to others is akin to the painter painting the painting, the sculptor chiseling the stone, the weaver threading the colored strings of wool through the loom.

The other exercises did what they were supposed to, prepared my body to sing and let my mind and spirit run wild. (1–103)

The rehearsal is more important than the performance. Every part of every rehearsal served to prepare me. I learned how to give (and accept) back rubs. I learned how to put energy and feeling into my music. I learned to rely more on the director than on a sheet of printed paper. I learned how to relax and be myself. I learned the importance of warming up. I learned how to apply knowledge from many areas to many [other] different areas. (1–11)

The warm-up exercises help us all to get focused . . . that is what I love about the activities such as the v, f, sequence thing that ends in the ah . . . I feel so relaxed and refreshed and know where I'm headed in the song. (1–22)

It [the warm-up exercises] makes us aware of our bodies and the uses for them even in everyday life. People are very disconnected from their bodies and don't realize that the mind *must* be connected to the body. (1–35)

The breathing procedures were helpful when we put our hands to one's back and one to ourselves. I then found how to sing through my diaphragm and within instead of through my throat as I had first started. With the touch I was able to feel how people used this and breath to create a variety of sounds. And also what everyone used within, especially in their spirit to make the words creative . . . I was helped with this to express my feeling and create the mood I needed as my support to form the song. In this I found my starting point from which I could really begin. (1–41)

We always had to feel the sound. If our bodies couldn't do it "the little piece of flesh in our throats the size of our thumbnail" won't be able to. (1–112)

I feel that the exercise of melting into the sand and using breath to lift and release our bodies has helped [me] to learn how to focus and how to become more in touch with myself and my feelings. If it weren't for these exercises I don't think I would know as much of who I am as I do today. (1–127)

I think the exercises that we did before our performances really helped us get a grip and got us to the point where we were totally able to go out there and sing. (1–143)

When we all lie down and you talk to us as if we were on the beach, drifting into the sand . . . when people came around trying to lift other people's heads off the ground – immature people might not understand the meaning or importance of these activities, but they, in their own way, moved me as much as the concert did. (1–145)

. . . the way we worked the singing in with the body movements made me realize that singing isn't just with the voice, but with the entire body. (1–202)

The exercise that comes into my mind is – you told us to let our minds wander, and to count how many thoughts ran through our minds. When you told us to concentrate on just one thought, I really had to try hard to not allow my mind to wander. So, in order to get the most I can out of an experience, I feel that I really need to be in tune and concentrate on what I am doing. (1–179)

C. Passion can be learned.

You pulled me into a whole different aspect of life through the years . . . Learning passion for what I sang also helped me have passion for the rest of what goes on in my life. (1–105)

I've learned to look deeper into songs instead of just listening to lyrics. I've learned to know the spirit, dedication and pure joy that goes into being a conductor, musician and even being human. (1–104)

A lot of times in our life (my life, everyday) people don't communicate – we tend to keep things to ourselves. You teach us to express what we feel. You're real and so human. You bring out a side in us, everyone normally wants to keep sheltered and hidden. That changes. (1–109)

No one says, "Can I trust you with this?" or "I'm baring my soul now, so be ready." We just give and take and share. . . . You developed a code with us. It was either words, or a face, or a flick of the wrist – but there was a motion for each sound. And with that simple motion we all clicked on to the same channel and it became music. It's like taking all the different energy particles and focusing them in through a big bright light. All it took was a wink or lift of your arm and BOOM. I can't even explain it . . . it was . . . MUSIC! (1–111)

I felt a sudden rush of many emotions just rush into me and suddenly for a brief moment everything made sense, everything that we have done, all the exercises, everything came together and made sense. (1–127)

People access passion when deep meanings are engaged; deep meanings, therefore, are at the core of intrinsic motivation. . . . In part, deep meanings provide a sense of direction

> because they govern what people look for and what they are willing to do, whether in sports, computing, music, finance, or writing poetry. And, in part, deep meanings are a source of the energy that people are capable of bringing to bear on a task or activity. (Caine and Caine, 97)

XIII. The performer comes to own the work through the perceiving/creating process.

> I felt that in a way the songs we sang became mine, and everyone else's who was a part of the production. (1–81)

> The music became mine, I sang it as though it came from my soul. (1–133)

> You had completed your job of giving us the tools to make beautiful sound. We made the evening our own. (1–151)

> It felt like we had a tiny secret that the audience was about to find out, a little surprise for them. (1–189)

XIV. Therapeutic or healing quality found in the perceiving/creating process.

> There are times when I can't open my eyes when we're singing because I don't want to see anything that will hurt . . . it smooths over all the rough edges in my life because it is such an inward personal spiritual experience pushing out all the demons and pain in my life. (1–100)

> It's such a good way to release the tension. Music is my medicine. (1–133)

> I saw myself flying over the world looking through wider and more "cleansed" eyes and I felt at peace both inside and outside my body. When I'm singing, that same feeling comes back and I feel as though I'm making a difference to myself and whoever is listening. It might be small and insignificant to one, [however] by singing I'm expressing myself in a way I usually can't. (1–102)

> The audience not only heard our voice, but our song, the song that each one of us holds in our hearts and reveals our most private joys, and heartaches. This concert allowed me to release all the pain and anguish that I had been keeping bottled up inside. (1–81)

CONCLUSION

In looking over my student's responses as they fit in with the outline of Csikszentmihalyi and Robinson, the one area in the visual aesthetic experience that wasn't addressed by the students was in the perceptual dimension and the area of skills. I believe there is reason for this.

As for the perceptual dimension, there were no questions given to the students that addressed the point. I did not ask them to write about the form of a piece or the relationship between one voice part and another and how the composer used this to create a certain affect because, though it is consistently addressed during rehearsals and provides the form out of which the rehearsal procedure develops form, it is also seen as the spring board for the pursuit of the other dimensions: a good diving board is good to have, but a good diver can dive with or without one. How a composer or painter or choreographer constructs a work *is* all-important: the brush strokes in Van Gogh's *Irises*, the sensual smoothness of *The Age of Bronze* by Rodin, the staggered entrances in the final moments of Charles Villiers Stanford's motet *Beati Quorum Vita*. Yes, this must be addressed when training the eye or the ear. But too many teachers, if they get that far in their instruction, go no further. As a result of Stanford staggering the entrances, from the lower voices upward, how does this act upon the resolution of the motet? With such a finish on the bronze, especially in comparison to Rodin's later works, what does this do to the viewer? Does it make the figure more lifelike? Does it make it more enticing to the senses? Does it embarrass the viewer? The construction is important but in its being put together in a certain fashion, something happens, something goes forth from it to those outside of it who are engaged with it. "It is safe to say that a philosophy of art is sterilized unless it makes us aware of the function of art in relation to other modes of experience, and unless it indicates why this function is so inadequately realized, and unless it suggests the conditions under which the office would be successfully performed" (Dewey, 12). Henceforth the emphasis on the other dimensions, the emotional, the intellectual and the communicative. Though the nuts and bolts of a work cannot be ignored, neither can it be assumed that from an awareness solely of the vocabulary will spring forth the poet.

In regards to the area of skills, an analogy might speak to this more directly. The analogy crystallizes when buying hiking boots so as to climb the trails of the Presidential Mountain Range in New Hampshire.

There are two groups of people at the base of one of these trails with their newly purchased hiking equipment ready to make the climb to the summit of Mount Washington. The first group is about ready to make the ascent when the trail guide tells them all to remove their hiking boots and undo the laces. "Now let us look at the patterns on the bottom of your shoes. Aren't they fascinating? Let's compare the patterns of everyone's shoes. In fact, count how many there of each pattern amongst yourselves. Can you design any other patterns? Why don't we take a few minutes to break up into groups and do this. And notice the laces. How many different ways can we lace these boots?" Etc., etc, etc, ad infinitum, ad nauseam. The second group, with new boots on their feet (where they belong) stands before their trail guide who says, "Let's go." *The boots were made to get you somewhere.* Yes, the design on the bottom is interesting, all of the questions put forth by the first guide could be answered and more. But while the first group is still looking at and discussing their boots and laces, the second group is well on their way up the mountain. Yes, the first group has discovered their boots and laces and are now aware of their equipment, but the other group is now clearing the treeline and getting on to the summit. The point of having some knowledge, the point of the skill is to take one to a place where one has not been before. Just as a child can be made into a pyromaniac by overemphasizing the dangers of fire, escape routes, don't do this and don't do that, so can a student come to be obsessed by skills to the point where the life in the art is totally lost. Many artists can testify to having teachers who were obsessive in their regimen and oppression in the area of skills to the point where their love of writing, singing, playing, painting was smothered. In some cases it took years for the person to recover what was lost.

> Parents who push their children to excel . . . succeed in perverting music into the opposite of what it was designed to be: they turn it into a source of psychic disorder. Parental expectations for musical *behavior* often create great stress, and sometimes a complete breakdown.
>
> Lorin Hollander, who was a child prodigy at the piano . . . tells how he used to get lost in ecstasy when playing the piano alone, but how he used to quake in sheer terror when his demanding adult mentors were present. When he was a teenager the fingers of his hands froze during a concert recital, and he could not open his clawed hands for many years thereafter. (Csikszentmihalyi 1990, 112)

It is an art in itself to be able to teach something to someone without that person even knowing she has been taught. Where the activity between the teacher and the student is a seamless current of information and action, where there is a pas de deux of minds and bodies wherein so little needs to be labeled and so much can be owned. The more developed a person is in a skill, the less she has to go back to square one and relearn it, relabel it, review it. Considering this is where my students are in their musical training, it is no wonder the area of skills is not more abundantly addressed. To ask my teenager what he does to get ready in the morning to go to school would elicit a blank look with a question behind it akin to, "Is this a trick question?" But to ask my five-year-old the same question, he would proudly sit up and begin to go through the long list of (boring?) details as to what *he* does to get ready in the morning.

The areas that were not in the model of Csikszentmihalyi and Robinson but found in the model based on my students' responses are unique to the musical, specifically choral, aesthetic experience because as I addressed earlier, the musician is in the unique position of being audience and creator. Therefore the musician's preparation for the performance or re-creation of the work, which would be akin to the painter's time of creating the painting, is like a second time of creating the work: the first being when the composer composed the work, and the second, when the musician rehearses it. Most performing, but certainly not all, involves performing along with someone else. Henceforth that communal experience that Johan Huizinga refers to in his book *Homo Ludens*:

> A play-community generally tends to become permanent even after the game is over. Of course, not every game of marbles or every bridge-party leads to the founding of a club. But the feeling of being "apart together" in an exceptional situation, of sharing something important, of mutually withdrawing from the rest of the world and rejecting the usual norms, retains its magic beyond the duration of the individual game. (12)

This may not necessarily be found in the experience of the viewer but it may possibly be found in that of the painter during the creation of the work: having friends, buyers, curators, etc. over to the studio to view the work in progress.

Also, there is the area of a relationship between the conductor and the singer found in the model I set out, but not found in the visual model. The person who acts as a camera taking in the light and colors; adjusting the amount of light let into the lens; determining the angle of the shot; judging how to frame the object; ascertaining what to focus on in the object before the camera – delivering this all to the "film" behind it: the audience. It is this relationship between the performer and the conductor, at least in this choral medium, that is vital to the overall process of creating the work: it becomes a co-creative act, a symbiotic relationship of creative energies. The scope of this relationship is beyond the bounds of this paper but it needed to be mentioned. In the visual experience the closest thing that can be thought of to compare with the role of the conductor would be that of the guide/teacher who illuminates the work for the viewer.

It becomes even more evident, once again: the extent to which the performer is the creator of the work. Are we hearing Bach or Glenn Gould? Or can we never hear Mozart in space, time and sound without hearing Alicia de Larrocha as well? Are creators destined to be grafted onto or woven into other creators, the performers, the interpreters, of their work?

MOZART: Look at us! Four gaping mouths. What a perfect quartet! I'd love to write it – just this second of time, this *now*, as you are! Herr Chamberlain thinking: "Impertinent Mozart. I must speak to the Emperor at once!" Herr Prefect thinking: "Ignorant Mozart. Debasing opera with his vulgarity!" Herr Court Composer thinking: "German Mozart. What can he finally know about music?" And Mozart himself, in the middle thinking: "I'm just a good fellow. Why do they all disapprove of me?" [*Excitedly to* VAN SWIETEN] That's why opera is important, Baron. Because it's realer than any play! A dramatic poet would have to put all those thoughts down one after another to represent this second of time. The composer can put them all down at once – and still make us hear each one of them. Astonishing device – a vocal quartet! [*More and more excited*] I tell you I want to write a finale lasting half an hour! A quartet becoming a quintet becoming a sextet. On and on, wider and wider – all sounds multiplying and rising together – and then together making a sound entirely new! . . . I bet you that's how God hears the world. Millions of sounds ascending at once and mixing in His ear to become an unending music, unimaginable to us! [*To* SALIERI] That's our job! That's our *job*, we composers: to combine the inner minds of him and him and him, and her and her – the thoughts of chambermaids and Court Composers – and turn the audience into God. (Shaffer, 57)

An audience turned into God. Not a passive God, who merely reclines and receives the reverberations of the creation, but an active God, the continual creator in relationship with its work. Persons in relationship with a creation from another person. A work set out before all who have come to see or hear in order to co-create and give rise to a new multi-individual work.

In the musical experience, the work's first audience, the first person to experience the work, is the individual singer, or pianist, or violinist – it is the performer who will translate the visual symbols into sound – who will at some further point in time re-create the work for his or her own audience – the second audience – to experience as sound.

Regardless of the performer's individual experience while performing the work (ref. to Section XI in the outline), and though the co-creating may be choreographed by the conductor, it is still the individual singer who walks away with her own, individual creation. It is not merely one grandiose work churned about and turned into a generalized blurb by a hundred troops and shot out of a canon into the open fields of the concert hall. Like anything found in nature, the one piece that makes up the whole experience and which is integral to the overall process of events also has its own, unique essence.

> In such experiences, every successive part flows freely, without seam and without unfilled blanks, into what ensues. At the same time there is no sacrifice of the self-identity of the parts. A river, as distinct from a pond, flows. But its flow gives a definiteness and interest to its successive portions greater than exist in the homogeneous portions of a pond. In an experience, flow is from something to something. As one part leads into another and as one part carries on what went before, each gains distinctions in itself. The enduring whole is diversified by successive phases that are emphases of its varied colors. (Dewey, 36)

It is the unique advantage of the musical experience that a person can participate in the creation of a work with other musicians while the work is passing through them to the listeners around them; for the musicians to receive and re-create within them the work for themselves. But the performer must not only parrot what lies before him or her in the score if the work is to take on a life of its own.

> Works of art are, visibly and palpably, human achievements, renderings of the ways in which aspects of reality have impinged upon human consciousness. What distinguishes one art form from another (music from poetry, say, the dance from painting) is the *mode* of rendering, the medium used, and the qualities explored. But all forms must be encountered as achievements that can only be brought to significant life when human beings engage with them imaginatively. (Greene, 163)

This is where most musical experiences break down and become hum-drum-going-through-the-motions-let's-put-on-our white-gloves-and-show-mom-and-dad-what-I-can-do type of experiences, regardless of whether it is in the concert hall or rehearsal room. The essence, the work's *is*-ness remains woefully dormant when the imagination of either the museum goer or the singer is left idle. Works of art cannot be engaged with imaginatively until the imaginative ability has been developed in the person doing the engaging. This is the whole purpose of the study conducted by Csikszentmihalyi and Robinson: to understand the form or nature of the visual aesthetic experience in order that museums and art educators might better grasp how to develop people's ability to engage with works of art. Formulating a model of the aesthetic experience is one matter, but drawing from that the wherewithal to create a pathway or curriculum for people, either students or the general public, to come to have significant encounters with works of art, is no small task.

A way of stimulating the imagination of the singer has been the life-long work of Kristen Linklater. The objective of her work has been and is

> a voice in direct contact with emotional impulse, shaped by the intellect but not inhibited by it. . . . To free the voice is to free the person, and each person is indivisibly mind and body. Since the sound of the voice is generated by physical processes, the inner muscles of the body must be free to receive the sensitive impulses from the brain that create speech. The natural voice is most perceptibly blocked and distorted by physical tension, but it also suffers from emotional blocks, intellectual blocks, aural blocks, spiritual blocks. All such obstacles are psycho-physical in nature, and once they are removed the voice is able to communicate the full range of human emotion and all the nuances of thought. Its limits lie only in the possible limits of talent, imagination or life experiences. (2)

But this involves the training of the performer's entire person, the whole instrument – the body, the psyche, the emotions – so as to penetrate the text or song more deeply and, in doing so, enjoin the listener in the same activity. But does not the performer need to have also trained the eye, the inner eye or the inner ear to hear what is going on within the text or the music of the song in order to be able to justly convey the work of the author or the composer? This is what Czikszentmihalyi and Robinson are addressing: what about the training of, not the painter, but the one who is at the other end of the artistic route, the final component of the aesthetic roundelay: the audience, the untrained viewer or listener?

To review the first question I had: isn't training a musician akin to training a painter to be a painter of paintings – to translate or transport and transmit the symbols set forth by someone else? Isn't it so that Csikszentmihalyi and Robinson are talking about how to go about training viewers to be painters? To train people not only to be able to translate what it is they see but to also be able to transport and transmit what it is they see to themselves if to no one else? To be able to articulate to their selves what it is they are seeing? To know how to unlock the doors not only for others to get in but for themselves to get in? Is it not enough for the singer to be able to sing well, to sing his or her own song, as a painter paints his or her own painting? Or must a singer also be able to not only sing well but also sing another's song? How is that done? Can it be done? There are many people who can analyze a song, its text, its history, its musical structure and all that can be said about it just as there are those who have well trained instruments, who sing beautifully. But the one who embodies both the bard and the song of Orpheus?

> A god can do it. But will you tell me how
> a man can penetrate through the lyre's strings?
> Our mind is split. . . .
>
> Song, as you have taught, it, is not desire,
> not wooing any grace that can be achieved;
> song is reality. Simple, for a god.
> But when can *we* be real? When does he pour
>
> the earth, the stars, into us? . . .
>
> True singing is a different breath, about
> nothing. A gust inside the god. A wind. (Rilke, 231)

I have had the experience of going into a museum like the Metropolitan Museum of Art in New York City and seen a painter sitting at a distance from a great painting, with his or her canvas, brushes and paints, repainting the painting. I would look back and forth from one to the other several times. I could not help but notice how the painter's work, as good as it might be, just did not make it. I was unable to enter into a dialogue with it that would in any way resemble the dialogue I would have with the original painting. Something was missing. Going beyond the many technical differences, there was a quality or entity in the master painter's work that could not seem to be found in the painter's painting. In reference to the master's painting, Dewey would assert that

> they are fine art only because the anonymous artist lived and experienced so fully during the process of production. An angler may eat his catch thereby losing the esthetic satisfaction he experienced in casting and playing. It is this degree of completeness of living in the experience of making and of perceiving that makes the difference between what is fine or esthetic in art and what is not. . . . Wherever conditions are such as to prevent the act of production from being an experience in which the whole creature is alive and in which he processes his living through enjoyment, the product will lack something of being esthetic. (Dewey, 26)

In *Artistry in Teaching*, Louis J. Rubin's assertion about teaching can apply to this and other areas as well: ". . . although we can show them gadgetry, tricks, razzle-dazzle devices, and clever maneuvers – all of these are the symbols, not the source, of artistry . . . . ordinary teachers may use skillful artistic procedures, but the end result is not authentic artistry. Hence, formula and rule have but limited utility" (160). In his book *Flow*, Csikszentmihalyi observes that "Even when children *are* taught music, the usual problem often arises: too much emphasis is placed on how they perform, and too little on what they experience" (112).

The painter in the museum performed. The master painter experienced. Is this not what needs to happen? Not having the student painter imitate what is seen but experience what is seen. Not for people to be solely exposed to the symbols of artistry but to be brought through the experience of the source of artistry. Not for the singer to perform what is on the musical score but to experience the music as it may have run through the composer's blood when the hand put pen and ink to paper. Not for people to look at Rodin's *The Burghers of Calais* but to experience *The Burghers of Calais*. Wouldn't this complex and multidimensional quality of teaching bring a student to the doorstep of the mysterious?

> The most beautiful thing we can experience is the mysterious. It is the fundamental emotion which stands at the cradle of true art and true science. To know that what is impenetrable to us really exists, manifesting itself as the highest wisdom and the most radiant beauty (quoted in Jones, Albert Einstein, 42).

If it is the composer's job to "turn the audience into God," and if the performer and the audience are taken to be co-creators of the work at hand, and the performer is both performer and audience, then it is not only the audience that is turned into God but everyone involved in the entire process: all participate in this God-likeness. All are transformed.

> Music: breathing of statues. Perhaps:
> silence of paintings. You language where all language
> ends. You time
> standing vertically on the motion of the mortal hearts.
>
> Feelings for whom? O you the transformation
> of feelings into what? –: into audible landscape.
> You stranger: music. You heart-space
> grown out of us. The deepest space *in* us, which, rising above us,
> forces its way out,
> holy departure:
> when the innermost point in us stands

outside, as the most practiced distance, as the other
side of the air:
pure,
boundless,
no longer habitable. (Rilke, 147)

REFERENCES

Bernstein, Leonard, *Findings* (New York: Anchor Books, 1982).

Caine, Renate Nummela and Geoffrey Caine, *Making Connections: Teaching and the Human Brain* (Alexandria, Virginia: Association for Supervision and Curriculum Development, 1991).

Cogan, Robert and Pozzi Escot, *Sonic Design: The Nature of Sound and Music* (Englewood Cliffs: Prentice-Hall, 1976).

Csikszentmihalyi, Mihaly, *Beyond Boredom and Anxiety* (San Fransisco: Jossey-Bass Publishers, 1975).

Csikszentmihalyi, Mihaly, *Flow: The Study in Optimal Experience* (New York: Harper & Row, Pub., 1990).

Csikszentmihalyi, Mihaly and Rick E. Robinson, *The Art of Seeing: An Interpretation of the Aesthetic Encounter* (Malibu, California: J. P. Getty Museum, 1990).

Csikszentmihalyi, Mihaly, Kevin Rathunde and Samuel Whalen, *Talented Teenagers* (New York: Cambridge UP, 1993).

De Bruyne, Edgar, *The Esthetics of the Middle Ages* (New York: Frederick Ungar Publishing Company, 1969).

Dewey, John, *Art as Experience* (New York: Capricorn Books, G. P. Putnam's Sons, 1934).

Feldman, David Henry, Mihaly Csikszentmihalyi and Howard Gardner, *Changing the World* (Westport, Connecticut: Praeger Publishers, 1994).

Fowler, Charles, *The Crane Symposium: Toward an Understanding of the Teaching and Learning of Music Performance* (Potsdam, NY: Potsdam College of the State University of New York, 1988).

Getzels, Jacob W. and Mihaly Csikszentmihalyi, *The Creative Vision* (John Wiley & Sons, 1976).

Greene, Maxine, *Landscapes of Learning* (New York: Teachers College Press, 1978).

Harnoncourt, Nikolaus, *Baroque Music Today: Music as Speech* (Portland, Oregon: Amadeus Press, 1988).

Huizinga, Johan, *Homo Ludens: A Study in the Play Element in Culture* (Boston: The Beacon Press, 1950).

Jones, Dewitt and Eleanor Huggins, *What the Road Passes By* (Portland, Oregon: Graphic Arts Center Publishing Co., 1979).

Jones, Frank Pierce, *Body Awarness in Action: A Study of the Alexander Technique* (New York: Schocken Books, Inc., 1976).

Kagan, Sergius, *On Studying Singing* (New York: Dover Publications, 1950).

Kivy, Peter, *Sound and Semblance* (Princeton, NJ: Princeton UP, 1984).

Kivy, Peter, *Music Alone: The Musical Experience* (Ithaca, NY: Cornell UP, 1990).

Langer, Susanne K, *The Philosophy in a New Key: A Study in the Symbolism of Reason, Rite, and Art* (New York: The New American Library, 1942).
Langer, Susanne K., *Feeling and Form: A Theory of Art* (Charles Scribner's Sons, 1953).
Leppard, Raymond, *Authenticity in Music* (Amadeus Press, 1988).
Lewis, C. S., *Reflections on the Psalms* (New York: Harcourt, Brace & World, Inc., 1958).
Linklater, Kristin, *Freeing the Natural Voice* (New York: Drama Book Publishers, 1976).
Lippman, Edward, *History of Western Musical Aesthetics* (Lincoln, Nebraska: Nebraska Press, 1992).
Meyer, Leonard, *Emotion and Meaning in Music* (Chicago: Chicago UP, 1956).
Nachmanovitch, Stephen, *Free Play: Improvisation and Life and Art* (Los Angeles: Jeremy P. Tarcher, Inc., 1990).
Reimer, Bennett, *A Philosophy of Music Education* (Englewood Cliffs, New Jersey: Prentice-Hall, 1970).
Rilke, Rainer Maria, *The Selected Poetry of Rainer Maria Rilke*. Trans. Stephen Mitchell (New York: Vintage International, 1989).
Rubin, Louis J., *Artistry in Teaching* (New York: McGraw Hill, Inc., 1985).
Shaffer, Peter, *Amadeus* (New York: Harper & Row, Publishers, 1980).
Shepherd, John, *Music as Social Text* (Polity Press, 1991).
Tovey, Donald Francis, *The Forms of Music* (New York: The World Publishing Company, 1972).

PART FIVE

THE EDUCATION OF THE EYE AND THE EXPERIENCE OF ART

ANDREA BOLLAND

THE EDUCATION OF THE EYE AND THE EXPERIENCE OF ART IN RENAISSANCE ITALY

For twentieth-century viewers looking backward across the artistic revolution effected by Cézanne, Matisse, and Picasso, Renaissance painting seems unproblematically naturalistic, a representation of figures and events in a legible approximation of the real, three-dimensional world. Indeed this notion that Renaissance art attempts – with progressively greater success – to faithfully transcribe the appearance of the natural world first finds expression in the Renaissance itself. We see this, for instance, in Leon Battista Alberti's 1435 comparison of painting to the reflective surface of the pool of Narcissus, or Giorgio Vasari's 1568 account of Giotto, the founder of "modern" painting, who is said to have learned his art from nature itself.[1]

If Renaissance art is a reflection of the world, modelled on the example of nature, then we might imagine that, unlike the "intellectualized" abstract art of the modern era, it was immediately accessible to a broad range of spectators: to all the people who inhabited that real, spatialized world it portrays. Yet, interestingly, some of the earliest appraisals of Trecento painting (and especially those addressing the art of Giotto) suggest that the new "naturalistic" art of the Renaissance was not so easily grasped by all viewers. For instance, in the *Decameron*, written circa 1350, Boccaccio praises the rather specialized appeal of Giotto's art by contrasting him with earlier artists who "painted more to delight the eyes of the ignorant than to please the intellect of the learned."[2] Some twenty years later, in his last testament, Petrarch echoes this distinction when, describing a *Madonna* by Giotto, he writes that "the ignorant do not understand the beauty of this work, yet masters of art are stupefied by it."[3]

What is it that the ignorant did not understand in a work by Giotto? We might suppose that it was the presence of learned doctrine, yet it can hardly be claimed that medieval art, with its intricate typological programs, had been lacking in complexity. It has been suggested that Petrarch's and Boccaccio's words merely demonstrate a generalized "contempt for the masses" that was characteristic of late Trecento humanism; to insist that the vulgar crowd does not understand Giotto's

A-T. Tymieniecka (ed.), Analecta Husserliana, Vol. LIII, 325–333.

art is simply to pay him a highly conventionalized compliment.[4] Yet I believe that these distinctions are not without meaning, and I would argue that what Petrarch and Boccaccio suggest is required of the viewer of a Giotto is not humanist learning, but a wholly new type of expertise: the visual skills necessary to discern and interpret the syntax and the rhetoric of illusionistic painting. This paper will be divided into two parts: first I will explore what types of skills are alluded to in the passages cited, paying special attention to the implicit relationship between pictorial and poetic language; and second, I will analyze selected paintings by Giotto in order to suggest that his visually complex art is, at a profound level, about vision itself.

Let us begin by examining in more detail the two passages in question. Boccaccio's statement follows directly on his praise of Giotto's ability to lead the visual sense into error, making it mistake for reality that which is actually painted. Petrarch, who claimed in his will that Giotto's artifice was best appreciated by artists themselves, makes this same association of illusionistic painting and learned viewers in his dialogue on overcoming good and bad fortune, the *De remediis utriusque fortunae*, written in 1354–66. In Chapter 41 of this work, one of the interlocutors – *Ratio*, or Reason – derides the other – *Gaudium* or Joy – for his simple delight in pictures; yet in doing so, he precisely enumerates the allures of painting. "You are fascinated," he explains,

> by the lifelike gestures, the movement in these inanimate and immobile pictures, the faces jutting out of posts, and the portraits that seem about to breathe and make you think that they might utter words. The danger here lies in the fact that great minds, in particular, are captivated by these things – and what a peasant will pass off with brief enjoyment, a man of intellect may continue to venerate with sighs of admiration.[5]

Petrarch, like Boccaccio, locates the learned viewer's pleasure in decoding illusionistic effects. Yet, if we take *Ratio* as an exemplar of the learned viewer, we should note that he also derives pleasure from the ability to work backwards from that illusion: to recognize that the moving and breathing body he sees is indeed immobile and inanimate, that the projecting face is indeed a flat post. Hence in Petrarch's will, it is the master of art – himself the creator of illusion – who is best suited to appreciate such an effect. Yet we must note that Petrarch is not leaving his Giotto *Madonna* to a painter, but to the ruler of Padua; he compliments this illustrious lay-viewer by suggesting that he, like a trained painter, appreciates both pictorial effect and the artifice used to create it.

The intellectually engaged viewer of illusionistic painting delights in moving from surface to depth – from the patchwork of color arranged on the panel or wall to the fictive, imaginative space that is implied beyond that support. In this two-tiered structure of painting, we might imagine that Petrarch and Boccaccio saw a structural connection with their own art. Both writers mounted defenses of poetry, in which they define the poetic art as the veiling of profound truths in a garment of delightful fiction.[6] The fable or myth which the poem narrates entices the reader, but it is only an integument through which he or she must pass in order to arrive at the deeper, allegorical meaning. Both writers, in statements echoing their claims for Giotto's art, write that such poetry is inaccessible to the common crowd.[7] The *vulgi* are forever detained at the poem's colorful surface; they are unable to probe its depths, because they cannot understand the poet's use of allegory, that artful trope which implies depth beyond surface, signified beyond signifier.

With this in mind, we may return to the question of what sorts of skills are required of the Renaissance viewer of pictorial fiction. Later Renaissance literature on the arts contains mention of spectators who, like the *ignoranti* of Boccaccio and Petrarch, do not understand the system of modelling in light and dark that creates illusion. In his 1548 *Dialogue on Painting*, the Venetian artist Paolo Pino tells of a woman who chided him for the spot he painted on her daughter's upper lip in his portrait of her; he had to explain to the simple woman that the dark mark on the canvas indicated the shadow cast by the nose.[8] In this anecdote, an ignorant viewer, who is perhaps accustomed to looking at less sophisticated images, is unable to distinguish the means from the ends of representation; this contains echoes of the ignorant reader who is unable to discern the structure of the allegorical trope, and thus misses the point of the poem. Pino's story illustrates an extreme, perhaps even fictive, example of "pictorial illiteracy"; we might imagine another class of slightly more sophisticated viewers, familiar enough with advanced art to differentiate a shadow from a stain, yet who might nonetheless be confounded when asked to read into the compositional spaces of a Giotto.[9]

If we compare a painting such as Coppo di Marcovaldo's thirteenth-century *Madonna* in Orvieto to Giotto's *Ognissanti Madonna* of around

1310, we see a transformation from a powerfully present figure whose form is co-extensive with the panel on which it is painted, to an image where surface serves as a transparent veil to an apparent world beyond.[10] Coppo creates an intricately ornate yet ultimately opaque surface, whereas the beauty of Giotto's image depends on the spectator inferring mass and volume from the visual cues that rest on the plane.

Both paintings present Christ incarnate, yet the demands each artist placed on the spectator to actualize this meaning are very different. In Giotto's painting, the discovery of meaning involves an imaginative journey through previously uncharted territory, in which our act of looking through the painting's fictive space is echoed within the picture itself, by figures who replicate our wondering gaze, tilting their heads back, adjusting their angle of vision to look over each other, and *through* windows opened in the sides of a fully spatialized throne. Giotto revolutionized the art of his day, not by simply creating the illusion of space, but by using that space – making it a medium through which vision may travel, and in which meaning may be created.

For modern viewers it is medieval space that seems difficult and ambiguous; yet we might imagine that spectators accustomed to the spatial conventions of thirteenth-century painting would have found Giotto's intricately constructed spaces much stranger to navigate. It is worth noting in this regard that the complexities of Giotto's *Madonna* may have been too much for Trecento church-goers. Originally made for the high altar of the Florentine church of the Ognissanti, it seems to have been replaced in the 1360s by a more traditionally composed altarpiece, in which the various figure groups occupy separate panels.[11]

The sophistication that characterizes the space in the relatively static form of the altarpiece is even more evident in Giotto's narrative art. In the frescoes of the upper church of S. Francesco in Assisi, Giotto (or perhaps an assistant following his design) depicts the *Crib of Greccio*.[12] The story being told is that of the first crèche scene – a commemoration of the nativity which St. Francis arranged to have made in the city of Greccio, near Assisi. On that occasion, a miracle occurred during the Saint's sermon, when a worshipper saw the effigy of Christ suddenly become flesh and blood, and witnessed Francis awakening the child with an embrace.[13] Although the actual event took place outdoors, Giotto sets the scene within the architectural space of a church choir, viewed from the unusual perspective of the apse. As in the *Ognissanti Madonna*, this framework organizes various types of visual access to the miracle.

The onlookers strain to take in the sight: monks and laymen peer through the altar canopy, leaning forward to see; the women's gazes through the choir screen provide them entry to a space from which they are physically barred; even the ass in the crèche scene cranes his neck to take a look. Giotto amplifies the medieval account of the miracle, adding to Francis's physical embrace of the Christ Child an act of psychological communion – an intense gaze shared by the two.

As scholars have noted, a piece of representational art – the crèche – provides the centerpiece for the miracle.[14] The miraculous transformation of lifeless doll into living Christ Child has a certain kinship with the pleasures of art that Petrarch would later enumerate in the *De remediis*. The inanimate object under the spectator's gaze becomes living and breathing, and it is tempting to imagine that Giotto mused on such matters in planning his composition. The remarkable substitution at the heart of the story is pointed up by the unusual point of view Giotto chooses. We see the backside of a pulpit at the left, and even more strikingly, we see the reverse of a wooden crucifix mounted on the choir screen – the wooden planks that provide the support for the painted scene on the other side. Without pressing the point, I would suggest that in this painting, Giotto makes reference to painting itself, alluding in the central episode to the miracles it is capable of producing, and pointing, in the flat boards that support the Crucifix, to the sleight-of-hand that lies behind those miracles.

The final group of works I will discuss provides perhaps the most nuanced example of Giotto's narrative art, and the strongest demonstration of what I would suggest is his thematization of vision. The frescoes narrating the life of the Virgin and the life of Christ in the Arena Chapel in Padua were executed around 1306. These paintings provide the fullest catalogue of glances, gazes, and stares imaginable. It must be said that in many examples of medieval painting, figures do indeed direct their eyes to one another and look attentively at miracles. Yet nowhere in previous art can one find an emotionally charged space such as that which lies between the disapproving temple priest and the embarrassed Joachim, whose sacrifice has just been refused in the depiction of the *Expulsion from the Temple*.[15] This same affective intensity – though with an entirely different meaning – is present at the *Nativity*, when Mary first looks into the eyes of her newborn son, God made flesh, and He returns the gaze with equal force.[16]

These one-on-one gazes are found in many of the frescoed scenes, and they constitute a kind of "language" through which the figures communicate. It is striking that in the *Birth of the Virgin*, the maidservants are shown washing out the baby's eyes – as if preparing her to inhabit the world Giotto has created.[17] I would suggest that in these forceful gazes, the artist establishes an analogue for direct speech, and that within the fresco cycle, he occasionally creates figures whose rhetorical power depends on the viewer's ability to recognize their deviation from this established norm. For instance, in the fresco immediately following the scene of the *Expulsion from the Temple*, Joachim meets up with two shepherds.[18] The absoluteness of his dejection is here measured by his *inability* to offer them his gaze. The shepherds, on the other hand, share a sidelong glance of suspicion – a glance that is striking in large part because of what Giotto does not show: the dramatic completion of the scene now depends on our ability to project the proper response onto the lost profile of the nearest figure.

Similarly, a departure from an implied norm constitutes the most moving feature in one of the best-known scenes from the Chapel: the *Lamentation over the Dead Christ.*[19] The scene is intensely expressive – unusually so for Giotto – as in the extreme gestures of Saint John and of the angels, who throw heaven itself into disorder by their grief. Yet the most moving element in the fresco depends on the spectator's ability to discern the implicit contrast with the loving gaze shared by mother and Son in the *Nativity* on the opposite wall of the Chapel. Here, the Virgin directs her eyes at her Son, whose head is being held up for her to contemplate; yet her imploring stare pathetically yields no response.

Giotto's mastery of the "language of vision" was doubtless one of the features that provoked interest in the Trecento poets who praised his art. Following the model provided by the *stilnovisti*, these poets – Petrarch, Boccaccio, and Dante – fashioned the language of love as a language of vision rather than tactility.[20] We may say that in so doing, they too incorporate an implicit space, in the sense of unbridgeable distance, into their love poems. In these poems, we find a catalogue of gazes: looks from the beloved which wound, which melt, which even kill the lover, as well as gazes by the lover which permanently imprint the beloved's form in his imagination.[21] Like Giotto's art, this poetry depends for its effects on the reader's possession of certain skills – for instance, the ability to decipher the visual language used by the poets, to distinguish the effect of the *mirar intento e fiso* from that of the *bel guardo sereno* or *soave* or *gentile.*[22]

It is tempting to speculate that Giotto's pictorial explorations of vision developed from a familiarity with earlier poetic treatments of the amorous glance, or even that the Trecento poets who praise Giotto may have been inspired by his depictions of the *mirar fiso* and its variants.[23] Such relationships can never be proven; yet a link between painting, vision, and the amorous gaze is in some ways certified by the fifteenth-century example with which this paper began. Leon Battista Alberti's assertion (taken, he claims, from the poets) that Narcissus was the inventor of painting derives, at a basic level, from the fact that a picture, like the reflection on a pool, is apprehended by sight alone. Yet the gaze of Narcissus was a gaze of longing, of desire for something physically untouchable (his own visage, intently staring back). In the story of Narcissus, canonically told by Ovid and reiterated by countless medieval poets, vision becomes a trope for touch; in gazing at his own reflection on the surface of a pool, Narcissus embraces with his eyes that which he cannot hold in his arms.[24] Painting, as Alberti suggests, replicates the world by means of that same trope – the tactility with which we experience the physical world is translated into pure visuality, and the spaces and objects of painting (as Petrarch and Boccaccio recognized) can only be traversed and embraced by those who can read its highly figured language.[25] Thus we might imagine that Petrarch's ingenious viewer in the *De remediis*, detained at length before the beauty of the painted panel (which he venerates "with sighs of admiration") is, like Narcissus gazing at the pool, refiguring desire.

University of Nebraska-Lincoln

NOTES

[1] Near the beginning of Book II of his *De pictura*, Alberti muses that Narcissus must have been the inventor of painting, for this art is nothing other than "the act of embracing by means of art the surface of the pool." See Leon Battista Alberti, *On Painting and On Sculpture*, ed. and tr. Cecil Grayson (London, 1972), 61–63. Vasari (*Le vite de' più eccellenti pittori scultori ed architettori*, ed. Gaetano Milanesi [Florence, 1906], 1: 370–371) writes of Giotto's "discovery" by Cimabue (who was to take him into his studio): when the older artist first saw him alongside a road, Giotto was making a drawing of a sheep from life on a rock, "without having being instructed in any way by anyone except

nature." Although Vasari notes that Giotto's skills were honed by his subsequent apprenticeship with Cimabue, he nonetheless claims that Giotto "deserves to be called a disciple of nature, and not of anyone else" (1: 378).

2 Boccaccio, *Decameron*, VI, 5.

3 Theodor Mommsen, *Petrarch's Testament* (Ithaca, 1957), 78–81.

4 Lawrence G. Duggan, "Was Art Really the 'Book of the Illiterate'?," *Word & Image* 5 (1989): 235–236.

5 *Petrarch's Remedies for Fortune Fair and Foul*, ed. and tr. Conrad Rawski (Bloomington, 1991), 1: 126–127. The Latin text may be found in Michael Baxandall, *Giotto and the Orators* (Oxford, 1971), 140–141.

6 A sampling of Petrarch's writings on poetry may be found in Wladyslaw Tatarkiewicz, *History of Aesthetics, Vol. 3: Modern Aesthetics* (The Hague, 1974), 9–11; Boccaccio's most complete defense of poetry is found in Book XIV of the *Geneologia deorum gentilium*; see Charles G. Osgood, *Boccaccio on Poetry* (New York, 1956), especially 39–40, for Boccaccio's explanation of how poetry "veils truth in a fair and fitting garment of fiction."

7 On the inability of the vulgar horde to penetrate the integumental surface of poetry, see Tatarkiewicz, 10 (Petrarch, *Invectivae contra medicum*), and Osgood, 44, 51.

8 See Paola Barocchi, *Trattati d'arte del Cinquecento* (Bari: 1960), 1: 134.

9 Petrarch's words in the *De remediis* would seem to indicate that his "peasant" viewer is not totally unable to understand pictorial artifice; though the clever man is so taken by the artist's illusion that he dwells for a long period before a painting, even the less sophisticated viewer experiences "brief enjoyment."

10 Coppo di Marcovaldo, *Madonna and Child*, Orvieto, S. Maria dei Servi; reproduced in Frederick Hartt, *A History of Italian Renaissance Art*, 4th ed. (New York, 1994), 50. Giotto, *Ognissanti Madonna*, Florence, Uffizi; reproduced in Hartt, 89.

11 Sandra Bandera Bistoletti, *Giotto: Catalogo completo dei dipinti* (Florence, 1989), 92. The polyptych that was placed on the main altar (parts of which are now in the Uffizi) was painted by Giovanni da Milano.

12 Reproduced in Bandera Bistoletti, 38. the cycle of paintings in the upper church at Assisi has been attributed to various masters. In the opinion of Miklos Boskovits, followed by Bandera Bistoletti (34), Giotto designed the scene, though it was probably painted by another artist.

13 For the story, which occurs in two separate *Lives* of St. Francis, see Erwin Rosenthal, "The Crib of Greccio and Franciscan Realism," *Art Bulletin* 36 (1954), 57–60.

14 Rosenthal, 57–58 (citing texts by L. White and L. Olschki).

15 Reproduced in Giuseppe Basile, *Giotto: The Arena Chapel Frescoes* (London: 1993), 32 (whole) and 37 (detail).

16 Reproduced in Basile, 122 (whole) and 128 (detail).

17 Basile, 74 (whole), 79 (detail). While the bath is normally included in such nativity scenes in early Renaissance art, this is the only scene I know of, of the eyes being washed.

18 Basile, 32 (whole), 46 and 48 (details).

19 Basile, 198 (whole), 248–249 (detail).

20 Dante makes reference to Giotto's fame in *Purgatorio*, XI, 91–95.

21 For examples of the gaze used in these ways, see Eugenio Savona, *Repertorio tematico*

del dolce stil nuovo (Bari, 1973), 170–171, 182–186; Petrarch, *Rime sparse*, poems XXX, L, CCLXX; Boccaccio, *Rime*, part one: poems XIII, XV, XVI.

[22] The most sustained example of the use of this language is found in Petrarch's *Rime sparse*; see, for instance, poem XVII, 8; LXXVII, 1; CCLXI, 3; CCCXXIII, 31; CCCLVI, 10; CCCLX, 140 (for *mirar fisso* and its variants); XXXVII, 83 (for *bel guardo sereno*); LXX, 40 (*bel guardo soave*); CXXIII, 12 (*bel guardo gentile*).

[23] For a discussion of the significance of vision in Trecento culture generally (with some discussion of the art of Giotto), see Margaret R. Miles, *Image as Insight: Visual Understanding in Western Christianity and Secular Culture* (Boston, 1985), especially 64–75. On the Trecento interest in the science of optics, see Graziella Federici Vescovini, *Studi sulla prospettiva medievale* (Turin, 1965), 137ff.

[24] Ovid, *Metamorphoses*, III, 407ff.

[25] The relationship between vision and touch in Renaissance painting, with reference to the erotically displayed female nude, is discussed by Mary Pardo in "Artifice as Seduction in Titian," in *Sexuality & Gender in Early Modern Europe: Institutions, Texts, Images*, ed. J. G. Turner (Cambridge, 1993), 55–89.

JORELLA ANDREWS

CUTTING THE LINE

The Late Works of Matisse Discussed from a Merleau-Pontean Perspective

CUTTING THE LINE

Although Henri Matisse first made use of the technique of cutting and assembling pre-coloured paper in the 1930s as a way of working out the composition of large-scale artworks,[1] commentators agree that it was during his last years of life as an invalid following a serious operation in 1941, that the creation of the cut-outs as independent artworks became his predominant artistic activity. "I now find the simplest and most direct way in which I can express myself is by cut-outs," he told Maria Luz in 1951.[2] Matisse spoke also of his pleasure in cutting: "A pair of scissors is a . . . wonderful instrument . . . working with scissors in this paper is an occupation I can lose myself in. . . . Why didn't I think of it earlier?"[3]

This prominence of the cut line for Matisse after 1941, and indeed his *pleasure* in cutting, raises questions concerning the relationship between this change of emphasis in the artist's working methods on the one hand, and, on the other, the alteration in his experience of the world due to disability. (He was now confined for long periods to bed-rest or to a wheelchair.)

Where this question has been touched upon in the literature on Matisse (I say "touched upon" because it does not seem to have been the subject of in-depth analysis), some have regarded the cut-outs as that to which Matisse had been reduced by his physical circumstances: at worst they have been regarded as his last hobby;[4] at best, as the most practical way in which, with the help of assistants, he could continue designing large-scale artworks. (This was very likely the case initially.) Other commentators, focusing mainly on the content or iconography of these works, and noting a predominance of mythical and pleasurable themes (for example, *Icarus*, 1943, *Creole Dancer*, 1950, *The Parakeet and the Mermaid*, 1952, *Memory of Oceania*, 1952–53 and *The Head of Hair*, 1952), have seen this as evidence either of the artist's escapism from

A-T. Tymieniecka (ed.), Analecta Husserliana, Vol. LIII, 335–343.

the circumstances of his life or, alternatively, as an expression of heroic domination over them.

The fact that such vastly different interpretations have arisen regarding the significance of the cut-outs supports the argument that it is ultimately fruitless to try and define in precise terms the relationship between a given artwork and the psychological or physiological condition of the artist who made it.[5] This is a point Maurice Merleau-Ponty insists upon in his various writings on art.[6] My own approach, therefore, taking the lead from Merleau-Ponty, is to point instead to a non-causal correlation between Matisse's experience of cutting (my emphasis is on process rather than product) and his experience of pain and disability. Specifically, however, I use to some degree as a model for my discussions, and certainly as a starting-point, Merleau-Ponty's commentary on drawings by Matisse in the much discussed fourth section of *Eye and Mind* (1961).[7] The point of focus for this text is the relationship between line and the paper surface with which that line interacts. He writes:

> And Matisse's women . . . were not immediately women; they became women. It is Matisse who taught us to see their contours not in a "physical-optical" way but rather as structural filaments [des nervures], as the axes of a corporeal system of activity and passivity. Figurative or not, the line is no longer a thing or an imitation of a thing. It is a certain disequilibrium kept up within the indifference of the white paper. . . . (PriP p. 184)

DETACHMENT FROM ILLUSIONISM

I begin with Merleau-Ponty's reference to the line in Matisse's drawings as "a certain disequilibrium kept up within the *indifference* [emphasis mine] of the white paper." Initially I was surprised at his use of this word "indifferent" – after all, as Mallin points out in *Chiasm, Line, Art*[8] (p. 248) paper is not usually regarded by artists as neutral or blank but as having its own texture and character. Furthermore, as Elderfield observes in his study on Matisse's drawings,[9] Matisse himself is known to have thought of the paper support both as "pure, living tissue" and as "a screen or field of light against which the graphic lines, marks and hatchings may act."

However, Merleau-Ponty's use of this term gained a certain suggestiveness when I reflected upon its wider context in *Eye and Mind*. I am referring in the first place to the fact that the section in which this text is found deals primarily with Merleau-Ponty's treatment of modern painting as an effort "to detach itself from illusionism and to acquire its own dimensions" (PriP p. 178). In the second place, his discussion of Matisse focuses on the way in which a specific element of visual representation, namely line, manages to find for itself this freedom by operating "no longer [as] a thing or [as] an imitation of a thing . . ." (PriP p. 184).

Where a drawing located within the illusionistic system of representation is concerned, the relationship between line and its paper support is necessarily in a state of imbalance. A drawing of this kind is regarded as successful only if the support has, for the viewer, the effect of being immaterial or transparent. It must be looked *through* so that the viewer has the impression of being led, by line, into a space on the other side of the paper which is considered continuous with the space in which he or she stands. That which has been drawn, in other words, must be drawn in such a way that it should seem to reside in that apparently enterable space behind the support.

If, however, we look at the Matisse drawing simply entitled *Dessin*, which is reproduced in the French version of *Eye and Mind*,[10] we see that far from describing discrete things (a woman, leaves, a river) such that a sense of their tangibility and solidity is evoked, his use of outline serves to undermine illusionism. Instead of cutting through the support, his lines (except perhaps for those lowest down) have the effect of spreading out across the surface of the paper, tangled and more or less undifferentiated in quality regardless of whether they refer, ostensibly at least, to flesh, foliage, hair or water. Indeed, the woman, flattened and emptied of her object-ness, is hardly more than a roughly drawn sign for a woman. She is not therefore the inhabitant of an apparently real world which we feel we could enter, nor can she be read as a clearly defined figure against a background (as would be the case in an illusionistic drawing). While it is certainly the case, then, that a woman, or at least the idea of a woman, has been represented, it is equally true that what is also depicted is line itself, which has now acquired "its own dimensions" in that it exists in a space that is appropriate only for it, and in which only it can move.

What is of course also evoked is the motion of the artist as his hand-cum-instrument move over the flat surface of the paper. The lines spread out before us reenact the artist's physical experience of drawing and it is this aspect of the work rather than its nominal subject-matter that strikes us with the greater force. Drawing as *practice* rather than drawing as representative of things seems to be of prime focus here. This is how I interpret Merleau-Ponty when he says of Matisse's women that they "were not immediately women; they became women" and of Matisse that it was he who taught us to see their contours "not in a 'physical-optical' way but rather as structural filaments, as the axes of a corporeal system of activity and passivity."

In this drawing, then, since the line no longer serves illusionistic purposes, the surface of the paper is no longer able to fulfil its traditional, self-effacing role. Indeed, Merleau-Ponty's reference to the line as "a certain disequilibrium *kept up within* the difference of the white paper" [emphasis mine] indicates that the surface is now no longer transparent but has gained a certain thickness and substantiality. Because the line spreads out over the paper instead of appearing to cut through it, it (literally) draws our attention to the surface. On the one hand, then, we are given a new awareness of its materiality, and of the way in which line and surface interact. But on the other hand, what now strikes us with disturbing insistence is the *passivity* of the paper support (a passivity that is unnoticed in sharply illusionistic drawings since the support itself is unnoticed). This passivity is perceived not only by the viewer of the artwork (whether the artist or we ourselves) but also by the artist as maker. For the artist cannot help but contrast the activity of his instrument (the pencil or pen) in the physical process of drawing, with the relative inertness of the paper surface upon which, or into which, the line is traced. It is from both of these perspectives that the ground may be described as "indifferent."

In the process of cutting, however – and here I begin to extrapolate from Merleau-Ponty's text – the status of the paper support is radically altered both for the artist as maker and for the artist as viewer. If we consider first the physical experience of cutting, we see that there is no longer an observable contrast between the paper as passive and the artist's instrument as active, as was the case in the production of *Dessin*, or indeed any other drawing. From the artist's point of view, both paper *and* scissors are held and manipulated like instruments. The paper is no longer a surface that is worked across, but a material that is worked *with.*[11]

Then, concerning our perception of the cut shapes themselves, the following observation may be made. We might think that since, in the process of cutting, the line now literally breaks through the surface of the paper support, a clear distinction would be created, comparable to that found in illusionistic works, between figure (the cut shape) and ground (the surface from which the cut shape has been taken). But this is not the case. Rather we see that the paper support, having now been cut out, takes on an ambiguous quality. Far from ceasing to operate as ground and now becoming figure, it clearly has the capacity to function as both. In the process of cutting, then, not only is our awareness of the materiality of the surface enhanced, but the distinction operating within the illusionistic system of representation with respect to figure and ground is also found to be inappropriate. Indeed, figure and ground now stand in a relationship of instability.

From this survey of image-making from illusionistic drawings, through the production of *Dessin* to that of the cut-outs, we find that once-accepted ways of defining not only the nature and functions of the constituent parts of a given image (here, figure and ground as well as line and paper support), but also the relationship between the artist and his materials, have in each case become inapplicable, or non-transferable. In each case, the artist finds himself in a position of having to reorientate himself, and allow new possibilities to suggest themselves.

It is within the context of this inapplicability of interactive factors once taken for granted, and the discovery of new ones, that we may discuss, in terms other than escapist, Matisse's statement concerning cutting being an occupation in which he could lose himself.

A PLEASURABLE LOSING OF SELF

Matisse said of the cut-outs that there was "no break in continuity" between them and his previous independent artworks, that is, his paintings.[12] But clearly, this sense of continuity was not based upon a standard of similarity between the various works in terms of the interaction of such elements as figure and ground, as discussed above, nor in terms of use of materials or formal appearance, and only superficially so in terms of iconography (that is, his continued representations, on the whole, of pleasure and ease). However, if we choose to regard as valid a statement made by Matisse two years before his death: "I create only with concern to draw closer to the absolute, with greater abstraction . . . ,"[13] we may define as the thread of continuity holding together the paintings and the cut-outs, the requirement, associated with this quest for abstraction, that Matisse purposefully and repeatedly let go of personally and traditionally accepted norms with respect to the making of images.

As an artist committed to challenge the constraints of the illusionistic tradition by, to paraphrase Merleau-Ponty, enabling such elements of visual representation as line and ground to acquire their own dimensions, Matisse repeatedly had to sensitize himself to, and explore, new material conditions, acknowledging that what might have been applicable in the production of previous types of imagery, was likely now to be inappropriate. His task, therefore, was to keep moving forward, with the resources available to him, through a process of shedding and inventing – or, to put it differently, through a process of losing himself in circumstances that were unfamiliar to him, and for which he had no ready-made rules of conduct.

Where the cut-outs are concerned, the practical changes to the making of images that he had to make due to his now restricted bodily existence may, initially, have felt more *imposed upon* him, rather than *chosen by* him. But what Matisse seems to have discovered in these circumstances – circumstances in which he might have expected little room for manoeuvre – was that the process of shedding and inventing both took on a new urgency *and* proved remarkably productive, as described earlier, where his overall artistic aims were concerned. This discovery may well be one of the factors underlying not only the importance of the cut-outs for Matisse but also his stated pleasure in cutting during this period.

Turning more directly now to the issue of Matisse's illness, and bearing in mind that it is my aim to identify correlations, rather than direct causal links between his experience of cutting and his experience of disability, I suggest that it was precisely this readiness, just described, to shed past modes of interaction and invent new ones, that characterized his response to the altered bodily condition in which he now found himself. Matisse's friend, the French writer Louis Aragon, suggests in his essay *A Character Called Pain* (1968)[14] that while Matisse's initial experience of illness was that of limitation, it did not remain so in any absolute sense. According to Aragon, illness became something with which Matisse negotiated and which he was able to dominate for quite some time. Another interpretation, however, would be to suggest that illness came to represent for Matisse a new structural relationship with the world which, far from bringing to an end life as he had known it, extended it in new directions, thus confirming to him, as possibly no other situation could have done, the degree to which personal continuity depended not upon retrospection, or upon attempts to reproduce past conditions, but upon the ongoing exploration of new possibilities. By regarding old habits and expectations as inapplicable to his now altered bodily existence, it was possible for previously under-emphasized or ignored capacities to come into sharper focus, particularly the more imaginitive and respective ones. But this is perhaps best discussed with respect to a concrete example.

Disability, by its very nature, made impossible for Matisse certain much loved modes of interaction with the world. One such activity was that of swimming in the sea. Matisse's response to this change, however, seems to have been to find a new form of participation in it by means of vision, imagination and memory as evidenced in 1952 by the creation of *The Swimming Pool*, a large environmental panel which covered the walls of his dining room, and in which he could become both physically and emotionally immersed. "I have always adored the sea," he is recorded as saying, "and now that I can no longer go for a swim, I have surrounded myself with it."[15]

While it may be valid to regard this expression of physical mobility and freedom as an imaginative source of release for the artist from the hard facts of reality, it is also possible to say that it celebrates Matisse's discovery that his relationship with an environment of great importance to him was *not* severed due to his inability to swim. Rather, his contact with it was now of a different but equally meaningful order, indeed, one in which the "connective" power of *vision* (no longer eclipsed by touch?) may have been impressed upon him with new force. For as Merleau-Ponty writes later on in *Eye and Mind* (p. 187), having now left behind his discussion of Matisse, "[W]e must take literally what vision teaches us: namely, that through it we come in contact with the sun and the stars . . . and that even our power to imagine ourselves elsewhere . . . borrows from vision and employs means we owe to it."

With these references to vision and imagination, however, I am moving into a new, and not unproblematic area which cannot be discussed here. Returning, therefore, to the issue of Matisse's interaction with illness, I conclude by suggesting that while there can be no doubt that Matisse struggled with his disability, the nature of this struggle may not solely have been that of trying to resist or dominate it. Rather, it may more accurately have involved the challenge of letting go sufficiently of old notions of self, so that new ones could be discovered. The realization, made possible only through the acceptance of new conditions, that he could experience his life as continuing, and continuous, despite the most radical of bodily alterations, may lie at the root of these words, written by Matisse in a letter to a friend:[16]

What I did before this illness, before this operation, always had the feeling of too much effort; before this, I always lived with my belt tightened. What I created afterwards represents me myself: free and detached.

Essex University

NOTES

1 For example, *Dance*, a mural commissioned by Dr. Alfred Barnes, Matisse's foremost collector in the United States.

2 Matisse in interview with Maria Luz in 1951, recorded in *Témoignages* (published in 1952) and reproduced in Jack D. Flam, *Matisse on Art*, Oxford, Phaidon Press Ltd., 1973, p. 137.

3 Quoted in: Gilles Néret, *Henri Matisse Cut-outs*, Köln, Benedikt Taschen Verlag GmbH, 1994, p. 85.

4 *Ibid.*, p. 7

5 The point is not that artworks should *not* be regarded as illustrative of an artist's life circumstances, but that it is not possible to identify in exactly what way they are illustrative of them.

6 In particular "Cézanne's Doubt" – an essay in *Sense and Non-sense*, Northwestern University Press, 1964, pp. 9–25.

7 "Eye and Mind," p. 184. Found in *The Primacy of Perception*, Northwestern University Press, 1964, pp. 159–190.

8 An essay in *Merleau-Ponty: Critical Essays*, ed. H. Pietersma, Washington DC, 1989.

9 John Elderfield, *The Drawings of Henri Matisse*, London, The Arts Council of Great Britain and Thames and Hudson Ltd, 1984, p. 12.

10 This illustration is found between pages 48 and 49 of *L'Oeil et l'esprit*, Paris, Gallimard, 1964.

11 Had it been possible, within the context of this paper, to discuss Matisse's cut-outs with respect to colour as well as line, this point concerning the materiality of the surface could have been expanded upon with reference to the following statement: "Cutting directly into colour reminds me of a sculptor's carving into stone" (*Jazz*, published by Tériade, Paris, 1947, the text of which is reproduced in Flam, *op. cit.* p. 112).

12 ". . . There is no break in the continuity between my previous paintings and my cut-outs." (Matisse in interview with Maria Luz in 1951, recorded in *Témoignages* (published 1952) and reproduced in Flam, *op. cit.* p. 137.)

13 André Verdet's interview of 1952 with Matisse reproduced in *Matisse – A Retrospective*, ed. Jack Flam, New York, 1988.

[14] An essay in Louis Aragon, *Henri Matisse: A Novel*, London, Collins 1972, pp. 166–188.

[15] *Yale Literary Magazine* special number, "Homage to Matisse" 123, Autumn 1955. Referred to by John Elderfield in *The Cut-outs of Henri Matisse*, London, Thames and Hudson, 1978, p. 30.

[16] Elderfield (1978) p. 9, and sourced in Dominique Fourcade, *Autres Propos de Matisse*, Macula I (1976), p. 114.

JILL MILLER

THE ADVENT OF THE MODERN CHILD AND STUDENT

François Bonvin's Seated Boy with a Portfolio

It is ironic that a sickly boy mistreated by his stepmother and who was denied his beloved schooling by his family[1] would paint a serene image of childhood laced with nostalgia, such as *Seated Boy with a Portfolio* (1857). François-Saint Bonvin (1817–1887) seems to have overcome the pain inflicted by his aloof parents who disapproved of his art career. The result was this peaceful vision of a child's discovery of art. Perhaps he saw his boy, surrounded by objects familiar to an artist's atelier, and felt sympathy and nostalgia. An intimate, familiar painting of an average boy, was not typical of most art on public view at this time. Official artistic taste was displayed yearly in an exhibition, the Paris Salon. The Salon and the state patronage that accompanied it favored large, highly finished paintings of historical or mythological themes that in many ways were polar opposites to this painting. Bonvin's sensitive image of a boy he saw every day was not a Salon painting, but a private and humble image. Is this adolescent, who seems to be in a studio, a representative of the mid-nineteenth-century child? Does the image say anything about perceptions of childhood in that era?

In the nineteenth century ideas about children, family and education received heightened scrutiny. Public devotion to and concern for the child was stronger than it had ever been before. This new attention to the child was reflected in the works of artists and literati. The education and training of young artists also underwent major changes during this time. The French Realists recognized these changes and created images of children appropriately new in mood and content. They did not, however, spontaneously invent the images of children they created. The Realists borrowed from older artistic traditions to portray young people in new ways and rejected many practices of child-painting left over from the *ancien régime*. The artists' new sensitivity to children stemmed from changing public perceptions of them; parents began to study child development and examine their own child-raising roles and techniques. Public concern brought about changes in child labor practices. Consciousness of hygiene became linked to the process of raising

A-T. Tymieniecka (ed.), Analecta Husserliana, Vol. LIII, 345–362.

Francois Bonvin, French, 1817–1887. *Seated Boy with a Portfolio*, 1857. Oil on panel. © Elvehjem Museum of Art, University of Wisconsin-Madison, Martha Renk Fund (Art Collections) purchase, 1982.57.

healthy children. The desire to educate boys and girls, both rich and poor, revolutionized schooling. By examining cultural and artistic issues of the day, it is possible to reevaluate Bonvin's *Seated Boy with a Portfolio* as an emblem of nineteenth-century youth that announces the arrival of the modern child and student.

The France of Bonvin's time has been termed a "child-centered cult."[2] According to social historian Philippe Ariès, the nineteenth century was the first time in history that raising children became the primary purpose of the family. Since his *Centuries of Childhood* (1962),[3] many of Ariès' colleagues, David Hunt, Rosemary Lloyd, and Lloyd de Mause among them,[4] have argued that Ariès packaged his divisions by century and development too neatly, among other issues. Most agree, however, that the nineteenth century was the first in which children were seen as people in a separate stage of development and not as small adults. Many of the developments that led to this attitude, however, took place in the real century of change for the child, the eighteenth century. To understand that childhood in 1857, when *Seated Boy with a Portfolio* was painted, was not an abrupt formation of a new "cult," it is necessary to examine the state of childhood and the changes that took place in the years preceding the painting.

The French child of the early eighteenth century was often subjected to what today would be considered abuse. Children of parents of any means were sent to a rural *nourrice* (wet nurse) for two to five years immediately after birth (to relieve the well-bred mother of the "disgusting" and "swinish"[5] practice of nursing). Most children were wrapped, or swaddled, for the first several months of their lives. Although this practice often led to deformity, chronic illness, or death (as did the neglect and filth the child received in the home of the already overburdened *nourrice*), the practice continued until late in the century.[6] In sociological literature it is often debated whether this treatment resulted from a lack of caring on the part of the parents numbed by a high infant mortality rate (suggesting that parents would emotionally remove themselves to avoid emotional pain caused by children's death). De Mause, however, sums up the situation convincingly when he states that parents have always loved their children (which Ariès seems to discourage one from believing to prove his "cult of child" thesis), but it was not until the eighteenth century that societal changes made it possible for parents to see their children as individuals separate from their own needs.[7] The child became a separate entity and a legitimate part of the family in the nineteenth century due to groundwork laid in the previous century.

The familial revolution began with the parents, specifically with the concept of "mother." As Carol Duncan discusses in her seminal article "Happy Mothers and Other New Ideas in French Art," the new family was no longer simply a "line" to be carried on by the oldest son, with all other children being incidental nuisances and burdens. Now children were individuals to be nurtured. As industrialization increased in France there was a more clearly pronounced split between the life of work and home (as opposed to the rural family work unit which had combined the two and had previously been the staple of the French economy). According to Duncan, the mother as a domestic goddess lovingly overseeing obedient children because the ideal. Loving physical contact was perceived as a practice of good parenting. Swaddling and the use of wet nurses slowly began to wane. Because society was seen to be based on the family unit, and adults came to be seen as products of happy childhoods, raising children to be good citizens was the most important job of the family. In other words, producing virtuous people became, for the first time, more important than raising good produce.

One of the individuals who was most responsible for changes in the construct of the child was Jean-Jacques Rousseau, who was responsible for changes in society's image of what childhood should be. The eighteenth-century philosopher Rousseau promoted many of John Locke's theories (a seventeenth-century English philosopher who believed that education, essential for civic structure and success, was gained through sensory experiences) and expounded upon them. Rousseau's most important contribution was convincing French society that childhood was a separate stage of development. He asserted that infants were born innocent, weak, and free of evil, rather than being agents of Original Sin from birth. The only individuals whose goodness was assured and admirable after society's corruption, however, would be the child, the peasant, and the savage. This philosophy influenced Realist goals and icons. The portrayal by Rousseau of the child as a malleable, worthwhile being, changed the way children were treated, viewed, and even clothed.[8]

The book which effected most of the changes for which Rousseau was responsible was the widely-read *Émile*.[9] Posing as the guardian of an imaginary boy with whom he has been entrusted, Rousseau makes recommendations for the raising of Émile on every subject imaginable:

discipline, education, breast-feeding, swaddling, and play, for example. Although he was one of the first individuals to scientifically study childhood (breaking childhood into developmental stages by a child's age, for example), Rousseau used the format of the novel to convince and instruct parents (with whom novels were popular) to take a personal interest in raising their children.[10] There were many ramifications of the book over a long period of time, including mothers breastfeeding their own children, efforts to make punishment of children humane and appropriate, and improvements within and democratization of primary education. Rousseau also asserted that fathers were the best teachers of their children, stressing the equal importance of domestic and public education as an element of a moral society. While Rousseau's disseminated influence is often difficult to trace as the nineteenth century progressed, there can be no doubt that the benevolence of Bonvin toward his student had some foundation in the philosopher's theories.

The nineteenth-century social scientist Frédéric Le Play contributed to the development of the modern child through his scientific analyses of French rural families.[11] This social analyst wrote several decades after the appearance of *Émile*, which influenced his vision of the special natures of children. Le Play contributed to change in the child's role in the family, as opposed to Rousseau's emphasis on the individual. Le Play's *Les Ouvriers européens* ("The European Workers"), stressed that working classes were the foundations of the society in which he lived. His hatred of the enslavement of children and of the working classes by industry was shared by many, particularly artists, for at least the rest of the century. Like Auguste Comte and the Positivists, Le Play considered the family to be the foundation of society. The family unit, of which children were an important part, was being threatened in his eyes, and thus so was France.[12] Le Play had many suggestions for changes to save his imperiled society. One of several theories he unsuccessfully attempted to develop into legislation was freedom of testation. By allowing a father to divide his estate as he saw fit (rather than it being automatically being received by the oldest son), Le Play insisted that the family could act in the best interest of the individual and the group, thus positively impacting society. This freedom would guarantee social mobility and happy, more functional families in which all members were valued. It is ironic that Le Play's concerns about societal moral crises stemming from working mothers endangering the family and fathers having little power or participation in the lives of their families, parallel similar concerns today. This connection indicates, however, that in the nineteenth century, the modern child, family, and their problems had arrived.

Assuredly these attitudinal changes were manifest in the work of nineteenth-century artists, like Bonvin. The large body of nineteenth-century literature that is concerned with children serves as evidence. "Christopher Columbus discovered America: I have discovered the child!" enthused Victor Hugo.[13] Indeed in works like *Les Misérables*, Hugo was a major force in the mid-century flood of child characters and related themes in French literature. Hugo and Alphonse Daudet (in *Jack*) lamented the condition of the working child, according to Clifford Stetson Parker. Parker unearths many other examples of the child in literature, such as Honoré Balzac's *Pierrette*, where a story of orphan exploitation was told. Chateaubriand and Lamartine showed that children had individual personalities with varying facets and that like adults, youngsters were subject to different moods. Autobiographies began to be filled with accounts of the authors' childhoods, as keys in the soul-searching process. Some authors, like Alexandre Dumas *fils*, took it upon themselves to rally for specific child protection laws to be enacted. More pedagogic in approach, Gustave Droz's *Monsieur, Madame, et Bébé* continued the Rousseauian tradition of instructing parents, specifically with pleas for hygiene and a compassionate rearing of children.[14]

Artists responded to the same social stimuli as their literary counterparts, and incorporated the new sentiments into their work. Bonvin's *Seated Boy with a Portfolio* can be compared to an anonymous work titled *A Father and his Children* (Musée Tessé, Le Mans). This is a surprisingly sensitive portrait of a father surrounded by his two adolescent sons, his younger daughter, and his very young son. Although the figures look toward the viewer for the most part, each of them looks in a slightly different direction or looks from a position or activity in which he or she is engaged. The children and their father are physically close, but the spontaneity of certain gestures and expressions give a sense of emotional closeness and individuality. This anonymous work was produced at the height of changes in the family which allowed the tender, thoughtful emotions in Bonvin's image to arise. A consideration of both works in relation to the changes discussed above reveals the special familial intimacy each artist recorded. *A Father and His Children*, circa 1794–1800, marks the pivotal time when the ideas of Rousseau were being absorbed and when parents were realizing they must love and raise their own children. The naturalistic grouping of the offspring around the benevolent father (who is not a smothering parent) is most significant. The relaxed arrangements and attitudes of the subjects suggest that this closeness is an everyday occurrence. Also of interest is the portrayal of the children with different facial expressions, clothing and interests (the daughter plays the piano, and the sons each seem to be absorbed in their own thoughts). Not only is each child an individual, but each child's age characteristics have been connected with his or her appearance and mannerisms, as discussed in the developmental stages of childhood promoted by Rousseau. In other words the teenagers, although close to their father, are in different moods and have different preoccupations. The youngest child, perhaps about six years of age, squirms during the long portrait sitting, as a child his age would. The father is obviously concerned about the betterment of his children (he holds a book and his daughter knows how to play the piano). It seems that he may also have the modern parent's desire to sacrifice for the well-being of his children (it could be that he has been working, judging from his rolled-up shirtsleeves and loose collar, contrasting with the more formal attire of the children). Perhaps most touching and instructive of all is the intertwining of the hands of the father and little son on the burly father's knee. Thus, *A Father and His Children* is an emblem of the changes toward children taking place.

The gentle observation in Bonvin's *Seated Boy with a Portfolio* is an emblem of the change fully manifested. Bonvin chooses to isolate one child; he considers the boy important enough as a subject to paint him. The children in the anonymous work seem to be thinking individuals, but the boy's consideration of and fascination with the portfolio are more complex. Bonvin indicates this specific and absorbed mental activity with the boy's downturned head and eyes, relaxed facial muscles, resting hands, and slightly parted lips. Bonvin's use of color and light are not even throughout the work, as they are in the family grouping. The boy, in his powdery blue smock and subtly highlighted face and hands, emerges from the darkness of the background. The whiteness of the paper in the portfolio is reflected on the boy's face – the spotlighting is symbolic of illumination through knowledge taking place. The placement of the boy on an overturned stool with items on the floor gives an informal quality to the image. We witness one of many moments of his childhood, in contrast to one summary image such as the Le Mans portrait. The earlier painting can be used to demonstrate how sensitivity to children as individuals took place at the dawn of the nineteenth century, but *Seated Boy with a Portfolio* shows that the child as an intellectual capable of independent thought and concentration was appreciated and even celebrated by mid-century.

Another image that demonstrates the compassionate humanity of Bonvin is *The Little Savoyard* (Boulogne-sur-Mer Museum). "Savoyard" (or one from Savoy) was an umbrella term used to describe children who seemed to pour into the city and try to scratch out a living.[15] Some were orphans, some worked for exploitative adults who were or were not their parents. Others tried to find small jobs, such as chimney sweeping, while many stole, begged, and panhandled. Although many people agreed there should be social programs for the unfortunate, it was the plight and the problem of these children that prompted public demand for help. Here the boy is surrounded by all that he has – an extra pair of shoes (a too-large pair of cast-offs), an apple, a piece of bread, and some coins. Bonvin intentionally forces the viewer to question where the boy will find his next meal and elicits pity from the recognition that the child must fend for himself. It is to Bonvin's credit that the picture does not descend into cloying sentiment. Many painters would allow the subject's face to be seen in this situation, which would interfere with the role usually taken by Bonvin. Rather than evoking pity from the viewer, Bonvin elicits a purer feeling of charity. He does this intentionally, positioning the boy so the viewer does not make eye contact. There is no direct link established between the viewer and the boy when the viewer cannot look into his eyes, allowing the viewer to contemplate charity and come to an individual conclusion. Such an anonymous image is not cloying, and may be a more potent form of propaganda. A faceless child can also better represent all children; individual features would remove the universality of the boy's desperate situation. Bonvin's success lies in his role as reporter and presenter, rather than commentator.

In addition to a more general concern for the welfare of street children like the Savoyard, Realist painters like Bonvin utilized newly-developed theories on children and education, giving them visualization in their art. In Bonvin's *School for Orphan Girls* (Cabinet des Estampes, Paris) the diligence, piety, and quiet dignity of the students and their teacher is emphasized by their austere surroundings. A group of small girls whose faces are not visible, are seated around the room. They gather around their teacher, a nun, in a large, spare classroom space. The girls and much of the classroom are painted in muted browns and grays and the sister is dressed in black and white. The smallness of the figures, the somberness of the hues, and the pious, obedient nature of the girls give a sense of hushed reverence. This mood is appropriate for a painting of a convent school. It also represents the seriousness with which many rural girls at this time approached their education; they were often the first women in their families to have the chance to go to school, and might only have a couple of years to attend school before being deemed adult enough to work on the farm for the rest of their lives. The recognition the painting immediately received was a bright spot in the often difficult life of Bonvin. The Salon jury awarded Bonvin a second-class medal for this piece, his highest lifetime honor.[16] The commendation this type of subject received is significant, as the jury was recognizing a small, contemporary painting amidst large history paintings. The award indicates another important shift: issues of children and education were pervading every level of society, and they were finally being deemed important enough subjects for state-sponsored art.

Because the Realists valued truth and empirical observation, the "pure" child made an honest subject and a malleable student. It was the concept of the child as student that was one of the primary issues in French politics throughout the nineteenth century. Free, compulsory education was the goal of many, and yet it was not until the 1870s that most children throughout France attended public schools. Schooling was important, not only for the welfare of the children, but "to fill the moral and intellectual void allegedly opening up in the lives of working-class children" and French society.[17] The system which was created to try to attract parents to send their children to school instead of to work was based on the Napoleonic "Imperial University,"[18] which was the first national primary and secondary school system used in France. Previously, churches had provided much of the education which did take place, often at a cost to the families. Napoleon recognized and objected to the control the church had over young minds, and saw the opportunity in widespread government-sponsored schooling for creating a love of the fatherland and his ideals. Education, however, did not make great strides during the Empire, because of an underlying lack of conviction on Napoleon's part. The structure his Empire conceived, the "Imperial University," nevertheless did survive, in one form or another, through the early twentieth century.

While attempts to institute public schooling did begin in the 1790s, the first half of the nineteenth century saw many definite civic commitments made to education. The first public kindergartens were established in 1827, and by 1857 (the date of our painting) there were over 400 in existence.[19] This indicates that there was concern for children of nursery school age during our era of study. In 1833 a landmark education act, the Guizot Law, was passed. Although this law, dedicated to the establishment of government-controlled primary schools, was rarely acted upon, the motives behind it indicate a change in French society. Under the provisions of the law, each commune was to found and maintain at least one elementary school. This indicates that poor and rural citizens were now seen as important resources, and there was much concern for their moral health and intellectual development as it would impact the nation. The Guizot Law may almost be considered a government not an observed phenomenon: there were about 23,000 primary schools in France in 1813, and about 45,000 at the time of the passage of the Guizot Law (and 60,600 by the 1850s).[20] Thus, the government was aware of the revolution of the child and the education of the child, even if it did not yet take an active role.

The quality of the schools established by the time of the passage of this law was deplorable.[21] If not located in a city, many students were not able to get to schools due to distance and a poor system of roads. The facilities themselves were often leftover community buildings, the conditions of which were not conducive to good health (they often were poorly heated, poorly lit, and unsanitary). It was not until the last half of the century that schools were provided with books or supplies, and until then many students did not have the financial means to provide their own as they were expected to do. The teachers themselves often had other occupations, were poorly educated, or were seen as morally or politically dangerous subversives. The schools communicated important messages and values on behalf of the government: *pays* (love of the share fatherland); the importance of the military; the virtue of the nuclear family as a social unit; the importance of speaking a common language (French, which was spoken by few in the outlying provinces); and the preference for an active, mobile workforce rather than a mass of manual laborers. Through the promotion of its own values, the government hoped to create a unified and patriotic nation through teachers and schools.

There were many obstructions in the way of this "Francization," however. The ethnic minorities could not speak to their children in the French or understand the use of the Latin taught in most of the schools, so the new mother tongue was not reinforced at home. They had a great mistrust for the individuals and motives connected with public education and often did not think attending school was important for their children. The people saw themselves as citizens of their community rather than of France, another factor hindering unification. Perhaps the greatest stumbling block in the process of educating every French youngster was the issue of child employment. Whether they were members of the old tradition of the working family unit in rural areas, or were part of the army of vulnerable factory workers created by the industrial revolution, most children did not have time to go to school. If not for this reason, many could not attend because their families could not afford to lose the child's income if they should quit work to attend school. Although most schools operated for children who were ages six or seven until they reached the ages of ten to twelve, the perception of children who were as young as six or seven years as being ripe for work left no time in their lives for school. A handful of employers tried offering education during breaks, lunch, or after work, but this was not the norm and many children were too tired to take advantage of what little was offered. In fact, it was late in the century before education came to be valued by most people. The 1841 child labor law was of some assistance, eliminating the possibility of work for the very young children and shortening the work day for those who were considered old enough.[22] Many activists and theorists were disgusted by the lack of enforcement or effective provisions of these laws. It was not until the 1880s, when school was made mandatory, that the work versus school controversy was settled. It was then that children all over France were guaranteed an elementary education. Secondary schooling was for only those who could afford it, until the arrival of the twentieth century.

Thus the boy in our painting may have received minimal schooling at a younger age, and the adolescent was probably finished with the formal education he was to receive before working as an apprentice with Bonvin. Since the boy is allowed to learn at his own pace, leafing through a portfolio, his education is continuing. Education was obviously important to Bonvin, indicated by his choice of subject matter in this image and scenes such as *The School for Orphan Girls. Seated Boy with a Portfolio* shows the impact of Rousseau, as this child is allowed to observe, which is appropriate to his stage of development. Rousseau's influence is also indicated by the innocent look of concentration on the boy's face – he has not been too corrupted by society to enjoy simply looking at a pile of drawings. Even the way the child is dressed shows remarkable concern for the child, prompted by Rousseau. Fifty years before, such loose fitting garments that moved with the active child were not used. A bit of influence from Le Play may even be detected, as Bonvin allows the boy to take a break, or at least a relaxed working pace, in contrast to many children who were slaving most of their waking hours in factories. The intense focus on the boy's reaction to the art indicates the artist's interest in the boy as a person – this is relatively new to art. The child's individual needs may be symbolized by the fact that he is the only individual in the painting; in the past the master had to be present in portraits with his unimportant apprentices. Perhaps the most important relationship the picture has to the changes in perceptions of childhood over fifty years might be the quiet, reverential treatment of the subject matter. There even seems to be a bit of pride in the sensitive rendering of a boy doing such a simple task. The boy does not look toward the viewer, as the *Savoyard* did not. This lends an even greater mood of concentration and absorption to the work. There is almost a fatherly appreciation for the boy, as if Bonvin feels responsible for him and realizes that he can provide a good atmosphere and kind direction to the boy, helping him to become a happier and more successful artist and adult.

It is possible that there is a preliminary drawing of sorts for *Seated Boy with a Portfolio* that tells us more about the education of an artist in the studio of a Realist. The black chalk drawing, *Boy Sketching* (Stadtmuseum, Frankfurt) may feature the same pupil in a different activity. A boy with curly hair sits on a woven stool in front of a drawing of a nude (both the boy and the stool appear to be the same as those in the painting in question). This is another example of an image where Bonvin documents the training process of the artist. A small boy sits on the same stool, with his drawing board resting on the wall just below the drawing of the nude. The very young boy is copying the drawing, which is probably one by his master, Bonvin. That the drawing is of a human figure indicates the boy has not yet reached the stage in his training when he can draw from the nude himself. He learns anatomy, foreshortening, and the master's style by copying drawings before he renders from life. This is an image of a step in the state of educating a young artist in Bonvin's studio in 1856. The Realists were among the first to suggest that all children could benefit from art education; they were also willing to take in pupils. As teachers they promoted the observation and recording of the observable world and the teaching power of images. Bonvin took on his first pupil in 1851, and accepted more in 1853 and 1856. This, coupled with the fact that Bonvin had no sons, may indicate that the boy in our painting is one who has been taken into his atelier for training.[23]

Serious reforms in the education of art began about the same time that *Seated Boy with a Portfolio* was executed. It was common for all education to be seen as an awakening; knowledge and ability were to carefully and gradually come to the child.[24] The proliferation of toys, children's books,[25] and public libraries, also in full bloom in the 1850s, can also be seen as evidence of the new desire to train the young mind. The training of the eye and the development of the awareness of life, however, were concepts crucial and nearly exclusive to art education. Painting and drawing therefore were a fairly exclusive and academic type of training until the early 1850s.[26]

Art education became more democratic at mid-century because peasants and children were considered natural artists and appreciators of art. This was prompted by the absorption of Rousseau's philosophies on the innate goodness and talent of those who were unspoiled by society. Gautier and Baudelaire were some of the first to connect and promote art education for children who were not artists' apprentices. Suddenly, there was merit in the art of children and value in their creative powers.[27] Certain individuals developed specific systems of art education to encourage and develop children's natural talents. Horace Lecoq de Boisbaudran, a revolutionary figure in mid-century art training, believed that a strong visual memory was the most important tool of an artist. While he did not want to replace academic traditions, Lecoq founded a school in 1848 based on "stored observation." He promoted the importance of preserving the individuality of young artists in his book, *The Training of the Pictorial Memory*. Lecoq's methods forced students to use their eyes, as they were required to draw figures and other subjects from memory after visual study. Many nineteenth-century artists were trained by Lecoq, seeing his method as a way to strengthen overall their personal vision.[28] Lecoq himself was always ready to see things anew, and enrolled in Courbet's atelier, despite his stature.[29] Bonvin and Courbet were close friends early in their careers. As Lecoq was a mutual acquaintance, it is possible that Bonvin and Courbet discussed Lecoq and his methods. Perhaps Bonvin read Lecoq's book or even met Lecoq. Whatever the nature of their connection, both Bonvin and Lecoq placed great store in young pupils and their visual training. Bonvin's teaching focused on having his students move quickly to studies from casts to studies from life, a teaching technique that rebelled against the academic tradition of having students spend a large portion of their training drawing from casts and still lifes. Bonvin also encouraged his students to draw from memory, a technique possibly influenced by Lecoq.[30]

Young men, such as the boy in the painting, were called *rapins*. The boys would assist in the preparation and clean-up of paints and other materials, copy prints of old master paintings and the instructor's own work, and sometimes would pose for the artist. The momentary quality of *Seated Boy with a Portfolio* leads one to imagine Bonvin, who is shown to be a kindly, contemplative man in the writings of others as well as in his own writings and artwork, walking in on his young assistant discovering art by perusing Bonvin's portfolio. The young man may have been taking a break or perhaps was supposed to be straightening up, judging from the brushes and sticks of drawing chalk on the floor. The stool, turned sideways on the floor, and the open mouth of the boy (as well as Bonvin's gentle spotlighting of him) indicate a moment of active, engrossed discovery. The momentary effect of *Seated Boy with a Portfolio*, then, is very much a part of the Realist belief in the importance of observation in children's art training and in the capacity for art to make an impression. It also plays into the new societal values discussed previously, as the child is an individual, independently developing his own interests and responding free of adult intervention.

In 1851, Flaubert said of artists, "between us and the crowd there exists no link, unhappily for the crowd, unhappily for us above all."[31] During the course of the decade, however, this lament would be proved false. Gustave Courbet and artists acquainted with him agreed that artists should record their own world, not an imaginary Romantic one. The Realists strove to find visual truth, boiling the subject down to its simplest essence. As outlined above, Bonvin and his circle thought peasants and children to be the truest expressions of the French citizen. Thus, there was a crucial point of correspondence between the commoners and the artists who portrayed them. Fortunately for many of the artists, the new middle class desired naive, soothing images for their homes. With the decline of the church and its patronage, the democratic tastes of the new patrons made secular genre scenes popular. Still lifes, images of people reading or otherwise preoccupied, and of course, scenes of children were now commercially viable.[32]

The Realists were one of the primary groups to paint children at this time. Contemplative images of a single child (as opposed to the children of the previous courts are who were usually depicted only with adults) are found in the work of virtually every Realist painter. The proliferation of this type of image can be easily explained. Realism is an art in which the point of view of the subject is considered. Because peasants and the downtrodden were often featured, artists saw things from a humbler perspective. By no means were they the first artists to experience poverty, but many of them (including Bonvin at times) were some of the first to both paint the unfortunate and to fall into their economic strata. Realist empathy could have lapsed into pity, but usually did not, due to the visual objectivity the painters felt they needed to have to see the subject truthfully. Qualities of spontaneity and intuitiveness, then, made children ideal as both subjects and students of Realism. To the Realists, children possessed many of the qualities they felt they personally had lost. The child was a symbol of their alienation in the changing world, which makes their images of pauper and orphan children particularly appropriate.[33] The Realists and writers with whom they corresponded were also some of the first to be aware of the loss of creativity the process of formal education brought about. Thus the naturally creative, innocent, individual child was an important emblem in the work of Realists like Bonvin. The importance of children as symbols of the future (especially the future of art), can be surmised by Courbet's inclusion of two boys, one sketching on the floor and one watching Courbet paint, in his philosophic visual manifesto, *The Artist's Studio* (Musée d'Orsay, Paris), 1854–1855.[34]

It is equally important, however, to remember that this image and the ideas behind it were not generated spontaneously. These social and cultural elements, combined with the very subtle, tender mood that makes this painting so unique and pivotal in the process of understanding the child, may signify the most important development of all: the modern parental adoration of the child. Finally the father/artist has put the child first in composition, iconography, and mood. The modern child is the purpose and center of the family's operations and Bonvin's *Seated Boy with a Portfolio*.

The painting of this solitary boy stands for a microcosm for the world of the young at that time. His appearance could be said to symbolize the new understanding for childhood's free spirit and the need for children to grow and develop. The slight shabbiness of the clothes and the stool, as well as the humble environment, signify that Bonvin was aware of class distinctions which often created suffering in children of that time. The child's engrossed expression suggests mental activity, to which in one form or another children were being exposed. The art in the portfolio is the symbol for new experiences and levels of learning, central to the code of the Realists.

In the nineteenth century "spouses and children came to be prized for what they were, rather than what they represented or could do. [This] is the essence of sentiment. . . ."[35] This feeling, brought about because of many changes in the eighteenth and early nineteenth centuries, is what we see in Bonvin's *Seated Boy with a Portfolio* that is so special. It is not that this feeling was new in 1857, but more that it reached a pinnacle after a long climb. Bonvin reached a high level of humanistic sentiment in most of his work depicting people. He seemed to have aimed for an individual understanding of his subjects, heroic in their simplicity and never maudlin in treatment. His sympathy with children in particular could have been created by a need he felt to protect them, his joy in seeing them have opportunities he did not, or nostalgia for those experiences he did have. Whatever the reason for his sensitivity to the children with whom he worked, *Seated Boy with a Portfolio* must be seen as Bonvin's response to historic changes in the state and perception of childhood.

University of Minnesota

NOTES

This study was conducted under Gabriel P. Weisberg, Professor of Art History at the University of Minnesota. Professor Weisberg's knowledge, patience, and motivation were inspirational and crucial to this study.

[1] See Gabriel P. Weisberg, *Bonvin* (Paris: Editions Geoffroy-Dechaume, 1979), *The Realist Tradition: French Painting and Drawing, 1830–1900* (Cleveland: The Cleveland Museum of Art, 1980), and "Francois Bonvin's *Seated Boy with a Portfolio*," *Elvehjem Museum of Art Bulletin* (1981–1983): 14–20.

[2] Carol Duncan, "Happy Mothers and Other New Ideas in French Art," *Art Bulletin* 65.4 (1973): 570–583.

[3] Philippe Ariès, *Centuries of Childhood* (New York: Alfred A. Knopf, 1962). Ariès

believed that the family began to center around the child in the seventeenth century, but uses the "cult" term to indicate that the purpose and intensity of child-raising took on a new character in the nineteenth century.

[4] See David Hunt, *Parents and Children in History* (New York: Basic Books, Inc., 1973), Rosemary Lloyd, *The Land of Lost Content: Children and Childhood in Nineteenth-Century French Literature* (Oxford: Clarendon Press, 1992), and Lloyd de Mause, ed., *History of Childhood* (New York: The Psychohistory Press, 1974).

[5] de Mause, 34.

[6] *Ibid.*

[7] *Ibid.*, 17.

[8] Hollister Sturges, III, *Angels and Urchins* (Omaha: Joslyn Art Museum, 1980), 57.

[9] Jean-Jacques Rousseau, *Émile*, Barbara Foxley, trans. (New York: Dutton, 1966).

[10] Colin Heywood, *Childhood in Nineteenth-Century France* (New York: Cambridge University Press, 1988), 220.

[11] Frédéric Le Play, *On Family, Work, and Social Change*, Catherine Bodard Silver, ed. and trans. (Chicago: The University of Chicago Press, 1982). Le Play was one of the first authors and scientists to value and write about his own childhood experiences.

[12] Carle C. Zimmerman and Merle E. Frampton, *Family and Society* (Cambridge: D. Van Nostrand Co., 1935), chapters 4–7.

[13] Clifford Stetson Parker, *Defense of the Child by French Novelists* (Menasha, WI: Banta Publishing Company, 1925).

[14] Lloyd, 17–18.

[15] *Realist Tradition*, 42.

[16] *Bonvin*, 44.

[17] Heywood, 202.

[18] Michalina Vaughan and Margaret Scotford Archer, *Social Conflict and Educational Change in England and France, 1789–1848* (London: Cambridge University Press, 1971), 181–188.

[19] Parker, 59.

[20] Roger Price, *A Social History of Nineteenth-Century France* (London: Hutchinson, 1987), 307–309.

[21] Eugen Weber, *Peasants into Frenchmen* (Stanford: Stanford University Press, 1976).

[22] Price, 212. The 1841 law prohibited work for children under age eight. Children eight to twelve were mandated to go to school instead of working (although this was largely ignored). Older children who did work were to have their schedules reduced to twelve hours per day and were only to work six days per week.

[23] *Seated Boy with a Portfolio*, 14. The likelihood that this is a studio apprentice is increased by the fact that these young men wore blue smocks, such as the boy in the picture is wearing. See also Sturges, 45. Blue was a common color worn by students, apprentices, and members of the working class.

[24] Ariès, 264. Before the Enlightenment, childhood was seen as a weakness to be "snapped out of" rather than as a gradual process of reaching adulthood.

[25] Lloyd, 74. The author describes the perception of books at this time as "magic carpets" to and from childhood.

[26] *Seated Boy with a Portfolio*, 14–20. Democratization, controversy, and development of art education in mid-nineteenth century France is discussed.

[27] *Ibid.*, 283.

[28] Petra ten Doesschate Chu, *The European Realist Tradition*, 281–282. Painters who studied with Lecoq included Realist painters Alphonse Legros and Henri Fantin-Latour. Sculptors Auguste Rodin and Jules Dalou were also in Lecoq's atelier.
[29] Chu, 242–244, 277–289.
[30] For Bonvin's teaching methods and theories, see *Seated Boy with a Portfolio*.
[31] F. W. J. Hemmings, *Culture and Society in France* (New York: Charles Scribner's Sons, 1971), 3.
[32] *Seated Boy with a Portfolio*, 17.
[33] Lloyd, 341.
[34] *Ibid.*, 93.
[35] Edward Shorter, *The Making of the Modern Family* (New York: Basic Books, Inc., 1975), 6.

ELIZABETH K. MENON

FASHION, COMMERCIAL CULTURE AND THE FEMME FATALE

Development of a Feminine Icon in the French Popular Press

Late in the nineteenth century in France, a peculiar combination of circumstances resulted in the visual manifestation of the dangerous woman dubbed the Femme Fatale. Usually associated with the period of the Belle Epoque (1890's), the Femme Fatale emerged from such symbolist literary sources as Baudelaire's *Flowers of Evil.*[1] An important new subject in symbolist art of the period, the visualization of feminine evil was part and parcel of a much larger cultural context which included a mass-produced consumer culture, continued interest in Japanese paraphernalia, a burgeoning high-fashion industry and changes in the private and public relationships between the sexes. It is through the examination of this larger context that the Femme Fatale can come to be understood as a volatile mixture of fashion and the feminine body; the icon embodies both an advertisement of sensuality and a warning against indulgence in physical pleasure. Perhaps most of all, the figure serves as a visual manifestation of conflicting masculine impulses towards women in general and the then-burgeoning women's rights movement in particular.

The French popular press provides a viable matrix for the evolution of this hybrid of fashion and danger. Journal illustrations appearing late in the 1870's and the 1880's will be primarily considered; this is the moment when Japanese-influenced works, scenes of elegant Parisian life and macabre intimations of the Femme Fatale exist simultaneously. Illustrations appearing in weekly and monthly produced journals such as the *Journal Amusant*, *Le Monde Parisien*, *Le Courrier Français*, *La Chronique Parisienne* and *Le Rire* serve as barometers of taste, fashion and humor as it gradually evolves among the bourgeoisie. Alfred Grévin's *La Première Traine* (Figure 1) demonstrates the symbiotic relationship between fashion and the periodicals.[2] Here the woman's parasol and the train of her dress have been created from a variety of popular newspapers. Quite literally the papers make the fashion and fashion makes the papers in this clever cover illustration from *Le Journal Amusant*. The positioning of the papers flowing from the woman's dress takes on a

A-T. Tymieniecka (ed.), Analecta Husserliana, Vol. LIII, 363–379.

Fig. 1. Alfred Grévin, *La Première Traine*. Illustration from the series "Fantaisies Parisiennes"; appeared on the cover of *Le Journal Amusant*, March 8, 1879. Photo by the author.

serpentine appearance – a significant formation which will be discussed in detail below.

Comparatively little is known about the illustrator Alfred Grévin, nor has another graphic artist discussed here, Henri Gray, been given a great deal of attention. Since these individuals rarely achieved the status of "artist" in the mind of contemporary critics, it is not surprising that they have practically fallen into oblivion. But their contributions during the nineteenth century are critical insofar as they quite literally were in charge of the visualization of contemporary society on a daily basis. Of the three illustrators considered in this study, only Henry Somm has received the attention of Art History scholars in recent years.[3] It is also in his art that Japonisme (the taste for things Japanese), commercial culture, changes in fashion and the masculine opinion of the women's rights movement combine most clearly to result in the iconic figure called the Femme Fatale.

François-Clément Sommier (known as Henry Somm) is perhaps best known as a "minor Impressionist," but he was actually an important transitional artist between Impressionism and Symbolism. In his art ideas of fashion, feminism and a type of romantic orientalism were developed over a twenty-year period that can be documented by his prodigious output in illustrated journals. Somm's writings (short stories, art reviews, cartoon strips and letters) are crucial to fostering an understanding of life during this period and the struggles faced by artists who were known primarily as "illustrators." Unfortunately, as is the case with Grévin and Gray, the private life of Henry Somm remains a mystery. Articles written during the artist's lifetime, some even written by close friends, fail to shed light on his personal nature; and a complete chronology of his life is unrecorded.[4] Somm's articles and short stories similarly fail to reveal anything about him personally.[5] Moreover, during the height of his career (1880–1890), Somm was not listed in the Didot-Bottin, where all popularly known creators were found.[6] These particular books listed addresses by name, by profession (where a great number of Somm's associates were listed under "Artists") and by the street address. It is clear that he sought and achieved a certain amount of anonymity; he further accomplished this by not revealing personal information to his associates and by not working extensively with one person or group. In the absence of documents clarifying any personal motives that any of these artists had in the creation of a Femme Fatale type, we must rely upon the content of the images themselves and deduce any deeper meaning from an understanding of the cultural and historical context in which they were created.

A work by the artist Uzès depicting Henry Somm (Figure 2) helps to reconstruct the world of the early 1880's into which he fits. Published in the popular journal *Le Courrier Français* in 1884, it shows the artist in a top-knot and dressed in Japanese garments.[7] This orientalism, termed Japonisme in 1872, grew in response to Commodore Perry's opening of Japan in 1853. The resulting flow of prints and art objects affected not only artists from Anquetin to Vincent van Gogh, but also the very nature of bourgeois society. Henry Somm's interest in Japan manifested itself in his art from 1871 until the turn of the century. Ultimately, Impressionism and Symbolism as artistic movements both depend heavily on this orientalism or Japonisme. In turn, Japonisme itself was a product of and indeed a form of commercialization. Like Philippe Burty, Siegfried Bing and others, Somm was a part of the very core of the movement

Fig. 2. Uzès, *Henri Somm*. Illustration from the series "Nos Dessinateurs"; appeared in *Le Courrier Français*, August 30, 1885, p. 4. Photo by the author.

and watched Japonisme evolve from private gatherings of aficionados to its complete pervading of everyday middle-class existence.

Included in Uzès' illustration are references to aspects of Somm's symbolism that must be further explained. The first is a string of figures in silhouette, that refer to Somm's participation in the *Chat Noir* shadow theater. The second is the small figure of a fashionably dressed Parisian woman, who because of her placement in the palm of his hand, seems to be under Somm's control. The chemicals at the lower left (specifically nitric acid) refer to the process of creating an etching and further suggest that the woman has been "fashioned" or "invented" by the artist. The proportions of the figures are really a reversal of a common situation in Somm's work, where the fashionable woman dominates male figures, who resemble puppets or dolls. As we will see, this type of composition will provide the genesis of his variety of the Femme Fatale.

Early works by Somm show not only his interest in Japan but also his fear of women. In the premiere issue of *Frou-Frou* (October, 1871)

Somm published a cartoon strip called *Légende Japonaise*, meant to be read in Japanese fashion – from the top down and from right to left. The cartoon describes a conflict between a certain Japanese siren, Havissavon, and Bavenavedavict of the literati (really the artist Somm). The captions explain that the woman has a voice that can lure innocent men to their doom. When the man of letters warns others to protect themselves from this dangerous woman, Havissavon summons her slaves to take her to France to confront her detractor. Yet the story remains unfinished, not naming the result of the conflict. Nevertheless, in this cartoon strip Somm combined his interest in Japan with a certain amount of ambivalence towards women, providing an important clue to the development of the Femme Fatale.

A turning point in Henry Somm's art parallels, intentionally, the moment of the popularization to the point of vulgarization of Japanese art. An illustration that Somm completed for the *Monde Parisien* shows the connection between Japan, the Parisian woman and current fashion.[8] Here the woman holds a print depicting a Japanese woman clad in a kimono. The accompanying poem, appropriately called "Japonisme," cautions the woman that she can search for the latest fashions, but her search will be in vain because today's fashion is Japan itself. Somm documented just how fashionable Japonisme had become in a print titled *Japonisme 1881*, which practically serves to advertise the movement itself. Here a fashionably dressed Parisian woman interacts with little Asian men who resemble dolls. One of Henry Somm's friends, Jules Adeline, possessed an extensive collection of art and art objects from Japan that included a large number of Japanese *poupées*, likely the models for this print.[9] Significantly, Somm did not use any of these comparatively rare and expensive dolls that were a part of Adeline's collection as the basis for his images. Rather he utilized the cheapest, most banal objects in the collection which would be an effective way to communicate vulgarization. In the print *Japonisme 1881*, the doll-like men seem to materialize from folders labeled "A. Cadart," which identifies them as booklets of prints from that publishing firm. Clearly they are infatuated with the woman, and the objects such as the fan and the teacup can be viewed in this context as a reference to their status as items of novelty. The disparate sizes of the men make them seem like novelties as well, which we will see is consistent with most of Somm's work.[10]

Significantly, the little men appear generically "Asian" rather than specifically Japanese. Somm had literally transcribed Japanese works and had seen examples of very specific Japanese dolls. In this context, his choice of a generic oriental model must be seen as deliberate. Somm used these generic figures to symbolize the status that Japanese art had following its popularization in France during the 1880's. In this way the artist becomes social critic, much in the same way that Japan-enthusiast Edmond de Goncourt functioned as a social critic when he verbalized his frustration at the current state of Japanese art in Paris: "The taste for things Chinese and Japanese! We were among the first to have this taste. It is now spreading to everything and everyone, even to idiots and middle-class women."[11] Grévin similarly poked fun at this mutual admiration society in an illustration for *Le Journal Amusant*.[12] Here a woman toys with an ornamental plant while her Asian visitor declares that he's developed "a taste" for her. What each of these artists document is middle-class women's surface attraction to Japanese art and culture.

But the interest in Japan held other important discoveries for Henry Somm, for in his oeuvre the Femme Fatale was compared to and contrasted with the Geisha – believed to be the ultimate woman of servitude by artists in his circle. Some of Somm's works indicate that the Femme Fatale may have evolved out of the fascination with the art and culture of Japan that grew in Paris from the 1850's to the 1880's. In many drawings and watercolors by Somm, Parisian women are compared to or are "disguised" as Japanese Geisha. This suggests a possible link between orientalism within the general population and the development of an anti-feminist movement within the male artistic community.

To be sure, it is difficult to gauge how accurate the French perception of Geisha was during the 1860's–1890's. However, the 1860's were, in Japan at least, the "Golden Age of the Geisha."[13] According to Japanese scholars, the Geisha provided men with the "deeper delights and torments of love-making."[14] Somm's interest in the Japanese women as suggested in his works is related to a maladjusted relationship to the "new" woman emerging during the Belle Epoque.[15] If the French believed that the Japanese Geisha was the perfect woman, unsoiled by the feminist events transpiring in Paris, this could prove to be a contributing factor in the creation of the Femme Fatale. Ironically, the Geisha and Femme Fatale have more in common than one might think, for "although Geisha can hardly be labeled feminists, ironically they are among the few Japanese women who have managed to attain economic self-sufficiency and positions of authority and influence on their own merits."[16] These were the same qualities that the "new woman" during France's Fin de Siècle possessed. Called the *femme nouvelle* by art critics Camille Mauclair and Marius-Ary Leblond, the characteristics of the potentially dangerous woman were three: she was independent, critical and mobile.[17] Victor Jozé, writer for *La Plume*, defined the separate spheres that men and women should inhabit. Man's world was the public, physical place of activity; woman's the private, emotional seat of maternity. "Jozé assumed that any movement of the female away from the private world would violate the natural order and yield an unstable and dangerous state of social inversion."[18]

Somm's attitude toward women can be gauged through his own writings, the content of the writings of others dedicating their work to Somm and the content of the texts that Somm chose to illustrate. Perhaps most important was his designing of the cover for J. Olivier's *Alphabet de l'imperfection et malice des femmes* (1876). This lengthy diatribe against women was originally published in the seventeenth century, then revived and reprinted during the period when women were more visible and more independent than ever. On the cover, Somm depicted the Devil himself as the puppetmaster controlling the movements of women. Olivier declared "man hath not in the world a more cruel enemy than Woman, and which more sensibly and apparently is hurtful to his life, his honor, and all his fortunes, nor which more maliciously opposeth all his designes."[19] Women were said in this text to be the ruin of houses and families and the health and life of men.[20] "She is the sworn enemy of friendship, an inevitable pain, a natural temptation, a desirable calamity, and domestic danger."[21]

In many other illustrations Somm would suggest that women (supposedly still under the influence of evil), aspire to take control of men. In 1879, in the journal *Le Monde Parisien*, a series of women appear who begin to exert unusual control over men. In one, a man is held on a string, and in a similar illustration (Figure 3) a woman has become an executioner – a little man kneels with his hands bound and awaits the final blow. These images did not accompany short stories about horrible women, rather they were done as breaks for the text and were repeated in many issues. Thus they have no relation to any specific text, which makes them all the more unusual. In a related drawing (Figure 4) a woman possesses at least four of these little men – one in a breast pocket, one in a hip pocket and one that strains on a string in front of her. The one she holds in her hand is screaming for help, but his cries fall on

Fig. 3. Henry Somm, illustration depicting a woman about to execute a puppet-man; appeared in *Le Monde Parisien* in 1880. Photo by the author.

the deaf ears of the real puppets in the toy store window. Here Somm has made a clear statement that although the men that the woman possesses are treated like pets or puppets, they are not really toys. For contrast, Somm has taken care to illustrate three types of puppets in the store window: the Pierrot, the Harlequin and the Marionnette. We can see the essence of the Japanese quality of "Mitate" or parody at work with the unusual puns created by the juxtaposition of the doll-like men and the toy-store dolls. Somm has here devised a European symbolist image which evolved from his earlier works based on *poupées*.

Quite clearly this woman treats men like playthings – and it is no coincidence that this drawing was made during the advancements in women's rights. Following early achievements during the French Revolution in 1789 and the subsequent suppression of feminist organizations by the Jacobins, the advocates for women's rights made progress throughout the 1800's. Specifically, the 1880's came to be known as the "epoch of suffragist activism" as well as the time period when women were actually

Fig. 4. Henry Somm, *Jouets* (Toys), ca. 1885, ink drawing. Private Collection. Photo by the author.

given the right to vote.[22] In 1879, when Somm's images of dominant women appeared, the government had just restored the right of women to hold public meetings. By 1881 women had also gained the right to publish feminist journals.[23] Aside from any personal fears that Henry Somm had where women were concerned, these images can be seen as relating to general societal implications of the "New Woman." The Chamber of Deputies made it public knowledge that French opinion was "not sufficiently prepared" for female equality, and of course here the words "French opinion" referred to the male population.[24] That Somm adhered to these feelings is shown by his penchant for collecting documents written by Jean-Jacques Rousseau. Rousseau worked against the equality of women and masked his prejudice in literary works that managed to reinforce the status quo while entrancing much of the female Parisian population.

Other drawings can be considered that document the prevalence of these ideas, such as a work entitled by the artist *The Rights of Woman*.[25] Here a young woman in a large-bustled costume (indicating that the work was completed between 1882 and 1885) holds the scales of equality in one hand; but the pans are not balanced. In her free hand she holds a pistol which she uses to systematically murder little men who appear in a crescendo at the right side of the composition. Her gaze is blank; she appears almost like a mindless zombie completing her predetermined mission. A woman with a similar blank stare had appeared in an 1880 illustration by Somm which accompanied a poem about an animal trainer, implying that a woman controlled all the men of Paris – he relied on his staple imagery of a woman dressed as a circus ringmaster with assorted puppet-men who perform for her.[26] The dress of this Dompteuse is elaborate and she holds a whip under her arm as she gesticulates at the puppet figures. The poem explains:

Pour dompter daims et canaris,
Trop petite serait ta cage;
Il te faudrait mettre un grillage
Tout autour des murs de Paris.

Henri Gray adopted similar imagery in a cover for the *Chronique Parisienne* (Figure 5) where an allegorical figure presides over the destruction of puppet-like figures. A monumental woman (likely an *Amazane*) dominates the foreground of the composition; the skyline indicates that the location is Paris. Clothed in a toga and seated in the pose of a classical ruler portrait, the short-haired *Amazane* cradles the 1885 volume of *Chronique Parisienne* in one hand; the other holds strings connected to smaller figures. A man sporting a monocle and top hat has been elevated on his string; he grins stupidly at the spectator like a child's toy. Another string controls a fashionably dressed Parisian woman; she sports the large bustle and carries a parasol. Importantly this is the only figure to share the pedestal on which the *Amazane* sits. Three doll-like male figures are sprawled lifelessly about the base of the pedestal; the only other female figure is nude and in a position of abandon just to the left of the *Chronique Parisienne* volume – indeed the string which manipulates her comes from the tome itself. Gray has appropriated the puppet imagery to suggest not only that there exists a new, dangerous woman afoot in Paris, but that current fashion and the popular press have a force in her production and dominion.

Fig. 5. Henri Gray, *Les Pantins de 1885*. Appeared on the cover of *La Chronique Parisienne*, December 27, 1885. Photo by the author.

Rosalind Williams, in *Dream Worlds: Mass Consumption in Late Nineteenth-Century France* has described the complex manipulation of men and women alike by the consumer culture of the giant department stores sprouting on the right bank of the Seine during the last quarter of the nineteenth century.[27] As Émile Zola demonstrated in *Au Bonheur des dames* (1883), these stores were carefully constructed to snare women whose fashions served to validate their husbands' status within society. Pierre Giffard described the process of a woman shopping in terms of original sin: "Eve's daughter enters this hell of temptation like a mouse into a trap. . . . In this abyss, whirlwinds are strewn with mirages each one more dangerous than the other. As if from Charybdis to Scylla, she glides from counter to counter, dazzled and overpowered."[28] Philippe Perrot, in *Fashioning the Bourgeoisie*, outlines how the middle classes, though they had assumed political control of the country, still relied on the remaining aristocracy to dictate trends in clothing.[29] Each of these aforementioned cases, two texts written during the period in question and two recent analytical studies of the Fin de Siècle phenomenon, have in common the notion of power and control issues centered on Parisian fashion. The women are manipulated by the fashions and the department stores where they are found; men alternately manipulate (the store owners) or are manipulated by (the husbands) these women. This phenomenon of controlling that borders on what can only be described as a zombie directed by some evil force is a fundamental component of the Femme Fatale, as is her contemporary, fashionable clothing.

As the Femme Fatale developed, she was given a variety of different attributes. Clearly the "woman as puppetmaster" represented one variation of an evil, controlling feminine influence. Several cover illustrations for the *Journal Amusant* by Alfred Grévin, however, indicate a somewhat insidious transformation of women's dresses into serpentine twists (Figure 6) that foreshadow the full development of the snake as a primary attribute of the Femme Fatale. In addition to the obvious connection to Eve's fall from grace, the serpent was also an important visual component in the story of the "fatal" Cleopatra. A depiction of events at the

Fig. 6. Alfred Grévin, *Avant la Fête*. Illustration from the series "Fantaisies Parisiennes"; appeared on the cover of *Le Journal Amusant*, December 6, 1879. Photo by the author.

Cirque d'Hiver published in *L'Illustration* showed that the tiny Asp had been replaced by a huge boa constrictor for a reenactment of the Egyptian queen's suicide. Serpents were also featured on the flowing costumes of Loïe Fuller, whose seductive nightclub performances were dubbed "la danse serpentine."[30]

An 1879 cover illustration for *Le Journal Amusant* by Alfred Grévin (Figure 7) and Henri Gray's 1885 cover illustration for *Le Courrier Français* (Figure 8) demonstrate an important iconographical shift in the depiction of the Femme Fatale. Connected to the devil directly by the presence of a serpent bearing an apple in its mouth, these women seem to experience a sensual abandon as their bodies are encircled by the snake's coils and they share a kiss mediated by the apple of "dangerous knowledge." Gray's title *Le Premier Lapin* suggests that it is the snake, not the woman, who is being seduced. Grévin's illustration, titled *Filles d'Eve* and proposed as a sculpture entry for the next Salon, adds further important elements. The central woman, stripped to the waist, holds a

Fig. 7. Alfred Grévin, *Filles d'Eve*. Illustration from the series "Fantaisies Parisiennes"; appeared on the cover of *Le Journal Amusant*, June 14, 1879. Photo by the author.

Fig. 8. Henri Gray, *Eve pendant le péché: le premier lapin*; appeared on the cover of *Le Courrier Français*, January 18, 1885. Photo by the author.

glass of champagne and a fashionable fan as she indulges in the serpent's kiss. Below, a second woman in similar *deshabillé* seems to have expired following a similar embrace – she clutches a lifeless snake. The fashion half-worn by both of the women has a serpentine appearance; loose about the hips, it is gathered in twists around the ankles. Henry Somm took this idea further in several illustrations where the woman's lower body becomes a snake. A page from *Le Rire* contains many of Somm ruminations about women and what he believed to be their inherent evil. His feminine concoction of fashion and serpent wears an elaborate hat and holds her hands in a muff; from under her skirt sprouts a snake tail that grabs its prey (a man) and begins to squeeze.[31] A variation of this iconography is based on Mermaids or, more precisely, the mythical Sirens who were suspected of luring sailors to their deaths. A haunting watercolor by Somm (New Brunswick, Jane Voorhees Zimmerli Art Museum) suggests both the allure of a mermaid and the danger of a sea-serpent that seems to accompany her.[32]

Examination of a cross-section of illustrations in popular periodicals shows the discovery and eventual assimilation of Japanese art, its association with fashion and consumer culture and its connection to the creation of the Femme Fatale in opposition to the exotic Geisha. The fashionable Parisian woman had, in the 1880's, chased after Japonisme to enhance her own dress, her own chicness. The emergence of the Femme Fatale in early works influenced by Japonisme suggests that this tendency affected more than just the spatial construction and subject matter of contemporary prints and paintings. The influence of the art and culture of Japan reacted with the Feminist movement, which having just begun to make substantial progress for women's rights, was met head-on by a wave of interest in a country where male domination was the traditional, accepted practice.

At the same time, other illustrations reflect a complicated relationship evolving between French men and women, and suggest the imminent development of the Femme Fatale as a new visual symbol. Drawings show little men performing tricks, held on strings, and larger men subject to evaluation by fashionable women. In the works considered, a male/female relationship is established with the woman frequently appearing larger than the man and functioning as a dominitrix in one way or another, and as such she appears as a prototype of the new "modern woman" as envisioned by the male illustrators of the popular press. The Femme Fatale is intimately connected to the phenomenon of mass production and exists, at least in part, to advertize the products of mass consumption. Thus with this icon the lines become blurred between Impressionism and Symbolism, between orientalism and anti-feminism, between commercialization and illustration.

Mankato State University

NOTES

[1] See Bram Dijkstra, *Idols of Perversity: Fantasies of Feminine Evil in Fin-de-Siècle France* (Oxford University Press, 1986).

[2] The image appeared on the cover of the March 8, 1879 issue of *Le Journal Amusant.* It belongs to a series called "Fantaisies Parisiennes."

[3] See Elizabeth K. Menon, "Henry Somm: Impressionist, Japoniste or Symbolist?" *Master Drawings.* vol. 33, no. 1 (Spring, 1995), pp. 3–29.

[4] Consider the following passage from Melandri in "Album du Chat Noir, Henry Somm,"

– "... Je crois de mon devoir avant de continuer, de faire ici un aveu dépouillé d'artifice. La description ci-dessus m'a été dictée par Somm lui-même. Mais, par les cendres de mes aïeux, elle est plus fausse qu'une pièce du pape! ..." See Achille Melandri, "Album du Chat Noir; Henry Somm," *Le Chat Noir*, 12 February 1882, p. 2. See also J. L. H., "L'Exposition Henry Somm," *Le Travailleur Normand*, 14 July 1907, p. 2 and especially Georges Dubosc, "Les «Légendes» d'Henry Somm," *Journal de Rouen*, 22 July 1907, p. 4.

[5] Including "Hu-yo-katzi," *Le Chat Noir*, 9 March 1889, p. 2; "Expo des Artistes Indépendants," *Le Chat Noir*, 7 April 1888, p. 2; "Contes pour rendre les petits enfants fous," *Le Chat Noir*, 20 January 1882, p. 4 and "Contes pour rendre les petits enfants fous; histoire de Mme. Lachemise et des quatre petits boutons," *Le Chat Noir*, 3 March 1883, p. 4.

[6] See Didot-Bottin commerce catalogs, 1880–1890. Located at the Bibliothèque de la Ville de Paris.

[7] The image appeared in *Le Courrier Français*, August 30, 1885, p. 4.

[8] *Le Monde Parisien*, April 17, 1879 (cover).

[9] *Actes du Musée de Rouen*. T. XIII, 1910, p. 2.

[10] For a more thorough analysis of this particular Somm print, see Elizabeth Menon, "Henry Somm's *Japonisme 1881* in Context," *Gazette des Beaux Arts*, February 1992, pp. 89–98.

[11] Reprinted in *Paris and the Arts, 1851–1896: From the Goncourt Journal*, trans. George Becker and Edith Philips (Cornell University Press, 1971), p. 106.

[12] The illustration appeared on the cover of the April 26, 1879 issue.

[13] Liza Crihfield Dalby, *Geisha* (University of California Press, 1983), p. 62.

[14] Kikou Yamata, *Three Geishas*, trans. Emma Craufurd (London: Cassell & Co., 1956), p. 14.

[15] See Bram Dijkstra (cited above) and Edward Marsicano, "The Femme-Fatale Myth: Sources and Manifestations in Selected Visual Media 1880–1920," Emory University Ph. D. Thesis, 1983.

[16] Dalby, p. xiv.

[17] See Debora Silverman, *Art Nouveau in Fin-de-Siècle France: Politics, Psychology and Style* (Berkeley: University of California Press, 1989), Chapter Four: "*Amazone, Femme Nouvelle*, and the Threat to the Bourgeois Family," pp. 63–74.

[18] *Ibid.*, p. 72. Jozé penned "Le Féminisme et le bon sens" in *La Plume*, September 15, 1895.

[19] J. Olivier, *A Discourse of Women, Showing Their Imperfections Alphabetically*, English translation (London, M. DC. LXII), p. 21. Somm's illustration appears on the 1876 French edition.

[20] *Ibid.*, p. 103.

[21] *Ibid.*, p. 158.

[22] Steven Hause and Anne Kenney, *Woman's Suffrage and Social Politics in the French Third Republic* (Princeton University Press, 1984), p. 9.

[23] *Ibid.*

[24] Claire Moses, *French Feminism in the 19th Century* (State University of New York Press, 1984), p. 197.

[25] Current location unknown. Photographs of this work exist in the documentation files at the Musée d'Orsay, Paris.

[26] "Une Dompteuse," *Le Monde Parisien*, March 29, 1880, cover.

[27] Rosalind Williams, *Dream Worlds: Mass Consumption in Late Nineteenth-Century France* (Berkeley: University of California Press, 1991).

[28] Pierre Giffard, *Les Grands Bazars* (Paris, 1882), p. 62. Translated and reprinted in Philippe Perrot, *Fashioning the Bourgeoisie: A History of Clothing in the Nineteenth Century*. Translated by Richard Bienvenu (Princeton: Princeton University Press, 1994), p. 63.

[29] See Perrot, cited above.

[30] "Au Cirque d'Hiver," center illustration for "Mort de Cléopâtre," *L'Illustration*, February 14, 1891 (vol. 97, p. 161); "La Danse serpentine – Mlle. Loïe Fuller et ses transformations," *L'Illustration*, November 12, 1892 (vol. 100, p. 392).

[31] "La Femme et un serpent," on a page titled *Faux Sceptique* which appeared in *Le Rire* on April 6, 1901.

[32] Henry Somm, *Mermaid*, ca. 1895. Jane Voorhees Zimmerli Art Museum, Rutgers University, New Jersey.

This book is a sequel to:

Volume LII, Phenomenology of Life and the Human Creative Condition, Book I

and to:

Volume XII: The Philosophical Reflection of Man in Literature. Includes the Essay by A-T. Tymieniecka *Poetica Nova*
Volume XIII: The Unhappy Consciousness. The Poetic Plight of Samuel Beckett. An Inquiry at the Intersection of Phenomenology and Literature (by E. F. Kaelin)
Volume XVIII: The Existential Coordinates of the Human Condition: Poetic-epic-tragic. The Literary Genre
Volume XIX: Poetics of the Elements in the Human Condition. Part I: The Sea. From Elemental Stirrings to the Symbolic Inspiration, Language, and Life-significance in Literary Interpretation and Theory
Volume XXIII: Poetics of the Elements in the Human Condition. Part 2: The Airy Elements in Poetic Imagination. Breath, Breeze, Wind, Tempest, Thunder, Snow, Flame, Fire, Volcano . . .
Volume XXIV: Logos and Life: Creative Experience and the Critique of Reason. Book I of "Introduction to the Phenomenology of Life and the Human Condition"
Volume XXVIII: Poetics of the Elements in the Human Condition: Part 3: The Elemental Passions of the Soul
Volume XXX: Ingardeniana II. New Studies in the Philosophy of Roman Ingarden (Ed. by H. H. Rudnick)
Volume XXXII: Phenomenology and Aesthetics. Approaches to Comparative Literature and Other Arts. Homages to A-T. Tymieniecka (ed. by M. Kronegger)
Volume XXXIII: Ingardeniana III. Roman Ingarden's Aesthetics in a New Key and the Independent Approaches of Others: The Performing Arts, the Fine Arts, and Literature
Volume XXXVII: New Queries in Aesthetics and Metaphysics. Time, Historicity, Art, Culture, Metaphysics, the Transnatural
Volume XXXVIII: The Elemental Dialectic of Light and Darkness. The Passions of the Soul in the Onto-poiesis of Life
Volume XLI: Allegory Revisited. Ideals of Mankind
Volume XLII: Allegory Old and New. In Literature, the Fine Arts, Music and Theatre, and Its Continuity in Culture (ed. by M. Kronegger and A-T. Tymieniecka)

A-T. Tymieniecka (ed.), Analecta Husserliana, Vol. LIII, 381.

INDEX OF NAMES

A-T. Tymieniecka (ed.), Analecta Husserliana, Vol. LIII, 383–389.